THE IMPACT OF AIR POWER

The Impact of Air Power on the British People and their Government, 1909–14

Alfred Gollin

Stanford University Press
Stanford, California
1989

Stanford University Press
Stanford, California

© 1989 Alfred Gollin

Originating publisher: The Macmillan Press, Ltd.,
 London

First Published in the U.S.A. by Stanford University
 Press, 1989

Printed in the People's Republic of China

ISBN 0-8047-1591-2

LC 88-63914

This book is dedicated to
my dear wife Valerie

Contents

Acknowledgements and Preface ix

1 Introductory: Britain in the Air Age 1

2 The Advisory Committee for Aeronautics 24

3 The Phantom Airship Scare of 1909 49

4 The Beginning of Air Power Politics in Britain 64

5 The Forward March of Aeronautics 89

6 The Origins of British Air Defence 109

7 The Paris Conference and its Consequences 134

8 The Agitation for a National Air Force 159

9 The New Arrangement and its Lapses 186

10 Home Air Defence and Lamp-posts 204

11 The Air Panic of 1913 230

12 Preparing for War with Germany 261

13 Winston Churchill Takes Charge 291

Notes and References 321

Index 343

Acknowledgements and Preface

I have to acknowledge the gracious permission of Her Majesty The Queen to publish a document from the Royal Archives reproduced in the text. In this connection I must also express my thanks to the Librarian at Windsor Castle, Sir Robin Macksworth-Young. He and his staff were generous in the assistance they gave.

Thousands of documents in the Public Record Office in London were examined and I wish to record my appreciation of the help provided by the staff of that great institution.

The Library of Congress in Washington, DC, furnished me with microfilm of the late Octave Chanute's Papers.

In the British Library I examined material in the Balfour Papers, and in Lord Northcliffe's Papers. The late Sir Geoffrey Harmsworth assisted me with the latter collection.

Dr John N. Howard, Chief Scientist, Department of the Air Force, Hanscom Air Force Base, Massachusetts, kindly provided copies of papers from the collection of the famous British scientist, Lord Rayleigh. The depository of the Rayleigh Papers is the Research Library of the US Air Force Geophysics Laboratory at the Hanscom Air Force Base.

In the Royal Air Force Museum several collections were exploited for this history. They include the Papers of Lord Trenchard, Sir Frederick Sykes, and Sir David Henderson. I wish to record my thanks to the staff there and also to Lord Trenchard, Mr Bonar Sykes, and to Ms Devina Stewart Peter, Sir David Henderson's granddaughter. It should be mentioned that for a time the collection of Sir David Henderson's Papers was considered 'lost'. However, the Hon. John Grigg, who has shown a most generous interest in this history, introduced me to Ms Devina Stewart Peter, his cousin. In her turn, she kindly presented me with her grandfather's papers. With her permission, I then gave them to Dr John Tanner, the Director of the Royal Air Force Museum at Hendon, and the collection is now preserved there.

In the House of Lords Record Office (HLRO) several collections of Papers were examined. For the purposes of this volume material in the Bonar Law Papers was exploited.

The Papers of R. B. Haldane, later Viscount Haldane of Cloan, were examined in the National Library of Scotland. In Edinburgh, Mr Alan Bell of the Department of Manuscripts was most helpful.

The Papers of J. E. B. Seely, later Lord Mottistone, were examined in the library of Nuffield College, Oxford, and I wish to thank the authorities there for their assistance.

The Papers of H. H. Asquith were examined in the Bodleian Library at Oxford and I must express my thanks to the staff there who helped me and also to Mr Mark Bonham-Carter for his generosity in allowing me to consult this collection.

The Papers of Reginald McKenna were studied in the library at Churchill College, Cambridge. The staff members there were extremely courteous and helpful and I wish to record my thanks to them.

In the Courtauld Institute in London I consulted the privately printed book by Arthur Lee, later Lord Lee of Fareham, *A Good Innings*. There, Mr. Michael Dorran showed me every consideration.

Documents from the private correspondence of J. L. Garvin, in my personal possession, were also used in the work.

Mr J. W. S. Brancker lent me some of the papers of his father, Sir Sefton Brancker, who served in the Military Aeronautics Directorate in the War Office in the period covered by this present history. He also sent me a manuscript autobiography written by his father. This was published by Norman Macmillan under the title *Sir Sefton Brancker* in 1935. I found, in a few instances, that Brancker's holograph differed from Macmillan's published version and this has been indicated in my text, below, where appropriate. I wish to record my thanks to Mr Brancker for his kindness.

Shortly before his death I spent a day with the late Lord Beaverbrook in his country home, Cherkley, near Leatherhead in Surrey. He told me a great deal about his public career on that occasion including his time as Minister of Aircraft Production. At the end of the day I wrote down all he said and reference to these notes is made in Chapter 10, below.

I have also to thank the officials of the British Library, those of its Newspaper Library at Colindale, and the librarians in the University of California, Santa Barbara.

The use I made of all this material is indicated in the footnote references in the text. My object, with the footnotes, has been to demonstrate exactly how these sources were exploited and each document reproduced or referred to in the text has its appropriate footnote, without exception.

Group Captain E. B. Haslam, formerly Head, Air Historical Branch, Ministry of Defence, London, gave help and encouragement in several ways and I wish to record my thanks to him. Dr John Tanner, Director of the Royal Air Force Museum, offered splendid assistance and encouragement and I wish to acknowledge it here.

I wish also to acknowledge the warm help and support offered to me throughout by the eminent biographer of Lloyd George, the Hon. John Grigg. He gave me valuable advice and comment at every phase of the work. The enthusiasm of this distinguished scholar was a tonic and it is a pleasure to recall it here. My dear friend Mr Maurice H. Smith, formerly of HM Customs and Excise, provided magnificent help. He commented on each chapter in very great detail and I am indebted to him for all he did. My dear wife, Valerie, helped in every possible way and I cannot express adequately my appreciation for all that she did.

* * * *

In preparing this study I have tried to show how the development of aircraft in the period before the First World War affected the British people and their Government. In my view this is an important theme of recent British experience which has not received the attention it merits.

My concern, of course, has not been limited to technical military and naval development. I have also tried to touch upon the political, diplomatic, and social aspects of the theme because they are significant features of British life in the present century.

The present volume is one of a planned series and my hope is I will be able to carry the story forward in later volumes.

It seemed appropriate to end this volume when a major change took place in Britain's defensive arrangements. This occurred when Winston Churchill, in his capacity as First Lord of the Admiralty, agreed to assume control of the home air defences in September 1914.

The War Office had regularly demanded 'sole responsibility' for these defences until the outbreak of war when it at once found it could not carry out the task. Churchill accepted the challenge but lived to regret his decision.

Department of History ALFRED GOLLIN
University of California
Santa Barbara

1
Introductory: Britain in the Air Age

Wilbur and Orville Wright Fly in Public – Count Zeppelin – Britain's Aeronautical Situation – Two Schools of Thought – Lord Northcliffe – His Concern With Flying and his Fear of Germany – The Aerial League of the British Empire – The Mansion House Meeting – Lord Montagu of Beaulieu's Idea – The Parliamentary Aerial Defence Committee – Attitude of R. B. Haldane – A School of Generals – Privy Councillor Rudolf Martin and his Plan for an Air Invasion of England – Asquith Decides to Act – The Aerial Navigation Sub-Committee – Can Towns Be 'Set on Fire?' – A Thousand Tons to be Dropped on London in One Night – Conclusions of the Sub-Committee – Lord Esher and C. S. Rolls – Attitude of R. B. Haldane – Visit of the Wright Brothers – A Revolutionary Plan.

I

The year 1908 has been called the *annus mirabilis* of flying. In that year Wilbur and Orville Wright, the great men of genius who invented the aeroplane, flew in public for the first time. Their earlier experiments with their flying machine had been carried out in relative secrecy in the United States. In 1908, however, Wilbur revealed the capabilities of his aircraft before admiring and enthusiastic crowds at various places in France while Orville flew near Washington, DC, in demonstrations that astonished the spectators who were stunned by his unprecedented and brilliant performances in the air.

The French pioneers of aviation were quick to benefit from Wilbur Wright's example. They seized upon certain technical novelties and innovations of his aircraft and sought then to improve upon it by producing more efficient machines of their own. Before the year was over Henri Farman and others in France were striving to develop aeroplanes that were more capable than the Wright 'Flyer', as their inventors called it.

There were also significant aeronautical developments in

Germany in 1908. After years of experiment Count Ferdinand von Zeppelin achieved several notable triumphs with his gigantic rigid dirigible airship, known as LZ4.

The Count was a lively patriot who feared the French might overtake the Germans in exploiting lighter-than-air craft for military purposes. He resolved to perfect a dirigible that would guarantee Imperial Germany mastery of the air. On 1 July 1908 his LZ4 managed to travel a distance of 240 miles while it remained aloft for a period of twelve hours. The German people, wild with enthusiasm, now began to adore the Count and his creation as symbols of Teutonic superiority. Eventually, they called him the 'saviour of his country'.

In August Zeppelin attempted a flight of 24 hours' duration. During the course of this excursion the LZ4 landed at the village of Echterdingen near Stuttgart so that one of its engines could be repaired. While the huge vessel was on the ground a sudden storm blew up and the ship's moorings were pulled out of the ground. LZ4 drifted for perhaps half a mile, bumped into some trees, and instantly exploded in flames. Count Zeppelin was almost broken-hearted when he contemplated his beautiful airship, now reduced to a charred and useless wreck. Although his spirits were crushed by this disaster there now occurred what has been called the 'miracle of Echterdingen'.

People from every part of Germany, rich and poor alike, began to send voluntary contributions of money to Count Zeppelin so that he could continue his work. In their minds the Zeppelin had become a symbol of national greatness. Millions of marks were delivered in a fervour of patriotism to the Count's headquarters in Friedrichshafen. Buoyed up by this remarkable outburst of popular enthusiasm the Count instantly resolved to build more and better Zeppelin airships for the future.

In this early period of aeronautical history contemporaries could not be certain which class or category of aerial vehicle would determine the future of flying. Many Germans, and others also, believed the best hope for further advance lay with the huge lighter-than-air rigid dirigibles of the Zeppelin type. Others concerned with the subject argued that while the aeroplane was still in a primitive condition it was bound to be improved with the passage of time and that eventually it would dominate. The advocates of each type became partisans who disputed and bickered about their favourites with a fervour sometimes rising to passion.

For British contemporaries the situation was somewhat different. It was clear enough that Americans, Frenchmen and Germans were making considerable forward strides in the field of aeronautics in 1908. It was also obvious that the British lagged behind in each category of the new technology. The result was that some people in Britain became uneasy and depressed at the manifest failure of their engineers and scientists.

The British always liked to believe themselves in the van of technological advances. With respect to aeronautics they could no longer comfort themselves with this reassuring notion.

This distress about the air blended in with other negative themes that afflicted British opinion in the later Edwardian period. The age was dominated by a growing sense of national insecurity, by the conviction that the golden years of the nineteenth century had come to an end, by the idea that Britain was no longer capable of competing with her foreign rivals as she had done in the past.

Traditionally and historically the British were often nervous about a possible invasion of their island home from the European continent. This ancient anxiety was now linked to the immediate and terrible menace of Imperial Germany which had already given clear signs of an intention to supplant Britain and the British Empire as the foremost power on the earth.

In the sphere of aeronautics two schools of thought now appeared in Britain. One of these held that the British Government's first and urgent task was to acquire by purchase a successful aeroplane and a successful dirigible airship and so at last attempt to overhaul the French and the Germans in this developing area of technology.

A second school proposed a completely different solution to the problem of creating a British Air Service. According to this analysis it would be unwise to begin so significant an undertaking by purchasing foreign aircraft, however successful they might be. Genuine progress could be secured only by systematic, organized, scientific research, research directed and controlled by the British Government. The leader of this second school was Richard Burdon Haldane, the famous Secretary of State for War. It was his purpose to begin by creating what he called in later years a 'real scientific Department of the State for the study of aerial navigation'.

II

One man who believed the British Government should at once purchase a Wright aeroplane and so learn all its technical secrets was Lord Northcliffe, the great newspaper proprietor. In 1908 he was a power in the land. His genius enabled him to see that Britain's strategic position would be affected more radically than that of any other country when aircraft were more fully developed. He had already resolved to employ all his influence and his various newspapers in order to make the nation 'air minded', to warn his countrymen that the insularity upon which Britain had relied for centuries past might soon disappear as aircraft became more capable and more efficient.

Lord Northcliffe realized the recent technical advances in aeronautics were not an immediate cause for national concern. But he was also convinced that the responsible British authorities were not acting boldly enough in the new sphere. He decided, in 1908, that a successful flight across the English Channel, from France to England, would alert his contemporaries to Britain's vulnerability in the air. It would make strikingly clear to them the altered conditions in which they now lived.

Shortly after Wilbur Wright astonished the world with his achievements in France Northcliffe proposed to him that he should attempt such a flight across the channel. Northcliffe's *Daily Mail*, Wilbur was informed, would pay a handsome prize for the feat. Although Wilbur Wright was tempted he eventually refused to take up the proposal. In the autumn of 1908 the *Daily Mail* publicly offered a prize of £500 for the successful completion of a cross-channel flight in an aeroplane.

Northcliffe was fascinated by the Wright brothers' accomplishments. As a patriot he wanted to force the British Government to purchase one of their machines. Early in 1909 he travelled to Pau in France so that he could there observe at first hand Wilbur Wright's exploits in the air. He took with him, as his guest, Arthur James Balfour, the powerful leader of the Conservative Party. He planned to employ Balfour as an ally in his campaign. Together, after their visit, each would press R. B. Haldane, the responsible Minister, to secure a Wright aeroplane for the British Government.

If the significance of aeronautical progress was one of his great concerns in 1908 Northcliffe was also seriously worried by the menace of Imperial Germany. He paid close personal attention to

Introductory: Britain in the Air Age 5

the reports and dispatches of the *Daily Mail* Berlin Correspondent, a brilliant American journalist named F. W. Wile. For his part Wile was convinced that the German people, their leaders, and their Emperor were all full of hatred for Great Britain.

These separate themes were fused together in the *Daily Mail* in July 1908. In that month F. W. Wile interviewed Privy Councillor (*Regierungsrath*) Rudolf Martin. The result was an astonishing report published in the *Daily Mail* on 11 July.

Rudolf Martin, the subject of Wile's interest, was a retired German official who believed that the future of Germany 'lies in the air'. He wrote books about Germany's international position. In these publications he was especially concerned with Anglo-German relations. In his opinion Count Zeppelin's recent advances provided the Imperial German Government with a new weapon that might alter the course of history, in Germany's favour.[1]

When Wile saw Rudolf Martin in Berlin he discovered that the Privy Councillor was not, at that time, thinking about employing Zeppelin airships to bomb targets in England. Martin entertained an entirely different plan. He believed Zeppelins could be employed in order to carry out an invasion of England.

He told Wile that if the Germans built a sufficient number of 'motor-airships' in advance they would be able, when war came, to transport 350 000 soldiers from Calais to Dover in a single night. A long account of this 'Teutonic Vision' was at once published in the *Daily Mail*.

Lord Northcliffe decided a British expert should be asked to comment on Martin's remarkable idea. He decided to interview Major B. F. S. Baden-Powell, Scots Guards, a prominent member of the Aeronautical Society of Great Britain, and invite him to discuss the invasion plan and to explain its significance for the readers of the *Daily Mail*.

Baden-Powell's statements, published in the *Daily Mail* on 11 July, were as sensational as Rudolf Martin's proposals. He declared that in time of war 'we should no longer be an island'. He believed the British fleet would be helpless if it were attacked by foreign airships. He demanded that the Liberal Government at once set aside millions to be expended in the construction of British airships and special guns that could fire projectiles high into the air.

The first leading article in the *Daily Mail* of 11 July offered more sober and more reasonable opinions than these. The leader reflected Lord Northcliffe's view of the situation.

The article criticized both Martin and Baden-Powell for their 'alarmism'. It pointed out that the Zeppelin airship did not yet possess the lifting power to carry large numbers of men or explosives of a weight sufficient to cause serious damage to dockyards, industrial centres, or ships at sea. However, the article did warn the British Admiralty and the War Office, in the sternest tones, that a Zeppelin airship of the future might indeed be capable of carrying out such tasks as these. These departments of the State, the *Daily Mail* concluded, 'would be well advised without further delay to appropriate money to enable us at least to keep abreast of Continental enterprise'.

In 1908 Lord Northcliffe was well aware of his great political power. His uncertain temper was made even worse at this time by a disease of the eyes which afflicted him in this period of his life. He was determined to see to it that Britain, as a result of aeronautical developments on the European Continent, began at once to prepare its own air defences.

He had co-operated with R. B. Haldane in the past. He hoped to work with him in the future, for the benefit of their country. However, he was not a man who readily tolerated opposition to his ideas. If the Liberal Ministry did not pay sufficient attention to his suggestions about aeronautics he was prepared to cause as much trouble as he could for it.

III

Others in Britain were also concerned by the developing aeronautical situation. In January 1909 a society that called itself the Aerial League of the British Empire was formed in London. The declared object of the society was to stimulate British interest in aeronautics. The members of the group proclaimed themselves patriots and their society a 'strictly non-party organization'. They proposed to convince their countrymen of the 'vital importance to the British Empire of aerial supremacy'.

In April 1909 the society took a major step forward toward the achievement of its goals. A large meeting was held at the Mansion House in London on 5 April. The object of this gathering was to promote a great movement for the aeronautical development of

Introductory: Britain in the Air Age

the country by stimulating the support and interest of the British public.

At the meeting, attended by prominent politicians and high-ranking military and naval officers, Lord Montagu of Beaulieu proposed a resolution which was later carried unanimously by the assembly. The resolution reflected the uneasiness of many people in Britain distressed by their country's aeronautical situation. It declared: 'That this meeting of the citizens of London . . . regards with considerable anxiety the rapid development of the science and practice of aerial navigation by other nations and deplores the backwardness and apathy shown by this country. . . .'

Lord Montagu of Beaulieu was a distinguished motoring authority who was also very interested in the new science of aeronautics and the effects it could have upon the future of his country. On 21 April, shortly after the Mansion House meeting, he read a paper to the members of The National Defence Association, a group composed of civilians interested in military matters and of officers in the army and navy. In this paper he offered certain shocking conclusions.

He declared that London lay open to a strategic attack from the air that might produce terrible and even devastating results. His views were more sophisticated than those of Privy Councillor Rudolf Martin and Major Baden-Powell. They were published in *The Times* of 26 April and thus gained a wide publicity.

According to Lord Montagu of Beaulieu's analysis there were, in a highly civilized country like Great Britain, certain very important places he referred to as 'nerve centres'. These 'nerve centres', in the British case, included government buildings, the Houses of Parliament, the railway stations, telegraph and telephone headquarters, the Stock Exchange, and also the main communication systems of the country. He argued, in his paper, that a single powerful blow from the air aimed at these 'nerve centres' could instantly cripple the country.

He suggested that airships would be able to attack so swiftly that the nation would be paralysed before the army or the navy could come to its aid. In his view 'war airships' were not to be feared because they could land an invasion force upon the British shore. They were to be feared because they could almost instantly destroy 'the nerve centres and all the main communications of this country. . . .'

He argued that Britain was in very serious danger because it possessed no means of disrupting or even hampering an air attack upon these places. He further explained that London was an almost unique target for such a bombardment from the air because so many of the country's 'nerve centres' were concentrated there where they could be attacked with relative ease by a determined enemy.

Lord Montagu of Beaulieu and his friends in Parliament now calculated that further organization was required if they were to carry their agitation forward. The politicians therefore decided to form a group to be called the Parliamentary Aerial Defence Committee.

While these political arrangements were being put in train there were further developments in the world of British aeronautics. There were in Britain at this time three principal organizations concerned with the subject. The first of these, the Aeronautical Society of Great Britain, a prestigious scientific body composed of engineers and scientists who had long believed in the possibilities of mechanical flight, had been founded in 1866. The second was the Aero Club of the United Kingdom. The members of this society were amateur balloonists and those interested in ballooning and other forms of flight. The third society was the Aerial League of the British Empire which had recently attracted so much attention to itself because of its Mansion House meeting. The leaders of these three bodies came together in April 1909 in order to work out an agreement defining the nature and scope of the activities of each of their societies. They had other great objects. They hoped to induce the government to spend large sums of money in order to advance the cause of aeronautics in Britain. Finally, it was decided a Parliamentary Committee should be formed 'with the strict understanding that the whole question is one entirely outside party politics'.[2]

Eventually, the leaders agreed upon the following demarcations: the Aeronautical Society would be looked upon as the 'paramount scientific authority on aeronautical matters'; the Aero Club would serve as the 'paramount body in all matters of sport, and the development of the art of aeronautics'; the Aerial League of the British Empire would be accepted as the 'paramount body for patriotic movements and for education'.

In order to assist the Parliamentary Committee that was being formed at this time it was further agreed that each of the societies would nominate three of its members to advise and assist the politicians whenever they were requested to do so. All concerned in drawing up this agreement were convinced they had taken a positive step that would benefit their country as it began to face the perils and the opportunities of the new air age.

Early in May 1909 at a meeting in the House of Commons attended by several Members of Parliament the Parliamentary Aerial Defence Committee was formally organized. The new body was established on strictly non-party lines for the 'purpose of cooperating with the Government on aeronautical questions affecting the defences of the country'.

Arthur Lee, the Conservative Member for the Fareham Division of Hampshire, was elected Chairman of the Parliamentary Aerial Defence Committee. This was a curious choice. Lee was a bitter Tory partisan. He despised the Liberal Government of the day and was looked upon with contempt by many of his Liberal contemporaries. He would find it very difficult to co-operate with the government in any sphere of Parliamentary activity. Moreover, Lee had already formed a close association with Lord Northcliffe so that the two could co-operate in the field of aeronautics.[3]

The other officers of the Committee were less truculent than their Chairman. Cecil Harmsworth, Lord Northcliffe's brother and Liberal Member for Droitwich, was elected Vice-Chairman. The Secretary was Arthur Du Cros, Unionist Member for Hastings and a founder of the Dunlop Rubber Company. He was passionately interested in the subject of aeronautics. He could be depended upon to force the government into positive courses whenever aeronautical questions were discussed in the House.

IV

R. B. Haldane at the War Office was determined to carry out his own aeronautical policies. He had shown a keen interest in the subject for a considerable period of time. After several experiences with aircraft and airmen he had come to certain particular conclusions of his own. He would not lightly abandon them.

In taking up the task of creating a British Air Service Haldane did not act alone or upon his own initiative. In a debate in the House of Commons on 2 August 1909 which dealt with the subject of 'Naval and Military Aeronautics' he revealed that the Prime

Minister and the First Lord of the Admiralty had requested him to embark upon this important national project.

On that occasion the House sat as the Committee of Supply and Haldane said:

> The Vote is a civil one, but it touches the Departments of the Navy and the War Office. . . . The Committee will be interested to know what progress has been made in aeronautics as applied to war purposes. . . . The Prime Minister and the First Lord of the Admiralty some time ago asked me to take in hand the general consideration of the principles which underly this Vote and the devising of the machinery which should be called into existence.[4]

In 1906 the Liberal Party won a great electoral victory. It was at that time that Haldane was selected to become Secretary of State for War. His assigned task was to reorganize the War Office and the British Army so that they would be able to perform with more efficiency than they had demonstrated in the recent fighting in South Africa during the Boer War. The subject of military aeronautics soon attracted his attention.

Haldane had been educated in Germany and as a result of his experiences in that country he developed a great admiration for German efficiency and German technological ability. He was fascinated by Count Zeppelin's early experiments with his rigid dirigibles. Later, Haldane also became very interested in heavier-than-air machines, aeroplanes.

In the British Army in this general period construction of aeronautical devices was carried out at an institution known as the Balloon Factory. Practice with these devices and field training with them were the tasks of a military unit in the Aldershot Command. This unit was variously known as the Balloon Sections or the Balloon School.

The most prominent British officer in the field of military aeronautics at this time was Colonel John Edward Capper, Royal Engineers. He served variously as Commandment of the Balloon Sections and the Balloon School and also as Superintendent of the Balloon Factory. He applied himself to his work with an almost passionate devotion. He was interested in gliders, captive observation balloons, man-lifting kites, dirigible aircraft, and in aeroplanes. In 1904 he visited the Wright brothers in their home in

Dayton, Ohio, and thereafter he regularly called the attention of the authorities in the War Office to their accomplishments. The great object of Capper's life was to prepare the British Army so that it would be able to fight a war in the air whenever it might be required to do so.

At his headquarters in Farnborough Colonel Capper experimented with semi-rigid dirigibles and also with aeroplanes. In this latter sphere he relied upon the technical abilities of two subordinates. One of these was his favourite, Lieutenant John William Dunne, the Wiltshire Regiment. In 1907 and 1908 Dunne and a party of Royal Engineers took his Farnborough designed aeroplanes to Blair Atholl in the highlands of Scotland where highly secret flight trials were undertaken. Unfortunately, the Dunne aeroplane could not fly at this time. Colonel Capper's second aeroplane expert was an American civilian, Samuel Franklin Cody. In October 1908 he made the first sustained and successful aeroplane flight in Britain, in a machine of his own design. While this feat has won Cody an immortal place in British aeronautical history, contemporaries and later authorities have sometimes questioned the quality of his work.

Haldane, as Secretary of State, exercised a general control and supervision over these activities. He was so fascinated by the subject that in 1907 he travelled to Blair Atholl in order to observe Lieutenant Dunne at his work. He was very disappointed by what he saw in Scotland. Eventually, he lost all faith in Dunne and in Colonel Capper. In a celebrated and remarkable passage of his autobiography Haldane later wrote:

> A problem which confronted me at the War Office at this time was the commencement of the Air Service. Numbers of inventors came to see me as the then responsible Minister, including the brothers Wright, and I examined many plans and specifications. But I saw that those whom I interviewed were only clever empiricists, and that we were at a profound disadvantage compared with the Germans, who were building up the structure of the Air Service on a foundation of science.[5]

A school of generals in the War Office took an even more dismal view of the subject of military aeronautics. This group was led by a brilliant military administrator, General Sir William Nicholson, Chief of the General Staff. Nicholson, a powerful and influential officer, believed that no type of aeronautical device – dirigible, balloon, glider, or aeroplane – would be of much use to the British Army. He never explained this singular prejudice but he argued his point of view forcefully and with effect in the highest councils of the State. Known as 'Old Nick' by his enemies, Nicholson would not tolerate anyone who disagreed with him. Norman Dixon has not assigned Nicholson a place in his remarkable book, *On The Psychology of Military Incompetence*, but the general certainly merits inclusion in this catalogue of British officers who were responsible for some of the most terrible disasters in the military history of their country.

On 23 July 1908 the War Office issued a *communiqué* to the Press. This document reflected Nicholson's views. It stated:

> In the highest military circles in Great Britain it is accepted that so far airships are a failure. The military authorities have had experts employed in watching the flights of the various airships and aeroplanes and the impression is that for a long time to come there is nothing to be feared from them. . . . When it is possible to cross the Channel, say with a party of excursionists, the War Office may be prepared to regard recent experiments seriously.[6]

The argument of this War Office *communiqué* was quickly overtaken by aeronautical developments in different parts of the world. About a fortnight later, on 8 August 1908, Wilbur Wright astonished the civilized world when he flew his aeroplane in public for the first time at the Hunaudières race-course near Le Mans in France.

It was the quality of Wilbur's flying that stunned his French contemporaries and rivals. Their machines could lumber clumsily in the air. His flew with the grace and competence of a bird. The newspaper press in France and also in Britain announced that a new age in the history of mankind had arrived.

Early in September it was Orville's turn to take up the challenge. He flew at Fort Myer near Washington, DC, in order to satisfy the terms of a contract the Wrights had worked out with the United States Army Signal Corps. He at once began to attack the height and endurance records established earlier by the French pioneers. On 11 September, for example, Orville remained in the air for one hour and ten minutes. Wilbur, still in France, wrote to his sister about the effect of these flights upon European opinion:

> Tell 'Bubbo' that his flights have revolutionised the world's beliefs regarding the practicability of flight. Even such conservative papers as the London *Times* devote leading editorials to his work and accept human flight as a thing to be regarded as a normal feature of the world's future life.[7]

At this stage Privy Councillor Rudolf Martin sprang into renewed activity in Berlin. In the *Daily Mail* of 25 September F. W. Wile reported that Martin now advocated the establishment of a German aeroplane fleet. According to Wile's account Martin proposed to travel to France in order to meet Wilbur Wright and induce him to attempt a flight across the English Channel from Calais to Dover. If Wilbur Wright succeeded in this adventure Martin was certain that public opinion in Germany would force the authorities there to purchase an 'enormous number of Wright aeroplanes for military purposes'.

A few days later on 5 October 1908 the *Daily Mail* published its offer to pay the sum of £500 to the first aeroplane pilot who flew across the English Channel, from France to England or from England to France.[8] The paper took care to inform its readers that the shortest distance across the channel was 21 miles and that both Wilbur Wright and Henri Farman had already exceeded this distance in their flying machines.

Privy Councillor Martin, despite his new interest in aeroplanes, still advocated the construction of a fleet of Zeppelin airships that could be employed in an invasion of England. On 9 October the *Daily Mail* reported that at a public meeting in Berlin Martin had called the attention of his audience to

> a plan for the conquest of England by airships. He asserted that the principal duty of aerial navigators was to induce the combined Continental Powers[9] to construct a fleet of 10,000 'Zeppelins', each to carry twenty soldiers, which should land and capture the sleeping Britons before they could realise what was taking place.

The *Daily Mail* account, written by F. W. Wile, made it clear that many of Martin's listeners laughed at his bizarre theories and suggestions.

On 13 October the *Daily Mail* informed its readers that 100 Wright aeroplanes were to be built in France. Fifty of these machines were to be acquired by the French Army. The paper quoted a French official who said: 'There is no time to waste, no reason to wait. The Wright aeroplane is sufficiently practical to be used by the Army. . . .'

It was in these circumstances that the British Prime Minister, H. H. Asquith, decided to act. As a result of his initiative the British Government now undertook its first serious examination of the entire question of 'Aerial Navigation'.

V

In order to carry out the survey of aeronautics that seemed to be required, Asquith, in October 1908, appointed an 'Aerial Navigation Sub-Committee' of the Committee of Imperial Defence. The membership of this body reflected the enormous importance the subject had acquired in the official mind by this time.

The Chairman of the Sub-Committee was Lord Esher, a distinguished military strategist of the day and also an intimate friend and adviser of the King, Edward VII. Among the politicians who were selected to serve with the Committee were Lloyd George, Chancellor of the Exchequer; Reginald McKenna, First Lord of the Admiralty; and R. B. Haldane. The service representatives were of an equally high calibre. Captain Reginald Bacon, the Director of Naval Ordnance, one of the most brilliant officers in the navy, represented the Admiralty. General Sir William Nicholson, Major-General Sir C. F. Hadden, and Major-General Spencer Ewart were the military representatives. The Secretary of this Sub-Committee was Rear-Admiral Sir Charles Ottley, a greatly respected officer who was also Secretary of the Committee of Imperial Defence.

The appointment of this Sub-Committee was an event of very great importance in the history of British military aeronautics. The Prime Minister, in his instructions to its members, pointed out that the success attending recent aeronautical experiments 'in France, Germany and America has . . . created a new situation, which appears to render it advisable that the subject of aerial navigation should be investigated . . .'.[10]

Asquith desired the Committee to report upon three separate but related themes. First, he wanted to be informed about the dangers to which Great Britain might be exposed at sea or on the land by any reasonably probable aeronautical developments that might occur in the near future. Secondly, he wanted to be told of the naval or military advantages that could be expected from the employment of airships or aeroplanes by the British. Finally, he wanted the Committee's opinion about the amounts to be allotted for aerial experiments, and the Department to receive such sums.

The Aerial Navigation Sub-Committee held four formal meetings. These began in December 1908 and concluded in January 1909. At these meetings several expert witnesses who supplied technical information and opinion were interrogated. When its deliberations were over the Sub-Committee produced a Report which contained recommendations and conclusions. These views influenced British strategic thought for a considerable period of time thereafter.

The final decisions of the Sub-Committee were determined by the fact that only two of its members knew exactly what they wanted it to recommend. The two members were Captain Bacon, the Admiralty representative, and Sir William Nicholson, the chief spokesman of the British Army.

Sir William Nicholson believed there was no valid reason why the British Army should involve itself in aeronautical experiments. His was a powerful personality. He impressed his opinions upon his colleagues very forcefully and they were reflected in the final Report of the Sub-Committee.

Captain Bacon, however, represented an entirely different outlook. He had been hand-picked for service with the Sub-Committee by Admiral Sir John Fisher, the famous and powerful First Sea Lord of the British Admiralty. At this time Fisher wanted to secure for the navy a large rigid dirigible airship which could remain aloft for several days during a single flight. He believed such an airship would be able to serve as a scout for his fleets at sea, thus rendering important and invaluable service. Captain Bacon carried out his instructions to the letter. In its final Report the Aerial Navigation Sub-Committee recommended that funds should be included in the Naval Estimates for the provision of a dirigible airship of the rigid type.

Although Lord Esher was Chairman, it was R. B. Haldane who made the opening statement at the first meeting of the Sub-Committee. The statement was nothing less than a declaration of principle.

In his public remarks about the future of British military aeronautics Haldane was sometimes equivocal and ambiguous. It was not the case on this occasion. He made it clear to his colleagues that in his opinion 'real progress' in aeronautical development could be secured only if 'systematic scientific guidance' were obtained for the purpose. He suggested the creation of a commission or committee, presided over by an eminent scientist, to provide guidance of the kind he desired. Despite this firm declaration, however, Haldane chose not to disagree with Sir William Nicholson whenever he offered his own aeronautical opinions, which were invariably negative ones.

At its second meeting on 8 December two expert witnesses appeared and answered questions put to them by the Sub-Committee's members. The first of these experts was a son of Lord Llangatock, the Hon. C. S. Rolls, a co-founder of the famous Rolls-Royce motor-car firm and an amateur balloonist of considerable experience. He soon revealed he knew a great deal about the activities of the Wright brothers and something of the work of the French military authorities in the sphere of aeronautics.

Rolls told his audience the French Government was very interested in Wright aeroplanes and was about to purchase some from the company which controlled them in France. 'The rights for England', he explained, '. . . have not been taken up by anybody, and the Wrights are wishing to sell these patents for England for a sum of money which they have told me about. . . .'

At this Lloyd George enquired if Wright aeroplanes could be employed to bombard naval bases and other targets in England. Rolls suggested that Wright machines, at this stage of their development, would serve most effectively as aerial scouts. However, airships of the Lebaudy type, already in service with the French Army, were capable of lifting 1300 pounds of explosives, and the French were hard at work training men to drop these explosives on enemy installations.

Rolls revealed also that the French were thinking of dropping petrol bombs and bombs containing inflammable liquid. When Lloyd George enquired if these bombs could 'set towns on fire?' he received an affirmative reply to his question.

When R. B. Haldane, in his turn, examined this witness a fundamental difference in outlook became obvious. Haldane again made it clear that he hoped the British Government would not proceed further in the field of aeronautics until its theoretical and scientific arrangements were completed and in place. Rolls proposed an entirely different course of action. He suggested the British Government should immediately purchase a Lebaudy airship and a Wright aeroplane. These proven devices would enable the British to overtake their French and German rivals more quickly than would a series of scientific and theoretical experiments of the kind Haldane envisaged and advocated. In the course of their exchanges at this time the two men were unable to resolve this basic difference of opinion.

The second witness on 8 December was Major B. F. S. Baden-Powell of the Aeronautical Society of Great Britain. He suggested to the Committee that thousands of Wright aeroplanes, each carrying three or four men, could be employed by an enemy nation in order to carry out an invasion of England. He thought also that the large rigid dirigibles of the Zeppelin type would prove more effective than the smaller Lebaudys used by the French. Earlier in the year H. G. Wells had published his famous novel, *The War in The Air*. In this book fictional German airships, 'big as the biggest mammoth liners afloat', bombed and destroyed cities in England and in other countries. Lloyd George enquired if this was the size of vessel Baden-Powell had in mind. Baden-Powell replied that he did contemplate airships of this size. Later, he suggested that rockets should be developed that could ignite the gas in dirigibles and so destroy them whenever they attempted to approach a target, either on the land or at sea.

The third meeting of the Sub-Committee was held on 15 December when Colonel Capper was interviewed. It is clear from the minutes of the meeting that Capper was not well received on this occasion. Sir William Nicholson, as might be expected, was very hostile and opposed almost all the conclusions Capper offered. However, Haldane and Lord Esher were also critical of Capper's work in his two roles as superintendent of the Balloon factory and Commandant of the Balloon School. Eventually, as a result of this attitude, it was decided to dismiss his two aeroplane experts, Lieutenant Dunne and Samuel Franklin Cody. For his part, Capper was allowed to retain his place for a while longer and he was permitted to continue his experiments with some small semi-rigid dirigibles he had helped to design but it was also clear that his days in British military aeronautics were now approaching their term.

Originally, it had been planned to end the Sub-Committee's interviews after this third meeting. However, Winston Churchill, an ardent member of Asquith's Liberal Government and keenly interested in the developing technology of aeronautics, insisted that one further authority should be examined. This was Sir Hiram Maxim, a distinguished engineer who, among his other achievements, had experimented with an aeroplane of his own design many years earlier and was now looked upon by some of his contemporaries as a technical expert of proven quality.

Maxim, originally an American, had long since transferred his allegiance to England. The great successes of Wilbur and Orville Wright convinced him that a new age had dawned. Even before he was invited to meet the Sub-Committee on 28 January 1909 Maxim had advanced the theory that an invading army of 100 000 troops could be carried to England in a single night 'by a kind of aerial ferry of 5,000 aeroplanes'.[11]

When he met the Sub-Committee Maxim told its members about another possible use for aeroplanes of the future. He said:

> If you were going to bombard a town, you might have a thousand of these machines, each one carrying a large shell, because it is the large shell that does the business. If a thousand tons of pure nitro-glycerine were dropped on to London in one night, it would make London look like a last year's buzzard's nest.

VI

When it had finished gathering evidence it was the duty of the Aerial Navigation Sub-Committee to prepare a Report which contained its conclusions and recommendations; and this would be submitted to the full Committee of Imperial Defence for approval.

The first conclusion in the Report addressed the problem of the air defence of the British Isles. The Sub-Committee decided that in the existing state of aerial navigation the country was in little danger of being attacked from the air; but they pointed out it would be 'imprudent and possibly dangerous' to assume there would not be grave risk in the future, as aircraft were developed and improved. They dismissed the idea that Great Britain could be invaded by airships but suggested these machines might be employed to land a small raiding force to assault an arsenal or a dockyard. Air bombardment of warships and dockyards could not, in the opinion of the Sub-Committee, 'be dismissed as an impossible operation of war . . .'.

A second conclusion dealt with the military advantages Britain might secure if airships and aeroplanes were employed by the country's armed forces. The Sub-Committee decided it was very important to construct a rigid dirigible airship for the navy. Vessels of this type would be able to serve as scouts for the fleets and also might be employed in the future to attack 'foreign warships, dockyards and canal gates and locks'. It was recommended that the War Office should be granted funds to continue its experiments with small non-rigid airships because the army would be served best by this class of aerial vehicle.

A startling conclusion was advanced when the Sub-Committee discussed aeroplanes. The Report declared: 'There appears to be no necessity for the Government to continue experiments in aeroplanes, provided that advantage is taken of private enterprise in this form of aviation . . .'. In a later section which summarized these various conclusions it was further stated: 'The experiments carried out at the military ballooning establishment with aeroplanes should be discontinued . . .'.

Sir William Nicholson had scored a brilliant victory for the generals of his school.

A third conclusion dealt with the sums to be allocated to the two services for their aeronautical experiments. This conclusion was a triumph for Captain Bacon. The Report recommended that a sum of £35 000 should be included in the Naval Estimates 'for the purpose of building an air-ship of a rigid type' and that a sum of £10 000 should be included in the Army Estimates for 'continuing experiments with navigable balloons of a non-rigid type . . .'.

VII

The full Committee of Imperial Defence took up this Report at a meeting held on 25 February 1909. The Prime Minister presided at the meeting but it was Lord Esher who introduced the subject. He did so in a remarkable way.[12]

Lord Esher did not begin by discussing his Sub-Committee's Report. Instead, he informed his colleagues that he had received a letter from the Hon. C. S. Rolls. The letter revealed Rolls had purchased a Wright aeroplane and that he hoped the government would provide him with 'facilities for experimenting with it on Government ground'. In return for this favour Rolls stated he was prepared to offer his services to the authorities 'in the event of their wishing to benefit by the experience that he gained'. Lord Esher approved of this suggestion by Rolls and urged his colleagues to accept it.

Years later when Percy Walker, the great British aviation expert, analysed this development he was shocked. He wrote:

> Now Rolls, despite his many excellent qualities, had a reputation for extreme closeness in financial matters and had never been known to have paid for anything he could get for nothing. In Lord Esher's vicarious generosity there seems just a hint of influence in high places, and we are left wondering whether the proposed facilities are to be granted to Rolls the famous aeronaut or to Rolls the wealthy son of Lord Llangattock.[13]

For some months past Rolls had diligently courted the Wright brothers. He promised if he was allowed to acquire one of their aeroplanes he would do all that he could to preserve its secrets from everyone, even from the British military authorities. Eventually the Wrights agreed to sell him one of their machines.

Now, however, Rolls offered to supply information about the Wright aeroplane to the War Office provided that he was given facilities to experiment with it, at no cost to himself. Later, Rolls went even further in this direction. In March 1909 he wrote to Lord Esher to say he was prepared to travel to France to see the Wrights '. . . on behalf of the W.O. . . . This would enable me to draw them & get definite information . . . which it must be important for the W.O. to have'.[14]

Lord Esher's idea was accepted by the entire Committee of Imperial Defence save for one of its members. Winston Churchill objected as vigorously as he could on 25 February 1909. He said the proposal could be considered '. . . too amateurish. The problem of the use of aeroplanes was a most important one, and we should

Introductory: Britain in the Air Age

place ourselves in communication with Mr. Wright himself, and avail ourselves of his knowledge.'

This advice was rejected by the Committee of Imperial Defence. When it discussed the Sub-Committee's Report the members decided to approve it and also to accept Rolls's singular proposals.

In this way the War Office abruptly abandoned its own experiments with aeroplanes but continued to maintain a vague and tenuous link with the new technology in the person of C. S. Rolls. Unfortunately for the arrangement Rolls was killed in the following year when his Wright aeroplane crashed at an air show in Bournemouth.

The Admiralty followed an entirely different policy. Even before the Committee of Imperial Defence had decided upon its final course of action representatives of the navy contacted the directors of the great armament firm of Vickers Sons, and Maxim. It was proposed that this firm should build a large rigid airship of the kind Admiral Fisher and Captain Bacon contemplated. Eventually, a tender by the firm was accepted and in May 1909 construction of Britain's first rigid dirigible was started, in the gigantic Cavendish dock at Barrow-in-Furness. The task was begun and carried out in the greatest secrecy.

Admiral Fisher's first idea was to place Captain Bacon in charge of the work at Barrow-in-Furness but the Captain, for pertinent reasons of his own, chose to resign from the naval service at this time.[15] He was instantly replaced by an aeronautical enthusiast, Captain Murray F. Sueter, who was appointed 'Inspecting Captain of Airships'. Sueter and his staff were assigned to HMS *Hermione*, a cruiser that was despatched to Barrow to serve as a depot ship for the entire project. Unlike the authorities in the War Office, Admiral Fisher did not propose to mark time in developing an aerial component for the Royal Navy.

For his part R. B. Haldane knew exactly what he wanted to do. His future course was fixed firmly in his mind. He already possessed the confidence of the Prime Minister who had invited him 'to take in hand . . . the devising of the machinery which should be called into existence . . .' in order to create a British Air Service.

In these circumstances Haldane could permit General Sir William Nicholson to disparage the military value of aeronautical devices as much and as often as he desired. He could allow the Hon. C. S. Rolls to challenge his own opinion that progress in aeronautical development could only be achieved if 'systematic scientific guidance' for the purpose was obtained as a first or preliminary step. In the end as he well knew he was the Secretary of State, the responsible Minister, who would decide the matter as he wished.

In the months after the meeting of the Committee of Imperial Defence Haldane worked out a new aeronautical plan of his own. However, before the plan could be explained to the nation a problem sprang up. In May 1909 the Wright brothers arrived in London. They received a tremendous welcome.

Wilbur and Orville Wright were living symbols of the new air age. Their British contemporaries were at this time distressed by the fact that Britain had fallen far behind France and Germany in aeronautical achievement. It was obvious that the Liberal Government's aeronautical policy was inadequate for the purposes of national defence. Now, with the Wrights in London, many began to argue that the situation could be restored if the government at once came to an agreement with them to purchase large numbers of their aeroplanes. The Conservative Press, in particular, urged such a course upon Ministers, and in the most enthusiastic tones.

However, there was no place for the Wright brothers in Haldane's new arrangements. In his mind they were not scientists. They were only clever 'empiricists' who had stumbled by chance upon a solution to the problem of mechanical flight. Moreover, the War Office had already come to its arrangement with the Hon. C. S. Rolls. He would divulge all the secrets of the Wright aeroplane as soon as he learned them. Despite the public clamour for him to do so, as Haldane saw it there was simply no reason for him to treat with the Wright brothers.

Haldane solved his political problem in a devious and cunning way. He invited the Wrights to visit him at the War Office. When the visit was over the brothers were taken to a formal luncheon at the Carlton Hotel in the company of: Sir Charles Hadden, the Master-General of the Ordnance; Sir Edward Ward, the Permanent Under-Secretary for War; and General Gerald Ellison, one of Haldane's most important advisers. It is scarcely surprising that the entire British Press came to the conclusion that the Wrights had been asked to the War Office in order to discuss an arrangement for the sale of Wright aeroplanes to the British Government.

Haldane's object, however, was simply to restrain his critics until he was ready to reveal the details of his own 'revolutionary' new plan. Asquith, the Prime Minister, had already approved of the arrangement. It was decided that an announcement explaining it to the country would be made in the House of Commons on 5 May. In this way the government's detractors would be foiled and the Liberal Ministry's response to Germany's challenge in the air made clear to friends and foes alike.[16]

2
The Advisory Committee for Aeronautics

Haldane's Classic Problem – Its Political Aspect – Haldane's Analysis – His Relationship with Lord Rayleigh – Formation of The Advisory Committee for Aeronautics – Haldane and the Wright Brothers – He Explains his Plan to Lord Northcliffe – The Prime Minister's Statement in the House of Commons – Haldane lobbies the Press – Reaction of the Newspaper Press – The Response of Lord Northcliffe – The Opinion of Arthur Lee – Haldane Explains Again – The Co-operation of Arthur Balfour – Advisory Committee for Aeronautics – Haldane's Misfortune.

I

In the early spring of 1909 Richard Burdon Haldane realized he was confronted with a classic problem. The future of his country could turn on the manner in which he dealt with it. As Secretary of State for War he now had to make some response to the appearance of two entirely new types of weapon, the airship and the aeroplane.

The Wright brothers and the great French pioneers of aviation had demonstrated that aeroplanes could fly successfully. Almost everyone concerned with the new invention realized it was not yet an efficient instrument of war. Nevertheless, contemporaries clearly understood its awesome potential.

A more immediate aspect of the problem was developing in Imperial Germany. There, the German military authorities were making impressive strides with the Zeppelin airship. Their ordered and carefully arranged experiments had already shown that the Zeppelin could fly distances that would enable it to appear in the skies over England. On such flights the Zeppelin airship was capable of carrying significant weights of explosive. In this instance also the new technology was still in an early phase of development. Nevertheless, the potential challenge to Great Britain's insular position was obvious.

Haldane would not shrink from a challenge of this order. He appreciated and took pleasure in his high position in the councils of the State. He was a patriot who gladly accepted the burdens and labours of office. As he saw it he was now required to build up 'the Air Service' in the most efficient manner possible.

Other factors were involved in Haldane's calculations at this time. He knew that in the spring of 1909 fear of Germany dominated the minds of his countrymen. Already, the Liberal Ministry had been subjected to bitter attack by the Conservatives because of the supposed inadequacies of its naval building programme. The famous Lloyd George Budget of 1909 had reinvigorated the Liberal Party at Westminster and in the constituencies. Lloyd George's audacious proposals for new taxation of the rich enabled Liberals to snatch the initiative from their Tory enemies in the desperate political battles of this period. Nevertheless, a Liberal Ministry always lay open to the charge that it was neglecting the nation's defences. In these circumstances Britain's weakness in the air might be added to the list of complaints regularly brought forward by the government's critics.

Up to this point those who condemned the War Office's aeronautical policies maintained that their campaign was 'non-party'. However, the Conservative leadership could seize upon this developing agitation whenever they believed it would benefit them to do so. Here was a further source of political danger to the government that might become significant unless steps were taken to meet it at once.

Finally, Haldane believed the Germans were building up their Air Force on much more logical and efficient lines than those adopted in Great Britain. He welcomed the intellectual task of devising a scheme designed to give Great Britain 'the first place in the air'.

He was convinced that progress in aviation depended first of all upon systematic scientific research. He believed the principles of flight had to be discovered and then practical applications could follow. Only after these principles were understood could consideration be given to the special needs of the armed forces and to the requirements of other users of aircraft.

Haldane was particularly impressed by the fact that in Germany a chair of aerodynamics had been established earlier in the year by the Prussian Government. A distinguished engineer and physicist, Dr Ludwig Prandtl, had been appointed to the place, in Göttingen University. Haldane wanted to respond to this German initiative by creating in Britain what he later called a 'real scientific Department of the State for the study of aerial navigation'.

Haldane already possessed a considerable amount of experience that could help him in the creation of a 'Department of the State' of the kind he now desired to establish. Some years earlier he had served on an Explosives Committee whose chairmen was the eminent scientist, Lord Rayleigh. The knowledge he gained at that time helped to guide him in 1909 when he was planning his new departure for British military aeronautics.

During the Boer War Haldane, reflecting upon the disappointments of the campaign, came to the conclusion that the explosives employed by the British Army in South Africa were inefficient. He suggested to Lord Lansdowne, at that time the Secretary of State for War, that a scientific committee should be appointed to study the technology of explosives. Lansdowne was so impressed by his knowledge of the subject[1] that he asked him to become chairman of an Explosives Committee which would work under the direction of the War Office and the Admiralty. Haldane demurred. He told Lansdowne the post should be filled by a scientist who could direct the committee's activities with much greater effectiveness than could a layman. Lord Lansdowne accepted this advice but insisted that Haldane should be a member of the committee. When Lansdowne enquired about the scientist who might serve as chairman Haldane replied instantly that Lord Rayleigh was the most qualified man. As a result of these discussions a committee was appointed. Haldane worked closely with Rayleigh on the Explosives Committee for a period of four years.

John William Strutt, third Baron Rayleigh, was a scientist of the very greatest distinction. His researches were carried out in the fields of physics, chemistry, and mathematics. His most striking achievement was the discovery of argon, an inert constituent of the atmosphere. He received many awards, including the Nobel Prize for Physics in 1904. He was Secretary and later President of the Royal Society. He became Chancellor of Cambridge University. He served on a great number of scientific and governmental committees. In addition to these accomplishments he was also a considerable owner of land. His wife was Evelyn Balfour, Arthur

Balfour's sister. This domestic arrangement did not make him a political partisan. In 1908 he arranged that H. H. Asquith, the Liberal Prime Minister, should be elected to a Fellowship of the Royal Society.[2]

Lord Rayleigh was also interested in the problems of mechanical flight. He had written several papers on the subject in 1899 and 1900. In 1909, in addition to his other duties, Rayleigh was chairman of the National Physical Laboratory at Teddington. Here was the man and here the institution that could assist Haldane in the implementation of his great plan to develop a British Air Service 'on a foundation of science'.

The National Physical Laboratory was exactly the kind of establishment R. B. Haldane believed in, a scientific institution supported with funds provided by the State. Its task was to carry out scientific tests for theoretical purposes and also for the benefit of British industry. The Laboratory had been founded after several British scientists observed the way in which the Germans exploited scientific work for the benefit of German industry and technology in the period after the Franco-Prussian War. The director of the Laboratory in 1909 was R. T. Glazebrook, a distinguished mathematician and physicist who was quite prepared to assist the Secretary of State in working out a plan that could help him to achieve his object.

Haldane invited Rayleigh and Dr Glazebrook to talk over the problem with him. After discussion he asked the two officials to prepare a scheme for his consideration which could serve to secure the co-operation of the National Physical Laboratory and the developing Air Service.

After deliberation the two experts decided that the scientific needs of the army and navy would be served best if a committee were appointed consisting of representatives of each of the services and a number of scientists. Their plan was submitted to Haldane and was then approved at a conference held in the room of the First Lord of the Admiralty. It was next placed before the Prime Minister, H. H. Asquith, who agreed to act upon the recommendation. On 30 April 1909 letters were sent from 10 Downing Street to several distinguished scientists proposing that a scientific committee should be formed to advise the government on aeronautical matters. This was the origin of the famous Advisory Committee for Aeronautics.

Haldane was greatly excited by the innovation he had arranged to bring about. He believed if the best scientific talent could be brought to bear upon the problems of flight, rapid advances must follow, and his country benefit as a result. His intense interest in this aspect of his work as Secretary of State is reflected in a remarkable passage in his *Autobiography*.

We should observe it was in this passage that he mentioned seeing the Wright brothers and his conclusion, as a result of meeting them, that they were 'only clever empiricists'. However, it should be pointed out that R. B. Haldane's course was set long before his interview with the Wright brothers in the War Office on 3 May 1909. Haldane arranged their visit for political reasons. He was not interested in their technical advice. He had no intention of purchasing any of their aeroplanes for the British Army.

Years after the event Griffith Brewer, the Wrights' patent attorney and friend, the man who acted as their host when they visited London in 1909, wrote about their interview with Haldane. Brewer's account was based upon information furnished to him by Wilbur and Orville Wright. He explained:

> I met the Wrights in Paris and brought them to London, where they were my guests. . . . We had arranged a very busy programme, which included a visit to the War Office at the invitation of Lord Haldane. There Wilbur and Orville had an interview with the then Secretary of State for War. The details of it were not reported, but it turned out to be a social meeting, at which Lord Haldane did not discuss any of the technical aspects of flying.[3]

Before the Wrights even arrived in London Haldane had decided that the future of British military aeronautics would be developed in co-operation with the scientific authorities of the National Physical Laboratory at Teddington. His great plan to create the Advisory Committee for Aeronautics had already been launched. This was an arrangement in which the Wright brothers could play no part at all. Haldane wrote in his Autobiography:[4]

> I . . . took the matter . . . out of the departmental hands of the Master-General of the Ordnance, and going to the Prime Minister, got his authority to add a special section to the National Physical Laboratory at Teddington. There we installed a permanent Scientific Committee, paid for its work, including

our best experts, both theoretical and practical. We were so fortunate as to persuade the late Lord Rayleigh, of whom I had seen much as Chairman of the Explosives Committee, to preside. The Committee began from an early stage to furnish us with guidance of great value.

II

Haldane's first task, when the initial phases of the scheme were completed, was to make certain that his plan was received so enthusiastically that the government's aeronautical critics were silenced. Ministers had to regain the initiative or else they would expose themselves to further attacks in an area of politics that was beginning to attract more and more interest and attention.

It was arranged that a question about aeronautics would be asked in the House and that the Prime Minister himself would reply to it in order to explain what the government proposed to do. Haldane further arranged to meet journalists representing the Liberal and Conservative Press alike, so that he could provide even more information about the new aeronautical programme.

A vital element in these preliminaries was to reveal the new course of action to Lord Northcliffe and so win his support for it. On 4 May, the day before the Prime Minister spoke about aeronautics in the House of Commons, Haldane wrote to Northcliffe from the War Office. His letter was prepared with several objects in view. The first of these was to tell Northcliffe of the great forward stride that had already been taken. The Secretary of State explained it to Northcliffe in enthusiastic terms:[5]

War Office,
Whitehall, S. W.

4th May 1909

Dear Lord Northcliffe,

We have at last elaborated our plans for the foundation of a system of Aerial Navigation for the Army and Navy and constructed (what particularly engrossed me because no other foreign Government has got it) a real scientific Department of the State for the study of aerial navigation. Lord Rayleigh is to preside over a permanent Committee with its laboratory establishments in the National Physical Laboratory at Teddington.

Haldane next declared that he was prepared to make a special arrangement with the *Daily Mail* so that the paper's representatives could receive a fuller account of the new scheme than would be contained in the Prime Minister's answer:

> The Prime Minister will tomorrow answer a question which is to be put to him on the subject and announce the new institution. I have told my Private Secretary to telephone to the 'Daily Mail' offices to say that if they would like fuller information than the Prime Minister's answer will afford I should be here at a quarter to six tomorrow night and will give it to them. But I thought you might like to know that what I believe to be a real step forward in a subject in which you are much interested is now being taken.

Haldane knew that all those interested in aeronautics in Britain were very critical of the negative attitude adopted by General Sir William Nicholson and those members of the Army Council who agreed with him. Their views upon the subject of aeronautics, as we have seen earlier, had been made public when the War Office issued its *communiqué* of 23 July 1908. The failure of Britain to keep up with her German and French rivals in this field made it obvious that these military authorities had not yet changed their minds in any significant way. For this reason Haldane took care to point out that the 'new Department' would not be controlled by the War Office. He concluded his letter to Lord Northcliffe by making this point absolutely clear:

> The new Department which will be connected with the National Physical Laboratory will be under neither the War Office nor the Admiralty but the Treasury will serve as a connecting link between the two Departments each of which will have its own construction establishment.
>
> Believe me,
>
> Yours very truly,
> R. B. Haldane

Although Northcliffe was in Germany when this letter was written and despatched his organization was quite capable of dealing with the matter. On 5 May the *Daily Mail* contained a report: 'The following question will be asked in Parliament today: Mr. Lehmann. – To ask the Prime Minister whether he can make any statement as to the steps being taken by His Majesty's Government with regard to aerial navigation.' Directly below this announcement was an account of recent aeronautical activity in Germany. The *Daily Mail* revealed that a fleet of Zeppelin airships would inaugurate aerial passenger services in the spring, that thirty different German aeroplanes were approaching readiness for trial, and that the Wright patents in Germany had been secured by the semi-official Motor Airship Study Society.

Later on that day Rudolph Lehmann, the Liberal Member for the Harborough Division of Leicester, asked his question in the House of Commons. The Prime Minister replied with a long and formal statement.[6]

Asquith began by explaining that the government was taking steps to put its 'organisation for aerial navigation' on a 'more satisfactory footing.' He revealed to the House that a report prepared by the Committee of Imperial Defence had recommended that the work of devising and constructing dirigible airships and aeroplanes should be apportioned between the navy and the army. The Admiralty, he said, was building one type of dirigible while certain others of a different type would be constructed at the War Office Balloon Factory. Work with aeroplanes had been assigned to the War Office.

He proceeded to tell the House that the 'highest scientific talent' would be brought to bear on the problems the two Departments might encounter in carrying out these duties. In order to achieve this object the National Physical Laboratory had been requested to organize at Teddington a 'special department for continuous investigation, experimental and otherwise, of questions which . . . must be solved in order to obtain adequate guidance in construction'.

Asquith further revealed that he had appointed a special committee to supervise these investigations at the National Physical Laboratory. This committee had a second function, also. It would furnish 'general advice on the scientific problems arising in connection with the work of the Admiralty and War Office in aerial construction and navigation'. This was the Advisory Committee for Aeronautics.

The Prime Minister, in his statement, then named those he had invited to join the committee. The high quality of his selections became obvious when it was realized that seven of its ten members were Fellows of the Royal Society. Only the most eminent and distinguished scientists could hope to be elected to this ancient and celebrated institution. Lord Rayleigh was president of the new committee and Dr Glazebrook was the chairman. The army was represented by Major-General Sir Charles Hadden and the navy by Captain Reginald Bacon. This latter officer resigned shortly afterwards and he was replaced by Captain Murray F. Sueter. The other members were: Sir A. G. Greenhill, a former professor of mathematics at Woolwich; Dr W. N. Shaw, Director of the Meteorological Office; Horace Darwin, the son of Charles Darwin and an eminent civil engineer in his own right; Professor J. E. Petavel, professor of engineering at Manchester University; H. R. A. Mallock, a consulting engineer; and Dr F. W. Lanchester, the motoring expert.

When Asquith finished his statement Arthur Lee at once asked him if adequate funds had been placed at the disposal of the new committee and at the disposal of the War Office and the Admiralty so that the necessary work could be carried out without delay. The Prime Minister replied that such funds had been allocated. At this, there was cheering from the Liberal benches.

After these exchanges Arthur Du Cros organized a meeting in the House of Commons to consider the question of 'aerial navigation in connection with home defence'. About thirty members of the Commons attended. It was on this occasion that the Parliamentary Aerial Defence Committee was formed. The group approved of Asquith's statement made earlier in the day as a 'welcome instalment in the direction of a proper organisation of what was, in the unanimous opinion of the meeting, a movement of supreme importance'.[7]

R. B. Haldane was gratified by this development. Nevertheless, he was still nervous and uncertain about the way in which his new departure in aeronautics would be received. On the evening of 5 May he was very busy in the lobby of the House of Commons. He made it his business to talk with the correspondents of several newspapers there and urgently sought to convince them that the government's new arrangements would provide Great Britain with 'the most complete and scientific organization in the world for dealing with the question of aerial navigation'.

However, in order to find out if this diligent advocacy had been successful he would have to await the appearance of the next morning's newspapers.

III

In the days that followed the Press paid a great deal of attention to the Prime Minister's announcement. Long accounts of it, together with leading articles on the subject of aeronautics, were published in all the newspapers. As a result of R. B. Haldane's activities on 5 May these commentaries did not follow strictly party lines.

On the 6th the War Correspondent of the *Daily Telegraph*, a Conservative organ, published a long article about the government and military aeronautics. The article began by describing the departure of the Wright brothers from Waterloo station on the previous morning. The Wrights expected, the correspondent reported, that one of them would return to England before the year was out. The correspondent wrote next that he had had a meeting with Haldane on the 5th. The Secretary of State, in a long interview, had taken care to point out that the government could not be accused of excessively delaying the development of British military aeronautics. Haldane explained to the journalist that the Liberal Ministry, unlike some more extravagant foreign governments, had proceeded in this sphere 'with discreet forethought'. The British taxpayers' money, he emphasized, had not been squandered on premature and useless experiments. On 7 May the first leading article in the *Daily Telegraph* declared:

> Mr Asquith and his colleagues are unreservedly to be congratulated upon the vigour and judgement with which they have intervened in the international movement for the mastery of the air. Their action has been long deferred. . . . Now that it is taken at last, let us acknowledge that something has been gained by delay. The Ministerial scheme is well and resolutely planned and . . . is so far in advance of any central organization known to be possessed by other countries, that we are justified in looking forward with new hope to the future.

Haldane was very concerned about the charge that the government had delayed its aeronautical programme for too long thus allowing the French and the Germans to gain an insurmountable lead in the air. If he managed to convince the *Daily Telegraph* that this was not so he failed in another quarter. Alfred Gardiner, the famous Editor of the *Daily News*, was an intensely loyal supporter of the Liberal Ministry but he was also one of those anti-armament Radicals who opposed almost every kind of military expenditure. On 6 May he published a leader in the *Daily News* which included in it exactly those opinions Haldane had hoped would not appear in any segment of the Press. The *Daily News* declared:

> Little as we like the passion to develop airships for purposes of warfare, it is no longer possible for any Government to ignore the possibilities of this new terror. The Government has decided wisely in determining, late in the day, to take up this study seriously. The slowness of English inventors and aeronauts to follow the French, American, and German pioneers has been more than a little puzzling. In mechanical ingenuity we used to lead the world. The failures of our military experiments, when compared with the successes of the French and the Germans, suggest that in technical skill there must be something seriously lacking . . . the attempts made hitherto have been half-hearted. From the new departure more will be expected.

There was a very marked reaction to Asquith's announcement about the government's new aeronautical arrangements in the *Morning Post*. Here, Haldane could exercise no influence at all. Fabian Ware, the Editor, and his proprietors represented the extreme right wing of the Tory Party. In this period they were intensely interested in aeronautical developments. After the Prime Minister spoke in the House of Commons the *Morning Post* launched an entire campaign to attack the aeronautical policies of the Liberal Government.

A long article on the leader page of the *Morning Post* of 6 May began by saying: 'At last the Government seem to have realized that they can no longer afford to maintain towards the problem of the navigation of the air the attitude of the ostrich with its head in the sand.' The article pointed out that only a short time earlier the Earl of Crewe, in responding to Lord Montagu of Beaulieu's speech in the House of Lords, seemed to think that a 'policy of inactivity was right and reasonable'. The conclusion offered was straightforward: 'The development of aviation . . . will make an end of the insularity of Great Britain, and the danger of attack from the air can only be prevented by the creation of a strong fleet of airships.'

Two further articles in the *Morning Post* of 6 May were devoted to aeronautics. One, by the paper's Aeronautical Correspondent, was a brief history of recent developments. The writer recalled that only a few weeks earlier Arthur Balfour and then the King had observed Wilbur Wright's 'wonderful machine' in flight. Their excursions to Pau had heightened interest in the subject. More latterly Arthur Du Cros in the House of Commons and Lord Montagu of Beaulieu in the Lords had begun to lead a 'Parliamentary party' whose object was to advance the cause of aerial navigation in Great Britain. The writer made the political point that Balfour, Lord Montagu, and Du Cros were all members of the Conservative or Unionist Party which demonstrated that 'the Opposition is as fully alive to the importance of aeronautics in their national aspect as is the Liberal Party'. The second article, by the *Morning Post*'s Military Correspondent, said: 'We are a long way behind Continental Powers in aerial navigation and have much leeway to make up to overtake them.'

On 7 May the *Morning Post* hit upon a theme which became a very significant factor as interest in the government's aeronautical programme increased. The paper argued that in the government's new course 'too much value has been attached to the purely theoretical side, while no evidence is forthcoming that the practical side will be advanced at all'. A representative of the *Morning Post* had interviewed Lord Montagu of Beaulieu on the previous day. Montagu believed the government scheme was an excellent idea. Nevertheless he further declared:

> In my view, however, the Commission is composed of theoretical and official people as distinct from practical men . . . I do not recognize the name of any man on it of actual practical experience; nor do I see the name of any representative of any one of the three great aeronautical bodies which have just concluded so admirable an agreement.

R. B. Haldane's greatest victory in his efforts to influence the Press on the evening of the Prime Minister's announcement of 5 May lay in the response of the *Daily Mail* to the government's new plan. In an interview with the paper's Lobby Correspondent the

Secretary of State had explained his views at very great length and these were reported in the *Daily Mail* of 6 May in an article that was more than a column long. The article was a full and clear account of Haldane's analysis of the contemporary aeronautical situation and of his hopes for the future. It also contained a great deal of technical information about the government scheme that had not been included in the Prime Minister's announcement. Large headlines in the *Daily Mail* declared:

WAR IN THE AIR

CABINET ACTION

NEW GOVERNMENT DEPARTMENT

Directly beneath these headlines the *Daily Mail* printed the Prime Minister's announcement. The account of the Lobby Correspondent's interview with Haldane followed. It was described in the paper as the 'substance of the great development which the Government has in view'.

Haldane had explained to the correspondent that all kinds of experiments with aeroplanes and airships were being made in several countries but they were carried out haphazardly. Scientific knowledge had not been applied to the problems of aeronautics in a logical way. In order to place England 'in the forefront of the world in the matter of scientific experiments concerning aviation' it had been decided to make use of the facilities of the National Physical Laboratory at Teddington. There, a special department would be organized supported with funds drawn from the Civil Service Estimates, thus making certain it could not be controlled either by the army or by the navy. The department would carry out 'continuous research' into aeronautical problems and the results of its experiments would be confidential.

The 'permanent committee' named by the Prime Minister would work with the new department upon a regular basis. This committee was a 'body analogous to the old Explosives Committee, except that it is organized outside the War Office and is directly under the Prime Minister'. Most of its members were scientists of very great distinction while the service members, General Hadden and Captain Bacon, were 'as highly trained in the scientific side of warfare as any men who could be selected'.

This committee would form a link between the army and the navy upon the one hand and the new department upon the other. The *Daily Mail* Lobby Correspondent's article explained that the army and navy, constantly experimenting either with aeroplanes or airships, would submit their technical problems to the committee and the committee would refer these problems to the new department for solution.

R. B. Haldane entertained very great hopes for this new arrangement. The article accurately reflected his attitude when it declared: 'No foreign Government has so complete a scientific organization as this will be. It is hoped that a good deal of wasted energy and ingenuity will be eliminated and that the British Government will soon be able to make up for lost time.'

Haldane realized that many critics of the government's policy were demanding that the War Office should purchase Wright aeroplanes. Therefore, the article also stated: 'The War Office have been in both indirect and direct communication with the Messrs. Wright.'

The *Daily Mail* devoted its first leading article for 6 May to this subject. The leader was entitled 'The Struggle for the Air'. Haldane's triumph was reflected in the first paragraph of the editorial:

> The world stands on the threshold of a new era in war, and Britain cannot afford to lag behind her rivals. . . . The Government has made a wise and salutary departure by instituting a new scientific department to study the innumerable abstruse problems of aeronautics, and thus to give the Army and Navy and private inventors the guidance which they imperatively require.

Later, the great significance of recent progress in aeronautics was emphasized:

> The day has passed when the 'blue streak' gave protection. Our insularity vanished with the nineteenth century. We have before us nothing less than to gain in the air the same great position which we have held heretofore at sea.

The article ended with a warning about Germany's airships:

The times in which we live are times of bewildering change. A decade ago men would have scoffed at the idea of the command of the air. The airship was then a mere dream. The heavier-than-air flying machine was a scientific curiosity and no more. . . . The German airships of today are at least as manageable as the sailing ship when the wind is against them, and with the wind they can run at high speed. . . . A hostile airship, hovering over London, would be unassailable, and could inflict enormous damage, had we no British Dreadnought of the air to send against it. It is then a real mark of progress that the Government is 'thinking aerially'.

If Haldane believed he had now captured the support of the Northcliffe Press for his new policy he was mistaken. Lord Northcliffe was still in Germany where he could not exercise instant control over his newspapers when novel developments took place. His reaction to the government's aeronautical scheme was not a favourable one. He lost no time in letting the Secretary of State know of it.

IV

As soon as Northcliffe received information of the government's new plan for aeronautical development he instantly dictated a letter to Haldane in order to set out his impressions. Northcliffe was so concerned by the subject that he began his letter to the Secretary of State before his copy of *The Times* arrived at his hotel in the German resort town of Königstein in Taunus. He based his opinions of the government's policy upon the report of the Prime Minister's announcement published in the Paris edition of the *Daily Mail*. *The Times* arrived before the letter was finished but the information it contained only confirmed his negative attitude.

Lord Northcliffe greatly admired R. B. Haldane's accomplishments as Secretary of State for War. He still hoped to co-operate with him in order to advance the cause of aeronautical development in Great Britain. Nevertheless, his letter, dated 9 May 1909, began very firmly:[8]

Many thanks for your letter of the 4th May. But I hope you will not mind my saying, as one who has closely followed Aeronautics and Aviation for nearly twenty years, that in certain directions the composition of the Committee is one of the most lamentable things I have read in connection with our national organization.

I have as yet only seen it in the Paris edition of the 'Daily Mail', and it may of course be incomplete. (*Later.* It is complete according to the last 'Times' to hand.) As far as I can see, there is practically no practical aviator on the Committee at all, and the list of subjects is such as would have been excellently discussed fifteen years ago. I showed it this morning to a German friend who is engaged in the business of dirigibles and he smiled grimly.[9]

Northcliffe turned next to the composition of the committee named by Asquith in his announcement to the House of Commons on 5 May. He angrily denounced the selection of F. W. Lanchester:

I cannot imagine a better Chairman for the Committee than Lord Rayleigh. But the Committee should certainly include the names of some of the now numerous English practical exponents. There is put down a motor expert, a Mr Lanchester, who is one of those unfortunate people who gave themselves away about the Wright aeroplane. . . . The Wright aeroplane is a practical machine, the flight of which can be taught in three hours. He is the same Mr Lanchester, I understand, who is responsible for the Lanchester motor car, and a very few enquiries will show you that that is one of the most complicated motor cars we have ever had. There are a number of suitable men in England, and he ought to be reinforced by such men as Mr Royce, Mr Napier, and others.[10]

At this time one great hope or ambition of the Wright brothers was to write a book which would relate the history of their achievements and also explain the technical details of their invention. Unhappily, Wilbur and Orville were never able to carry out this project because business concerns consumed all their energies after their return to the United States. When they met Northcliffe at Pau they told him of these literary aspirations. He now used this information to continue his attack upon Haldane's committee:

Much of the work to be laboriously considered by this

Committee will be published when the Wrights issue their forthcoming book. Much could be ascertained by any intelligent observer who came to Germany for a few months and stationed himself fairly close to one of the numerous balloon sheds. I am sorry to be so pessimistic, but this proposed waste of money and time investigating that which is already known to foreigners is indeed lamentable. . . . I would suggest that a practical sub-Committee be formed and given authority (1) to test, say, the Wright and Voisin aeroplanes, which can be bought for a thousand pounds each, (2) to make as near a copy of a Zeppelin as possible.

Despite these harsh criticisms of the new organization Northcliffe took care to make certain his letter was not entirely negative. He wanted Haldane to understand that they could continue to assist each other in the future as they had in the past. One subject dear to the Secretary of State's heart was the Territorial Army, his own creation. Northcliffe now proposed that if they worked together his newspapers could help Haldane to win more recruits for service with the Territorials. He exercised a great deal of charm in this part of his letter. It was Lord Northcliffe at his best:

To turn to a more pleasant topic. I find that much more is known here about your Territorials than in England, and, apart from the artillery, which is laughed at, the rifle, which is poor, and the short time of service, they are more seriously regarded than in certain quarters at home.

I have been thinking for some time whether I should suggest to you some easily reported demonstration of Territorials this Summer, something that will give the newspaper men a chance to fan the increasing interest in the movement. . . .

You may remember that in a small way my 'Daily Mail', 'Daily Mirror', and 'Evening News' tried to help last Summer, but we began too late. The matter should be planned early in June, and the scheme should be put forward by the War Office so as to prevent newspaper jealousy.

Northcliffe ended his letter by proposing that he and Haldane should meet upon his return to London:

I am in the hands of oculists here and in Berlin until the end of the month, but could see you the first week in June. Meanwhile, I hope to be able to bring you some useful news with regard to dirigibles. I can at any rate bring all the information as to what has been done by private firms.

V

While he was in Germany Northcliffe maintained a steady correspondence with J. L. Garvin, Editor of the *Observer*. He read every page of the *Observer* each week and then sent Garvin criticisms and suggestions designed to improve the journal. Garvin, in Northcliffe's opinion, was the most brilliant Editor in his employ but his work was scrutinized as carefully and as regularly as that of anyone in the Northcliffe organization. Aeronautics were occasionally mentioned in their exchanges. On 7 May Northcliffe told Garvin: 'When I am free after my cure, I am going to Berlin for a few days, when I hope to learn a great deal about airships. Dr Solm is fairly pleased with my eyes, but says I shall not have much left in four or five years unless I am careful and leave off working so hard, which, as I told him, is much easier said than done.' When Asquith made his aeronautical announcement on 5 May, *The Observer* criticized the membership of the new advisory committee. Northcliffe at once told Garvin:

> I was very glad to see the article about that lamentable Aeronautical Committee. The composition of the Committee, in view of the practical work that has been done by France, Germany and the United States, makes one wonder whether we have not ceased to be an organizing nation.[11]

While Northcliffe was abroad he also took care to maintain a correspondence with Arthur Lee. They arranged to meet in London to discuss aeronautics upon his return from Germany. He made certain Lee quickly learned of his low opinion of the newly appointed advisory committee. On 13 May he told Lee: 'I am in correspondence with Haldane about his collection of primeval men'.[12]

In London, Arthur Lee was upset by the government's new plan for the development of aeronautics in Great Britain. He believed a gross error had been made when practical fliers were not included in the membership of the new committee. He explained to Northcliffe:[13]

> I agree with you that the Government's 'Scientific Committee' is almost ludicrous in its composition. I do not believe there is a single man on it who has either been in any form of flying machine or has probably even seen one fly. I will go further and express my hope, in the interests of humanity, that none of the venerable scientists who are to ponder over these problems will ever trust themselves to any locomotive machine more enterprising than a bath-chair.

By this time Lee had been selected as Chairman of the Parliamentary Aerial Defence Committee. He took his new duties seriously. He revealed to Northcliffe he had reason to believe Haldane was the one who chose the several members of the advisory committee. Indeed, Haldane had attempted to explain the entire government scheme to Lee but they had been unable to agree about it. They were divided by a basic difference of opinion. Lee's letter continued:

> I understand that Haldane is mainly responsible for the selection of this Committee and thinks it is admirably adapted for the process of 'clear thinking' to which he is so devoted. He attempted to explain the whole matter to me the other day, and claimed that as a result of this Government Committee we in this country should be 'incomparably better organized for aeronautical purposes than any other country in the world'. I said that I would not argue that point with him, but would like to know if it were not also the case that we were the only great Power which had not a single practicable airship or aeroplane. This he had to admit; but, as usual, he seemed to think that theory was of far more importance than practice.

In the War Office R. B. Haldane was convinced the course he had selected was the correct one. He was not prepared to alter it in any way. He believed a scientific enquiry designed to discover the great principles of flight was the paramount requirement if rapid and genuine aeronautical progress was to be achieved in Great Britain. On 18 May he replied to Lord Northcliffe's letter in detail. He began this response by attempting to explain yet again the nature of the government's new arrangement. The tone of his letter was cordial but the Secretary of State sought to make it absolutely clear that not one of Northcliffe's arguments had had the slightest effect upon his own analysis of the great problem he was attempting to deal with:[14]

> Now as to Aeronautics. I think that you have not quite appreciated the purpose of the new arrangement. The Admiralty and the War Office will be responsible for the working out of practical experiments: they will build and buy machines, and for this purpose will reconstitute their staffs in large measure.[15] The expert Committee is not intended to be and is not chosen with a view to being a committee of experts in the sense of constructors and inventors; these have all their own views and that element is only represented on Lord Rayleigh's Committee to the small extent that is necessary to make sure of getting the requisite technical information.

Haldane then touched upon one of his favourite themes. As a statesman who was responsible for the well-being and the future of his country he believed that the British did not approach any of their great national problems in a proper way. In his view the only correct and valid course was to master the scientific issues involved and only then, fortified by such knowledge, proceed to practical activity. He explained to Northcliffe, as he would explain again years later in his *Autobiography*, that the practical inventors of the day could make no contribution to the type of system he now desired to create for his country. He also pointed out that in their own aeronautical preparations the German General Staff had already adopted the methods he advocated:

> The advice which I have received convinces me, of what I was very ready to be convinced, that here as in other things we English are far behind in scientific knowledge. The men you mention are not scientific men nor are they competent to work out great principles: they are very able constructors and men of business. But in this big affair much more than that is needed. The Naval and Military experts have demonstrated to the Defence Committee that dirigibles and still more aeroplanes are a very long way off being the slightest practical use in war. Something very different is wanted and that something we have

got to produce. The scientific problems are not worked out. Your German friend simply did not know what his own Government was doing. At Göttingen, my old University, which I visited the other day, I found that the Prussian Government had set up a Chair of Aviation,[16] the directions for which were that no student was to be admitted who was not an expert with the differential calculus. The German General Staff knows how much has got to be learned before practical results can be obtained in this new region.

R. B. Haldane had a further important point to make in this letter. A fascinating development had taken place at Westminster.

It will be recalled that Northcliffe had invited Arthur Balfour to accompany him, as his guest, when he visited Wilbur Wright at Pau in February 1909. At that time Balfour had agreed to speak to Haldane, upon his return to London, and inform him of the technica capabilities of the Wright aeroplane. Northcliffe looked upon the Conservative leader as a powerful ally who might be able to exert genuine influence upon the attitude of the Secretary of State in this matter.

However, when the Prime Minister announced the government's new plan for the further development of aeronautics in Britain Arthur Balfour was so impressed by it that he was prepared to cooperate with the Liberal leadership in the House of Commons so that several of the details of the scheme which were somewhat unclear could be explained yet again.

Arthur Lee and many others believed that under the government plan members of the advisory committee would be required to invent or construct aircraft and that the work of private inventors would be neglected by the authorities. This was not the case. Balfour had agreed to assist the government by asking a question in the House of Commons which would enable the Prime Minister or Haldane to restate the duties and functions of the new advisory committee, and thus put the matter into a clearer perspective. Haldane ended his letter to Northcliffe by informing him of these issues and of Arthur Balfour's decision to help the Liberal Government:

> The Government here proposes to encourage the inventor in every way and to give him scientific assistance. They will also experiment and work for themselves not through the National

Physical Laboratory or through Lord Rayleigh's Committee but in the new Admiralty works and in the War Office establishment at Aldershot.

Arthur Balfour is as clear as I am on this point and is about to address a question to the Prime Minister or me to emphasize the distinction. You will see that what we have done we have done deliberately. Rightly or wrongly our principle is that without science progress in this difficult matter will be very slow and we want to make it real.

Two days later, on 20 May, Balfour asked his question in the House of Commons. It was designed to permit the Prime Minister to make it clear to the critics of the government's new scheme that the members of the advisory committee would not be employed to invent or construct aircraft. The committee's functions under the plan were quite different. Balfour 'asked the Prime Minister whether he could state the nature of the duties entrusted to the scientific Committee on Aerial Navigation, and explain the relation of the Committee to the executive officers who were understood to be designing balloons and aeroplanes for naval and military purposes'. In his reply Asquith declared:

> It is no part of the general duty of the Advisory Committee for Aeronautics either to construct or invent. Its function is not to initiate, but to consider what is initiated elsewhere, and is referred to it by the executive officers of the Navy and Army Construction Departments. The problems which are likely to arise . . . for solution are numerous, and it will be the work of the Committee to advise on these problems, and to seek their solution by the application of both theoretical and experimental methods of research.[17]

It may be that Arthur Balfour's co-operative attitude with respect to aeronautics helped to mollify the asperity of Lord Northcliffe's outlook. Years later, as we shall see, Northcliffe bitterly condemned Haldane's actions at this time. However, this was not the course he followed in this period of aeronautical history. On 4 August 1909 the first leading article in the *Daily Mail* was full of warm praise for the Secretary of State. The article said of Haldane: 'Happily, we have at the War Office, a Minister of an unusual type. Mr Haldane has frequently dwelt in his speeches upon the need for science in government, and he is not one of those who keep precept and practice studiously apart. . . . He is therefore dealing with the command-of-the-air problem in a thoroughly sound and scientific way.'

VI

The formation of the Advisory Committee for Aeronautics was Haldane's achievement. The committee began its work in 1909 and within a short period of time it began to make significant contributions to the development of aeronautics in Great Britain. These were continued for many years thereafter.

The key to the success of the committee was that it enabled the men and the organizations concerned with aeronautics in Britain to co-operate in a way that had not existed before. The scientists were able to offer their help and advice to the military officers who were experimenting with aircraft. The soldiers and sailors at work in the field presented those technical or theoretical problems they could not solve to the members of the committee. The new department at Teddington was organized by the National Physical Laboratory and equipped with all the devices needed for the investigation of aeronautical questions. The Advisory Committee for Aeronautics took care to co-operate with the Aeronautical Society, the Aero Club, and the Aerial League. These bodies were invited to make suggestions about the research that should be undertaken. In this way their varied experience was also brought to bear upon the problems of flight in Britain.

The Advisory Committee was a national institution. It was not controlled by either of the two great services. It assisted the army and the navy. Eventually, British industry benefited from its endeavour. It later became concerned with aeronautical science in the universities.

Men like Lord Northcliffe and Arthur Lee wanted Haldane to adopt an entirely different approach to his problem. In 1909 they believed there was no time for delay. In the face of Germany's aeronautical advance they urged Haldane to acquire successful aircraft in any way that he could and so build up at once a military aeronautical organization based on them. Haldane refused to entertain such suggestions. His object was to create a system. He allowed the War Office to exploit C. S. Rolls and his Wright aeroplane but this was merely ancillary and peripheral to his main arrangement. Sir Walter Raleigh surely had this basic difference of view in his mind when he wrote in his official history:

> Delay and caution are seldom popular, but they are often wise. . . . The rapid growth in power and efficiency of the British air force owed much to the labours of those who befriended it before it was born, and who, when it was confronted with the organized science of all the German universities, endowed it with the means of rising to a position of vantage.[18]

In the course of a long public career R. B. Haldane rendered great services. Among his many contributions to British life were his remarkable series of reforms at the War Office, and also his work in what he called 'the commencement of the Air Service'. As a result of what he did his country was better able to confront the German challenge when it came in August 1914.

It may be pertinent to point out, at this point in our narrative, that party politics became so bitter in England in 1915 that Haldane was forced to leave the government because he was falsely and ignorantly charged with being too friendly to the Germans. The Prime Minister, Asquith, his closest personal and political friend, refused to stand by him in the great political crisis of May 1915. He gave way to the Tory demand that Haldane should hold no post in the coalition Ministry that was formed at that time. Haldane was banished.

Haldane's persecution did not end when he left office. A further incident in what has been called 'this shameful and cruel exhibition of national ingratitude' occurred on 12 July 1916. On that occasion Haldane got up in the House of Lords in order to speak on the subject of educational training, a topic in which he was expert. Suddenly, the seventh Duke of Buccleuch, white with emotion, sprang to his feet. He said: 'Before the noble Lord directs your Lordships' attention to foreign policy I suggest he should explain his past conduct in misleading Great Britain upon the German danger, and in misleading Germany upon British policy.'[19]

The embarrassment of the House at this curious interruption of its proceedings was obvious. Haldane managed to maintain his composure. He told the Duke they were not assembled to speak about foreign policy. He added that he had been subjected to an extraordinary stream of misrepresentations, untruths, and inaccuracies. He declared no one would be more pleased than he when the facts could be revealed 'as to what was done before the war and the preparations that were made'. Haldane then proceeded to deliver his speech. When he sat down the Lords paid him the unusual compliment of clapping loudly in order to express their sympathy with him in the rude attack he had just sustained.

The incident was not over. The Editor of the *Morning Post* in 1916 was H. A. Gwynne, an embittered partisan who regularly and ferociously assailed all those who failed to agree with him. The *Morning Post* now declared in an article:

> We think it monstrous that, while Lord Haldane is conscious – for he must be conscious – of the way the nation regards him, he should go about lecturing on education or on any other subject. If he knows the feelings of his suffering countrymen, if his front were not brass, he would not desire to speak nor to see nor to be seen. He would relapse into silence and seclusion until such time as it was thought safe for the world to hear his defence. When the time comes his place will not be to lecture but to plead. In the meantime his appearance in public is an outrage.

Thus, despite his many contributions to the welfare of the State, Haldane was made to pay a bitter price because in the period before the war he often liked to express his admiration for German society and German culture. The great Halevy summed Haldane up in a brilliant phrase in his history when he wrote of 'the restless Haldane, who dressed his window with German goods but sold merchandise of a very different character . . . .'.[20]

3

The Phantom Airship Scare of 1909

A Period of Great Tension for the British People – Fear of Germany – A National Crisis – Mysterious Dirigibles in the Skies Over England – One is Seen at Peterborough – And at Clacton – And at Cardiff – The Press Reacts – The German View of the Scare – Lord Northcliffe and the Panic – He Condemns his Countrymen – 'England is Becoming the Home of Mere Nervous Degenerates' – R. B. Haldane's Analysis – A Novel Element in his Calculations – A Question in the House – Aeronautical Developments in 1909.

I

The years 1908 and 1909, when it was first realized that men could fly, were a period of great tension for the British people. They were told by some of their most respected leaders that their country lay open to a German invasion. As a result a widespread suspicion arose that German aliens in Britain were either spies or trained soldiers ready to spring into action as soon as a signal to do so was flashed to them from Berlin. The newspapers began to receive agitated letters from readers who claimed to have witnessed acts of espionage of various kinds, and in all sorts of places. It was sometimes stated that 80 000 German soldiers disguised as waiters, business men, barbers, and shop assistants were already in the country. It was charged that stores of German rifles were hidden at strategic places in London, ready for instant use by a merciless enemy. The Conservative leaders regularly argued that Liberal Ministers were unfit for high office because they failed to pay adequate attention to developments like these, developments that might result in a national disaster.

The naval crises of 1908 and 1909 served to make the situation even more tense. The people of Britain now began to fear that the navy, their first line of defence and the pride of the nation, might be overtaken by the Germans. Such a result could only be looked upon as a catastrophe. Winston Churchill explained the popular reaction in his *World Crisis*:[1]

> All sorts of sober-minded people in England began to be profoundly disquieted. What did Germany want this great navy for? . . . There was a deep and growing feeling, no longer confined to political and diplomatic circles, that the Prussians meant mischief, that they envied the splendour of the British Empire, and that if they saw a good chance at our expense, they would take full advantage of it.

There was another and very formidable aspect to the Anglo-German naval rivalry of these years. Many contemporaries believed that the gigantic costs of the new ships of this period could not be borne for long, even by the wealthiest of the European States. It was feared the result would be national insolvency, economic chaos, and revolution if the competition continued for any significant length of time. Considerations of this sort contributed even further to the general feelings of anxiety.

Some writers have argued that the introduction of the Lloyd George Budget in April 1909 helped to bring the naval crisis of that year to an end. However, it should be realized that the campaign to make the Budget popular in the country was fought out so bitterly at Westminster and in the constituencies that many contemporaries became even more uneasy than they had been before.

The politicians presented the Budget controversy to the people as a crisis of nation and Empire. The British future, it was argued, would turn upon the fate of the Budget of 1909.

Contemporary politicians had to deal with two novel problems in the new century. First, they had to secure for the State the funds necessary for the tremendous armaments of those days. Secondly, the General Election of 1906 had taught Britain's political leaders that if they desired to win at the polls the working men of the country could no longer be ignored. It followed from this that further huge sums would be needed for social reforms. The workers were now demanding such reforms as remedial measures to which they were legitimately entitled as British citizens.

The great issue of the day was: how could the State rake in these huge and unprecedented amounts? The Liberal answer was the 1909 Budget. However, the Conservatives objected to the new taxes contained in the Budget proposals as nothing less than confiscatory socialism. Since their defeat of 1906 the Tories had been thirsting for another trial of their policy, the policy of Tariff Reform and Imperial Preference.

The Liberal Free Traders told the people that if the Conservative programme of protection was adopted it would destroy the economy of the country. Britain had grown rich by adhering to the system of Free Trade which did not permit the imposition of many tariffs. A change of the kind the Conservatives proposed, the Free Traders argued, would result in nothing less than national ruin.

The Conservatives wanted to bind the Empire more closely together. They proposed to do so by levying tariffs upon items of foreign food imported into Great Britain. However, foodstuffs imported from the so-called Great Colonies, Canada, Australia, New Zealand, and South Africa would be excused from these tariffs. The Great Colonies would be granted an 'Imperial Preference' in the British home market. The Liberals were horrified by this idea. Everyone agreed that under such a system the British worker would have to pay more for his food. Moreover, the ultimate result of such a course, the Liberals declared, would be that the foreigners would retaliate with high tariffs of their own and Britain's position in the markets of the world would be gravely damaged.

The Conservatives, in this period, warned the people that the fate of the nation was now at stake. Britain could not hope to survive in the harsh conditions of the new century if she continued to clig to the outworn and outdated practices of the past. Britain, alone, was no longer able to maintain herself in the modern world. However, if she bound the other parts of the Empire more closely together in a new Imperial Union including Canada, New Zealand, Australia, and South Africa she would be in a position to defy any other Power upon the earth. By the terms of the Conservative plan the Empire would be unified in the economic sphere first; and when that was achieved, when Imperial economic co-operation provided a firm base for further development, the various parts of the Empire could proceed to closer military, diplomatic, and naval co-operation.

There was another very significant aspect to the Conservative plan. In order to protect the British worker from the competition of foreign industry the Conservatives advocated the imposition of yet a second set of tariffs. These tariffs would be designed to keep the manufactured products of foreign countries out of the British market. As a result the British worker, secure from the competition of the foreigner, would enjoy fuller employment at home and would thus be compensated for the higher cost of his food.

By the spring of 1908, after two years of political setbacks, the Tariff Reformers began to believe their hour had come at last. A severe depression in trade set in. This made a mockery of the Liberal plans for social reform. Working men in every part of the country began to appreciate the Tory goal of industrial protection as something more valuable and more concrete than the vague promise contained in Liberal schemes like Asquith's plan for old-age pensions. If the Free Trade system could not provide men with jobs the electorate was prepared to reconsider the programme of Tariff Reform and Imperial Preference as a more reasonable alternative:

In January 1908 the Tariff Reformers won an important by-election in Mid-Devon. This was merely a beginning. During the year they carried one constituency after another in an impressive series of by-election victories. The Conservative leaders hoped they were now in a position to reverse the great electoral decision of January 1906.

The Lloyd George Budget was the Liberal response. In order to halt this electoral trend the Liberals needed a gripping and dramatic scheme of their own. They believed they had found it in the revolutionary new taxes the Chancellor included in the 1909 Budget. These taxes would provide the funds for what Lloyd George called a 'war on poverty'. They would also force the rich to pay for the battleships needed in order to keep pace with the rapid growth of the Imperial German Navy.

The Budget campaign of 1909 was one of the most exciting in recent British history, and for good reason. The political leaders, in order to gain adherents for one side or the other, always argued in the most strident tones. They issued dire warnings. The electorate was told that the choice was its own. If the voters made a mistake in this crucial hour and chose the wrong course for their country it might never be able to recover. The campaign, we must be clear, began long before Lloyd George introduced his Budget and it became more ardent and more vehement from April 1909 onward.

This was the intense condition of British affairs when the Prime Minister, H. H. Asquith, made his announcement about the government's new aeronautical policy in the House of Commons on 5 May 1909. His speech produced a curious and remarkable result.

People in several parts of England now began to see airships in flight, in places where it was impossible for such aerial vehicles to be.

II

As early as March 1909 reports began to be received in police stations and in newspaper offices stating that mysterious dirigibles had been seen in the skies over England. Most of these reports came from the Eastern counties which bordered on the North Sea, the German Ocean as the British sometimes liked to call it. Then, a police constable in Peterborough declared he had seen a dirigible with a powerful light pass over the town in the early hours of the morning. When the police investigated the incident they decided the constable had seen a kite. In the next few days there were similar reports from March and Wisbech. Late in April a dirigible was seen at Ipswich. After that, the accounts of such observations became much more frequent.[2]

On 17 May the *Daily Mail* revealed it had received several reports of this kind but no satisfactory proof of the statements made had been furnished to it. The paper explained that ever since the policeman at Peterborough said he had seen an airship over the town the 'eastern counties have been buzzing with rumours of similar airships having been seen at night'. The *Daily Mail* stated it had investigated many of the rumours but it was found that not one of them could be confirmed. At Northampton a police officer and several other people said they had seen an airship with lights crossing the town on a Saturday evening. When the Chief Constable investigated he satisfied himself that the dirigible was 'really a fire balloon, carrying lighted Chinese lanterns, and that it was sent up by two young men with the object of hoaxing the residents of the locality'. The *Daily Mail* offered a conclusion about all the sightings: 'We are not completely satisfied that these observers have really seen what they believe they have seen.'

At exactly this time however this attitude of scepticism was rudely shaken. A panic began to sweep the country. As we have just explained the British people in this period were regularly threatened by the possibility of a German invasion of their homeland, by the deep menace inherent in the Anglo-German naval rivalry, and by the crisis in the affairs of their country symbolized and made immediately obvious by the battle the politicians were waging over the Lloyd George Budget. After Asquith spoke in the Commons on 5 May the Press issued further dire and novel warnings about the nation's vulnerability to an attack from the air. The *Daily News*, after the Prime Minister's speech, spoke of 'this added horror of warfare'. The *Morning Post* declared the development of aeronautics 'will . . . make an end to the insularity of Great Britain'. The *Daily Telegraph* wrote 'We have more at stake than any other nation'. The *Daily Mail* predicted that 'A hostile airship, hovering over London . . . could inflict enormous damage . . .'. All this produced a remarkable psychological reaction.

On 17 May the *Daily Mail* and other newspapers reported the observations of a certain Egerton Free who lived on the East coast, near Clacton. According to his statement to the Press he saw an airship near the town exactly two days after Asquith's announcement in the House of Commons:

> On Friday, May 7, he observed what he describes as a long, sausage-shaped, dirigible balloon, which manoeuvred at an apparent height of 600 ft. for some minutes above the cliffs. . . . The balloon was almost over Little Holland Gap, where the troops were landed by the combined British naval and military manoeuvres in 1907. The airship then passed away in a north-easterly direction towards Frinton-on-Sea and Harwich.

On the next day the vigilant Free visited the cliffs over which the dirigible had passed. There he found a curious object of steel and rubber, five feet in length and weighing 35 pounds. The curious object was stamped: 'Müller Fabrik Bremen'.

The matter was at once taken up in Parliament. On the day Free's account appeared in the Press, 17 May, Haldane was answering questions in the House of Commons about 'Dirigible Balloons'. Members pressed the Secretary of State to ask Parliament for powers that would enable him to construct airships like those in use in France and in Germany. Suddenly, the proceedings

The Phantom Airship Scare of 1909

were interrupted by Horatio Meyer, the Liberal Member for North Lambeth. He said to Haldane:[3]

> Will the right hon. Gentleman, in any Report he may circulate, tell us about a certain dirigible supposed to be hovering about our Coast?

The Speaker intervened at once. He refused to allow Haldane to answer this question and the House proceeded to other business.

This exchange was merely a precursor of what was to follow. The 'Phantom Airship Scare' of 1909 exploded into prominence on 20 May. On the morning of that day all the British newspapers reported the astonishing news that the mysterious dirigible had now appeared before witnesses on the other side of the country, at Cardiff in Wales.[4]

III

Two entirely separate incidents occurred in the vicinity of Cardiff. In the first a railway signalman working in the Queen Alexandra Dock at Cardiff declared that early in the morning of 19 May he had seen a 'weird object flying in the air. In appearance it represented a boat of cigar shape . . . It was lit up by two lights . . .'. Many other men engaged in loading a steamship in the dock also saw this airship. Their foreman said: 'There is no question about the reality of the mysterious airship. Too many of us have seen it to leave room for doubt . . .'.[5]

The second incident involved a traveller who was crossing Caerphilly Mountain, eight miles north of Cardiff. He also reported seeing an airship, only a few hours before the sighting at the Queen Alexandra Dock. His was a bizarre experience. In this instance the airship was lying on the ground. Two men, dressed in heavy fur coats and caps, made up the crew of the vessel. The traveller claimed that when the noise of his cart attracted their attention they 'jumped up and spoke excitedly in a foreign tongue'. The foreigners made their escape by leaping into a 'kind of little carriage' suspended from their airship and 'the thing went higher into the air and sailed away toward Cardiff'.

When, on the next day, a journalist was sent to the scene of this occurrence he found a number of newspaper cuttings strewn upon the ground. Almost all of these contained references to airships or to the German Army. One of these articles, the *Daily Telegraph* reported, contained the following headlines: 'War in the Air; Government appoints a Committee of Experts; Bid for Supremacy; Wright Brothers have a Conference with Mr Haldane.' The *Daily Telegraph* of 20 May also published a story that the captain of a Lowestoft smack had seen an airship over the North Sea 'with which he exchanged signals'.

The Press reacted vigorously to this intelligence. When the Tory *Morning Post* of 20 May reported the appearance of the 'Mysterious Airship' at Cardiff it included two other items of news in the same column. The first of these called 'The New Peril' was a report of a speech made the previous evening in the Hackney Town Hall at a meeting held under the auspices of the Imperial Maritime League. The League's speaker, Harold Wyatt, was a prominent writer on naval and Imperial affairs. He was a member of the Executive Committee of the Navy League and a sometime Unionist candidate for Parliament. Later, he served in the Naval Intelligence Division at the Admiralty. On this occasion he said:

> The existence of German airships similar to that (whether German or not) which had been hovering over the East Coast of England had already gone far to prevent our fleets from keeping the sea. . . . This was a point which had not previously been placed before the British public, and he submitted it now as one likely to be a serious and terrible factor in the naval warfare of the immediate future.

The second news item printed in the *Morning Post* was a telegram from Colchester, the important garrison town in eastern England. It stated: 'serious attention is being paid in that town to continual reports that are received by the police . . . as to the presence in East Essex of foreigners, whose sole business seems to be to take notes of cross-roads and buildings in the neighbourhood of Colchester'.

The Liberal Press responded somewhat differently to the airship scare. The Radical *Daily News* sought to make light of the mysterious airship. On 20 May the *Daily News* declared: 'There is panic in the air – literally in the air, for mysterious airships have been seen in different parts of the country. They carry German spies, of course . . .'. The reaction of the highly respected *Man-*

chester Guardian was more moderate. In an editorial of 20 May the paper remarked: 'There is a great deal of evidence for the actual existence of this airship, and we are in great hopes that the owners may, after all, be Englishmen – not of the War Office, for we have almost given up hopes of airships from there, but ingenious inventors experimenting in secret . . .'.

On 21 May the *Daily News* and the *Manchester Guardian* published a Reuter's telegram from Berlin. It struck a more serious note. According to it a distinguished German publicist, Friedrich Dernburg, father of the German Colonial Secretary, had written a 'striking' article upon the subject of the English airship scare in the *Berliner Tageblatt*.

Dernburg was well known as a writer who regularly advocated warmer and more cordial relations between his country and England. In his article he wrote:

> the lines upon which Anglo-German relations are developing cannot be regarded without grave anxiety, the danger lying not in any methodical hostility of the two countries . . . but in the continual accumulation of explosive matter and in the temperament of the two peoples . . . while the Germans may shrug their shoulders at the symptoms recently manifested of the state of the British mind towards Germany – namely, the invasion scare and stories of forty thousand spies disguised as waiters . . . and of a mysterious airship hovering over England . . . these are most serious factors in the situation, for when an external incident exciting the popular imagination occurs, even a peace-loving Government may be driven to most fateful decisions.

Accounts of airship sightings continued to appear in the British Press. Some railwaymen near Liverpool Street Station reported they had seen an airship in the night skies above that part of London. An airship was seen at Herne Bay. C. S. Rolls and other aeronautical authorities were interviewed by the newspapers and asked to offer some explanation. They suggested that hot air balloons, fire balloons and other devices had been employed by hoaxers. The experts agreed it was most improbable that a German dirigible was sailing over the length and breadth of England, flying at night and hiding by day. Nevertheless, reports of sightings continued to be made. On 21 May the *Daily Mail* revealed 'the phantom airship' had been seen afresh at Norwich, Southend, Birmingham, Pontypool, 'and many other places'.

IV

Historians, in writing of the naval and air panics of this period, have usually blamed the popular Press, and Lord Northcliffe in particular, as fomenters of the scares and exploiters of them. In a standard historical account one scholar explained: 'The cheap popular press . . . fastened on the issue, not because it was a good political scrap, but because their readers were interested. The Daily Mail . . . gave itself to the task of pure scare exploitation . . .'[6] A more recent authority observed:

> We have already suggested that Northcliffe's anti-German campaign in the *Daily Mail* . . . and . . . his strident leaders publicized the hostile feelings of the country. In later years the paper pushed the invasion scare, the spy menace, and the fleet rivalry. . . . When in 1913 Northcliffe urged the government to build dirigibles, *Daily Mail* readers saw zeppelins in the night sky over Yorkshire just as they found German spies under every hedge in 1909.[7]

The matter was not quite so straightforward as these comments suggest. We have already seen that when Privy Councillor Rudolf Martin made his bizarre suggestion that England could be invaded by a German Army carried across the channel in Zeppelin airships, the *Daily Mail* at once explained to its readers that such a military operation was not yet technically possible. When Major Baden-Powell warned his countrymen that the Zeppelin was an immediate menace to their insularity and their safety, in his comments on Rudolf Martin's notions, the *Daily Mail* criticized him also. Later, when F. W. Wile reported an account of one of Martin's speeches his telegram from Berlin, published in the *Daily Mail*, took care to point out that even Martin's German audience laughed at some of his ideas.

At the time of the Phantom Airship Scare of 1909 Lord Northcliffe was in Berlin. He deplored the excited reaction of the British people and their newspapers. He felt humiliated that his countrymen should behave in such a way. He decided to take a remarkable step. It was not designed to exploit the panic.

In May 1909 Northcliffe was so distressed by the airship scare he cabled a two column story to the *Daily Mail*. It appeared below large headlines and over his own name on 21 May:

THE PHANTOM AIRSHIP
GERMAN VIEW OF THE SCARE
THE HARM IT DOES

Northcliffe began his article by informing the readers of the *Daily Mail* that the accounts of phantom German airships flying over England placed their country in a 'ridiculous and humiliating light before the German people'. At first, he explained, the Germans looked upon such accounts as a joke but now responsible leaders of German opinion were giving expression to their feelings of 'disgust, astonishment, impatience, and some little alarm'. The belief prevailing in Germany was that practical jokers in England were sending up balloons with lights attached, but 'seen through the spectacles of a nation which has given itself over to panic, take the form of Zeppelins, Grosses, and Parsevals'.

Next, he produced a few comments from the German Press. The *Berliner Neueste Nachrichten* wrote:

> Things which are being said and done in England these days strike us Germans as magnificent material for farce and comedy. But the ridiculousness of it all is only one side of the matter. Madness is also dangerous. . . . What stands out conspicuously is the lack of any sturdy resistance to these hallucinations in the English public. . . . It is the mere spectacle which fills us with astonishment and regret.

Excerpts from an article in the *Hamburger Nachrichten* were also reproduced: 'We refrain for the present from guessing what aims England has in view in these machinations . . . people in England must not wonder if grave mistrust gradually begins to take root even in those German circles which otherwise lay the greatest weight upon good relations with the island Empire.'

Northcliffe urged the readers of the *Daily Mail* to give up the foolish fiction of imaginary German airships. Instead, he asked them to concentrate upon genuine dangers which were: the accelerated German naval programme; the construction of new docks in Germany; the strengthening of the alliance with Austria-Hungary and Italy; and fresh German efforts to become more friendly with the United States.

Northcliffe ended his article with a blunt and even brutal assertion:

> Germans, who have so long been accustomed to regard Great Britain as a model of national deportment, poise, and cool-headed men, are beginning to believe that England is becoming the home of mere nervous degenerates.
>
> NORTHCLIFFE

On the day after this article appeared Northcliffe wrote to J. L. Garvin in order to tell him about his impressions of Germany. He revealed both his fear of and his admiration for the Germans. He made it clear that the accounts of airships flying over the night skies of England had produced a more profound impression in Germany than in Britain. He explained also that his signed article in the *Daily Mail* had done something to reduce or ameliorate the feelings of contempt the Germans entertained for the English as a result of the airship panic:[8]

> I do wish you had been here with me. We are having a most tremendously interesting time, and are learning lots. We must come here together. I do not believe, however, that any visits or writings will affect the inevitable purpose of these people, though they may stimulate English activity. My Gracious! What a penance it is to be here amidst this marvellous organization and prosperity. . . .
>
> The phantom airship story has created much more stir here than in England, and the Germans are very cynical, contemptuous and unpleasant about it; even one's own German relations are – and I have many. . . .
>
> I do not often air my views publicly, but I was moved the other day over our humiliating position here at this moment, and sent a telegram to the 'Mail', which has been sent back to the Papers here and has done good.

V

R. B. Haldane, reflecting upon recent developments, could not be entirely content with the results of the Prime Minister's aeronautical announcement of 5 May. He was certainly pleased by the formation of the Advisory Committee for Aeronautics and the arrangements he had contrived that would enable it to co-operate with the army and navy upon the one hand and with the scientists and technicians of the National Physical Laboratory upon the other. This was exactly the kind of scientific organization or 'Department of the State' he had hoped to create; and now he had created it. Moreover, he had blunted the impact of the Wright brothers' visit to London by allowing the Press to believe that the War Office was planning to co-operate with them in the development of a military aeronautical service in Britain. Furthermore, his diligent lobbying of the Press had produced mixed results but he could, nevertheless, quietly congratulate himself that his deft touches in this delicate area had worked some of the effects he desired.

However, the airship panic of May 1909 introduced a novel element in his calculations. His object was to launch the new Air Service by slow degrees. It was to be based upon the conclusions of scientific research and study. He would permit it to go forward only when the required principles of flight had been secured by British scientists, working in a British laboratory. Quick or immediate results were to be sacrificed by his design for the sake of steady, definite, and assured progress. In order to proceed in this manner it would serve Haldane best if the aeronautical programme of the government did not become a matter of partisan dispute between the great political parties. The airship panic, for him, was a tedious aberration that threatened his entire concept since national spasms of this order might drag his programme into the political arena, whether the politicians desired it or not.

Oddly enough, this theme had been touched upon in the German Press. Northcliffe, in his signed article in the *Daily Mail*, had reproduced for his readers excerpts from the *Berliner Neueste Nachrichten*. In condemning the airship panic in England that journal had declared:

> Perhaps Englishmen will say this babble is only a Conservative manoeuvre against the Liberal regime. We reply that it is

dangerous to permit conscienceless party politics to proceed to a point which does not hesitate to drag foreign politics and relations with a great neighbouring country into the vortex of partisan strife.

R. B. Haldane looked at the matter in a slightly different way. He had already helped to secure Arthur Balfour's co-operation with the government in this sphere. On 20 May, as we have seen, Balfour, the Conservative leader, obligingly asked the Prime Minister a question in the House of Commons that enabled Asquith to explain certain points about the Ministry's aeronautical policy which had been somewhat unclear up to that time. Haldane had taken care to inform Northcliffe of this development before it took place. These were definite tactical victories.

Now, a further touch was required. Haldane's new object was to secure a general calming effect in the country. In order to achieve this goal it was not necessary for him even to mention the subject of aeronautics.

After the publication of Northcliffe's article in the *Daily Mail* the airship panic began to subside. On 24 May Haldane created an opportunity that would enable him to ridicule the alarmists in England in terms very similar to those Northcliffe had employed. It was arranged that a question should be asked in the House of Commons and that the Secretary of State for War would reply to it.

On 24 May Sir John Barlow, the veteran Liberal Member for Frome in Somerset, asked the Secretary of State for War:[9]

> whether he has any information showing that there are 66,000 trained German soldiers in England, or that there are, in a cellar within a quarter of a mile of Charing Cross, 50,000 stands of Mauser rifles and 7½ millions of Mauser cartridges, that is 100 rounds per rifle?
>
> Mr Haldane: My hon. Friend has done well in bringing before the House this illustration of a class of alarmist statements to which credence is too often given by thoughtless persons . . . Such statements tend to lower our reputation abroad for common sense, and my hon. Friend has done well in exposing this one to the ridicule which it merits.

In this way R. B. Haldane reinforced the impressions Lord

Northcliffe had sought to create in his signed article in the *Daily Mail*. However, the forward march of aeronautical development in 1909 could not be affected by a Parliamentary tactic of this kind.

At the end of May Count Zeppelin made another flight in his airship. Although his craft was damaged at the very end of its journey he established a new endurance record which attracted the attention of the civilized world.

Early in June the *Daily Mail* set out to awaken renewed interest in its prize for an aeroplane flight across the English Channel. This time very serious competitors came forward in order to accept the challenge.

4
The Beginning of Air Power Politics in Britain

Activity of the Prussian Army's Airship Battalion – Count Zeppelin's New Venture – Excitement of the German Emperor – Arthur Du Cros's Plan – The Clément-Bayard II – A Sad Chapter – The *Morning Post's* National Airship Fund – Its Unhappy Result – The Channel Flight – Lord Northcliffe's Preparations – Latham Fails – The Success of Louis Blériot – British Reactions – The Comment of H. G. Wells – The Distress of the *Morning Post* – Defiance in the *Daily Telegraph* – Comment of the *Daily News* – The Opinion of Field-Marshal Lord Roberts – The Initiative of Baron de Forest – The Great Aeronautical Debate – Haldane in Charge – The Comments of Arthur Lee – Arthur Du Cros's Proposal – Bombs Will not be Dropped on Cities – Reactions to the Debate – Arthur Lee's Opinion – The Opinions of Octave Chanute.

I

By March 1909 several soldiers and officers of the Prussian Army's Airship Battalion had been ordered to serve at Count Zeppelin's establishment at Friedrichshafen. The experts of the Zeppelin factory now undertook to train army personnel in the management of rigid dirigible airships.

Meanwhile, the Count began to construct a new airship for the military authorities. They had already agreed to purchase this vessel but Zeppelin wanted to demonstrate its capacities, before he delivered it, by making a flight of some 36 hours. It was thought he planned to fly from Lake Constance to Berlin, and back again.

On 29 May the new Zeppelin began its journey. At first all proceeded according to plan. The German Emperor was intensely excited by the exploit because it would demonstrate German mastery in the air to all those interested in the subject. He ordered detachments of troops to hold themselves in readiness at the great Tempelhof parade ground near Berlin so that they could greet the Count upon his arrival there.

The Emperor, the Empress, and the Crown Prince waited beneath the Imperial standard at the field for several hours. Unfortunately, adverse winds were encountered and Count Zeppelin found it was necessary for him to turn back when he reached Bitterfeld, only a short distance from Berlin.

The return journey to Friedrichshafen was almost completed successfully when the crewmen, having been at their posts for 36 hours, began to show signs of exhaustion and fatigue. Zeppelin decided to land. Owing to the extreme tiredness of the men the airship rammed into a tree and was badly damaged.

Nevertheless, the flight was a magnificent accomplishment. The new airship had maintained itself in the air for more than 37 hours and had covered a distance of over 700 miles. The damaged vessel was repaired and returned to its base on 2 June 1909. As the *Daily Mail* put it in an editorial published on 1 June: 'Germany has reason to be proud of this achievement'.

Those in Britain who were concerned about the aerial defences of their country could point to no similar accomplishments. The large rigid dirigible being built at Barrow was still not completed. Colonel Capper's efforts in the Balloon Factory at Farnborough could not compare with the remarkable results already achieved in Germany. On 21 June the *Morning Post* declared with respect to British aeronautical development: '. . . so far as any useful effort goes absolutely nothing is being done, and nothing will be done.'

This provoked an immediate reaction from Arthur Du Cros, Secretary of the Parliamentary Aerial Defence Committee. He wrote to the *Daily Mail* in order to inform the public that an arrangement had already been concluded whereby an attempt would be made to fly a large dirigible balloon of the most modern type from Paris to London. Furthermore, his Committee had secured an option for 'the purchase by the Nation of this airship'.

In his letter, published in the *Daily Mail* of 22 June, Du Cros further explained that there was no shed or 'garage' in England large enough to house a balloon of the size involved in the transaction. He therefore appealed for the intervention of some public-spirited person to remedy this deficiency by supplying the funds required to build such a shed, in the amount of £5000, so that the project could be completed. At once, the *Daily Mail* offered to contribute this sum if the French designer of the new airship would remain with it in England for one month after its arrival.

Arthur Du Cros was the founder and President of the Dunlop

Rubber Company. He was a pioneer in the pneumatic tyre industry. Among his professional friends was one of the proprietors of the Clément-Bayard Company of Paris, a firm of motor-car builders that had recently begun to construct airships for the French Army. The first of their dirigibles had proven to be a success. Clément now proposed to build a second to be called the Clément-Bayard No. II. It was this airship, a non-rigid,[1] that the Parliamentary Aerial Defence Committee hoped to acquire for the nation.

It was planned at first that the Clément-Bayard II should be delivered in England in September 1909 but there were delays and the dirigible did not arrive until 16 October 1910, after a flight of six hours from Compiègne. Meanwhile, the *Daily Mail* shed was constructed on a site provided by the War Office at Wormwood Scrubs, near London.

The airship, by this time, had made more than forty ascents and had also taken part in the recent French Army manoeuvres. When she was at last secured in the shed at Wormwood Scrubs it was discovered the envelope was in a very porous condition. It was further learned the French Government had refused to purchase the vessel when she was offered to the French Army. The War Office, for these reasons, refused to pay the price Clément demanded. However, Arthur Lee now intervened and secured by private subscription the difference between the amount offered by the War Office and Clément's selling price. When Lee informed the authorities in the War Office about the private contribution it was decided there to accept the money and so complete the purchase. On 28 October 1910 Clément-Bayard II, still in her shed at Wormwood Scrubs, was handed over to the control of the War Office.[2]

Later, the military authorities discovered that Clément-Bayard II regularly leaked so much gas it would cost a very great deal of money to keep her inflated. As a result, the vessel was dismantled and never used again.

This sad and ridiculous chapter in British aeronautical history was the result of blunders made by all concerned in it. Men like Arthur Lee, Arthur Du Cros, and Lord Northcliffe were determined to secure a successful dirigible aircraft for their country. They rejected R. B. Haldane's concept that progress in aeronautical matters must be slow and steady, the result of careful scientific work in the National Physical Laboratory at Teddington. They exerted a good deal of political pressure, in Parliament and in the Press, to force the Secretary of State for War to acquire the French airship. Reluctantly, Haldane acquiesced in their plans. Nevertheless, no one in the War Office sought to obtain any advice about Clément-Bayard II from the Advisory Committee for Aeronautics. Twelve thousand five hundred pounds of public money and 5500 obtained by private subscription were expended in this way, for no valid purpose.

The proprietors of the *Morning Post*, in their patriotic enthusiasm, decided they would not be outdone by the *Daily Mail*. This ambitious resolve resulted in yet another curious episode of early British aeronautical history.

At the end of June 1909 the *Morning Post* launched a National Airship Fund. The object of the Fund was to help to establish Britain as a leader in aeronautical development. In order to achieve this end the *Morning Post* asked members of the public to contribute the sum of £20 000 so that the 'finest foreign-built airship that is obtainable' could be purchased and presented to the War Office. The Earl and Countess Bathurst, controllers of the *Morning Post* and high Tory patriots, presented the Fund with the sum of £2000 so that it could start off in an appropriate way. At once, R. B. Haldane sent the paper a telegram expressing his satisfaction with the new departure.

The purchase of the foreign dirigible was to be preceded by a series of tests which were to be worked out by a committee of experts appointed by the *Morning Post*, and by the appropriate authorities in the War Office. The committee consisted of the chief officers of the three British aeronautical societies, the Aero Club of the United Kingdom, the Aeronautical Society of Great Britain, and the Aerial League of the British Empire. There were three further members of the *Morning Post* committee: Field-Marshal Lord Roberts, Viscount Milner, and Admiral Lord Charles Beresford. Of this trio only Lord Charles claimed any expertise in this field. Roberts, Milner, and Beresford were invited to serve on the committee because they represented 'martial patriotism, statesmanship, and sea-faring supremacy'.

In the months after the National Airship Fund was launched the *Morning Post* regularly printed articles about it while the names of all the contributors were published in the paper, together with the amounts they had donated. Eventually, an arrangement was concluded with the famous French firm of Lebaudy Brothers; and one of their dirigibles was selected for the appropriate tests and trials.

On 26 October 1910 the Lebaudy airship flew from its factory in France and arrived at Farnborough some five hours later. There, the War Office had prepared a shed to receive the vessel. Unfortunately, the constructors had increased the dimensions of the dirigible's envelope without informing either the proprietors of the *Morning Post* or the officials at the War Office. In consequence of this novelty in the arrangements when the landing party began to 'walk' the huge ship into its shed the envelope fouled the roof and was torn quite badly.

As might be expected, intricate discussions and conversations between the Lebaudy firm and the War Office now followed. After negotiation the military decided to raise the height of the shed's girders while Lebaudy Brothers agreed to rebuild the airship at Farnborough. This work was completed in May, 1911.

On her first trial flight after she was repaired the Lebaudy sailed aloft for a time but then collapsed on some houses in the nearby town of Aldershot. She was wrecked beyond repair and was never employed again. The householders of Aldershot demanded the sum of £4000 for the damages they had incurred in the incident but the Treasury refused to pay. This was the final episode in the story of the *Morning Post*'s National Airship Fund.

II

Meanwhile, Lord Northcliffe was determined to press forward with his plan for an aeroplane flight across the English Channel. He would not abandon the idea he had first proposed in 1908.

On 3 June 1909 a leading article in the *Daily Mail* reminded the paper's readers of the prize it had offered for the first successful completion of such a flight. The article pointed out that the *Daily Mail* had consulted the Wright brothers about the prize competition before news of it had been published. The paper further declared that Wilbur Wright could have made the flight on almost any day of the year but he and his brother were so preoccupied with their business affairs they had been unable to find any time for the proposed crossing. Nevertheless, the Wright brothers were convinced the journey, though dangerous, could be accomplished by a determined and capable aviator.

By 29 June the *Daily Mail* was able to announce that three competitors had declared themselves ready to seek the prize. The first of these was Hubert Latham, a young Frenchman of British descent who had flown a monoplane with much success for some months past. The second aspirant was the Comte de Lambert, one of Wilbur Wright's pupils who had made many flights at Pau under the critical eye of his master. He had already rented a large piece of ground at Calais and had constructed an aeroplane shed there, in preparation. He planned to make the attempt for the prize in a standard Wright machine. The third competitor was the famous French flier, one of the great pioneers of flight in France, Henri Farman.

Latham had travelled to Dover, in company with a *Daily Mail* reporter, in order to survey the ground. After a motor tour of the coast he decided he would try to land his machine on top of the Shakespeare Cliff near the town. At this early stage, he was looked upon as the most formidable of the entrants for the prize.

Northcliffe saw to it that the most modern and detailed arrangements were made so that every aspect of the flight could be reported upon the instant. Some *Daily Mail* reporters were stationed in the Terminus Hotel at Calais while others established themselves in the Lord Warden Hotel at Dover. These journalists were equipped with 'Marconi's wireless telegraph'. The permission of the French Government had been obtained to erect one wireless station near Calais while another was installed on the roof of the Lord Warden. From there, messages were to be relayed by telephone directly to the *Daily Mail* news-room.

As these preparations went forward genuine excitement was aroused in England and in France. It was argued that when the channel was crossed by a heavier-than-air flying machine the event would mark 'the dawn of a new age for man'. It would also make obvious Britain's vulnerability to an attack from the air.

All the observers concerned with the competition took care to point out that each of the fliers who hoped to secure the prize was a Frenchman. No English aviator possessed the skill required for a flight of this order nor was there in existence an English aeroplane capable of completing so hazardous a journey.

Hubert Latham attempted the crossing on 19 July in his Antoinette monoplane. Unfortunately, his engine failed in flight and his machine fell into the water. He was rescued by a destroyer the French Government had ordered to accompany him, as an escort.

Latham was preparing for another attempt when on 20 July a new and very formidable competitor entered the lists. This was Louis Blériot, a wealthy manufacturer of automobile lamps and a pilot of outstanding quality.

Blériot was a determined aviator. After years of preparation and practice he fixed upon the monoplane as the most efficient of the heavier-than-air flying machines. In the course of his flying career he suffered many accidents but none was serious enough to deter him from his objects which were to make a name for himself as a great pilot and designer of aircraft.

By 1907 he had begun to emerge as one of the leading fliers in France. Shortly before he attempted the channel flight he was badly burned on one of his feet when the engine of his aeroplane caught fire. He was reduced to walking with the aid of crutches but his resolution was unimpaired by the accident. Such was his fixity of purpose that he proposed to wear a boot on his uninjured foot and a carpet slipper on the other when he made his attempt to grasp the prize offered by the *Daily Mail*.

The weather at Calais was stormy early on the morning of Sunday, 25 July 1909. It seemed impossible for anyone to fly in the frail aircraft of those days. At half-past two in the morning while Hubert Latham, convinced that no flights could be attempted slept in his bed, Blériot arose and observed that a lull in the storm had occurred. With his friends he at once departed for his airfield at nearby Baraques and also ordered the torpedo destroyer *Escopette*, placed at his disposal by the French Government, to weigh anchor and start into the English Channel.

At 4 a.m. Blériot made a trial flight of fifteen minutes around Calais and its environs. This practice convinced him his machine was in order. He had only to wait until the sun came up, a condition imposed by the *Daily Mail*, and then he was ready for the flight. However, he had a question to ask. He requested some of the bystanders to point out the direction he should take in order to arrive at Dover. He was told, in reply, the destroyer would show him the way.

At 4.30 a.m. Blériot took off for the English coast and the *Daily Mail* wireless installation at once flashed the news to the Lord Warden Hotel. He flew at a height of 250 feet and soon overtook the *Escopette*, racing through the waves below. After ten minutes he was shocked to discover he had entirely outdistanced the ship. He could not see it nor could he see the coasts of England or

France. Since he flew without compass or maps he was lost, lost in the air above the sea. Later, he wrote in an article for the *Daily Mail*: 'I touch nothing. My hands and feet rest lightly on the levers. I let the aeroplane take its own course. I care not whither it goes.'

Twenty minutes after Blériot left the French coast he was able to observe in front of him the green cliffs of Dover, the castle, and far away to the west the Shakespeare Cliff where he had proposed to land. However, he was unable to reach his original destination. Instead, he brought his machine down in the Northfall Meadow behind Dover Castle. The journey had taken 37 minutes but in that time, many were convinced, the 'dawn of a new age for man' had been ushered in. On the next day Blériot was honoured at a luncheon at the Savoy where he was presented with his *Daily Mail* cheque in the sum of £1000.

Strangely enough, British aeronautical historians have been unable to agree about the impact of his flight upon the outlook of British society. Percy Walker declared: 'The shock to the British people was comparable to that produced in the United States by the Japanese attack on Pearl Harbor in December 1941. Britannia no longer ruled the waves, since an aeroplane could fly over them.' But Snowden Gamble wrote: 'The military importance of this flight was appreciated more by the Continental nations than by this country. . .'.[3]

III

There were, indeed, mixed reactions in contemporary Britain. The *Daily Mail*, in its leading article for 26 July, offered its readers some disturbing conclusions:

> The *Daily Mail* prize of £1,000 for the first crossing of the Channel by a heavier-than-air flying machine has been won. . . . The event, which stuns the imagination by its far-reaching possibilities and marks the dawn of a new age for man, took place in the early hours of yesterday morning. Thus have the sceptics been put to scorn. When, less than a year ago, we offered our prize there were many who declared it ridiculous to imagine that man would ever fly across the Channel. . . .
>
> Now, however, we have to reckon with the fact that a small and inexpensive machine, which can readily be multiplied by

the hundred, has bridged the Channel. . . . British insularity has vanished. We would not be understood to say that in a few weeks or months hordes of aeroplanes will follow where M. Blériot has led, but his example has shown the way. . . . Men who can navigate the air know nothing of frontiers and can laugh at the 'blue streak'.

. . . The British people have hitherto dwelt secure in their islands. . . . But locomotion is now being transferred to an element where Dreadnoughts are useless and sea power no shield against attack. . . . Such a dramatic moment in human history occurred yesterday.

On the next day, 27 July, the *Daily Mail* published a long special article by H. G. Wells. In it, Wells lamented the terrible failure of his country to keep up with the aeronautical development of other nations. His article was typical of the outlook of many in Britain in the Edwardian era who were convinced their country was falling into decay and would soon be required to pay a bitter price for its failure to maintain itself in the world of great nation-states. The *Daily Mail* identified Wells as the 'Author of The War in The Air'. In his article he explained of Blériot's flight:

It is not the first warning we have had. It has been raining warnings upon us – never was a slacking, dull people so liberally served with warnings of what was in store for them. But this event – this foreign-invented, foreigner-built, foreigner-steered thing, taking out silver streak as a bird soars across a rivulet – puts the case dramatically. We have fallen behind in the quality of our manhood. . . . We are displayed a soft, rather backward people.

Within a year we shall have – or rather they will have – aeroplanes capable of starting from Calais . . . circling over London, dropping a hundredweight or so of explosive upon the printing machines of *The Daily Mail* and returning securely to Calais for another similar parcel. . . .

The foreigner is ahead of us in education, and this is especially true of the middle and upper classes from which invention and enterprise come – or, in our own case, do not come. He makes a better class of man than we do. His science is better. . . . His schools are places for vigorous education instead of genteel athleticism. . . .

It is the vital riddle of our time. I look out upon the windy Channel and think of all those millions just over there, who seem to get busier and keener every hour. I could imagine the day of reckoning coming like a swarm of birds.

The *Morning Post* of 26 July devoted its first leading article to Blériot's great achievement. Its conclusions were almost as pessimistic as those offered to his readers by H. G. Wells:

> the news that the feat has been accomplished will come to most Englishmen as a shock. We are a cautious race, sceptical of all innovation, and not a few have at heart disbelieved the probability of success, at least for many years. But here it is . . . While fully and freely congratulating M. Blériot, it is impossible not to feel a touch of jealousy that this historic achievement has not fallen to the lot of an Englishman. . . . If the problem of flight was of scientific interest or even of commercial utility alone this country could afford to look with equanimity at the progress other nations are making, but since . . . the practical aspect of flight which absorbs attention and completely overshadows all others is the use of flying machines in time of war, no nation, and least of all the United Kingdom, can afford to be one whit behind the foremost of its competitors. It is difficult for the ordinary citizen to realize that the progress made in flight during the last year alone is revolutionizing many of the accepted ideas of warfare, and threatening the traditional safeguards upon which this island has relied in the past.

A leading article in the *Daily Telegraph* of 26 July recognized the gravity of the situation made obvious by Blériot's successful flight. However, the paper's attitude was not entirely negative. It offered sturdy defiance:

> Yesterday the Channel was crossed by monoplane from shore to shore in one brilliant flight. The island, after the lapse of ages, was, in a sense, reconnected with the Continent. . . . The beginning of the end of the old conditions has come.
> . . . We must compete, and compete in earnest . . . our insular immunity in the old sense is coming to an end. Air-power will gradually become as vital to us as sea-power has ever been. . . .

England must wake up in this matter; and of a certainty, she will.

The Times sought to react more placidly. On 28 July it explained to its readers that 'Since the discovery of the New World no material event has happened on this earth so impressive to the imagination as the conquest of the air which is now half achieved'. However, *The Times* emphasized that people were not yet genuinely over-awed by the achievements of those who could fly through the air. Such achievements were not as 'concrete as riches'. Speculation about the future was still idle. The conquest of the air might change the fate of nations but no one, at present, could predict the future with any certainty. Meanwhile, aviation was a 'mere topic of the newspapers . . . like the account of something that has happened at the other end of the world'.

The Radical *Daily News* of 26 July did not permit itself to show any excitement at all in its response to Blériot's flight across the channel. It lamented the fact that such an event should suggest 'all manner of fresh dangers'. A reaction of this kind, the paper declared, was simply 'one more evidence of some absurdity inherent in our civilization that our first thought about aeroplanes and 'dirigibles' should be their possible uses and perils in war'. The *Daily News* assured its readers that the nations would not give up building Dreadnoughts simply because one or two men could fly through the air.

This was not the impression of Field-Marshal Lord Roberts. Although Roberts was in his seventy-eighth year in 1909 his keen interest in military developments and his still lively intelligence enabled him to realize at once that something very important had occurred. When he learned of Louis Blériot's successful flight across the English Channel he sent the French aviator a telegram of congratulations. An account of the telegram was printed in *The Times* for 28 July: 'Lord Roberts telegraphed congratulating M. Blériot on the successful result of his splendid perseverance and pluck and added that it was impossible to imagine what far reaching effects it might have. M. Blériot might be leading the way to great changes in the conduct of future wars.'

There was one further result of Blériot's flight which we should, perhaps, notice at this stage of our account. A curious figure of Edwardian history, Baron de Forest, sought to associate himself, in a minor way, with the achievement of a successful flight across the English Channel.

Shortly before Blériot won the prize de Forest wrote as follows to the *Daily Mail*:[4]

> I think that the extraordinary progress accomplished in aviation during the last year places the practicability of an invasion of this country through the air within the bounds of a perfectly reasonable and not far distant possibility.
>
> I think that the feeling of apprehension increases when it is observed that the candidates for *The Daily Mail* prize at present in the field are all foreigners about to compete on foreign-built machines.
>
> While this should not be considered as an argument for Tariff Reform, but rather as one for protection in the better and more national sense of the word, I think that some good purpose may be served by a prize being offered to the first Englishman who successfully flies across the Channel on an English-built machine.
>
> I beg to subscribe £2000 pounds for that purpose, and I hope that further amounts may follow.

When Blériot succeeded in his attempt de Forest at once sent a telegram to the *Daily Mail* in order to state that he was doubling the amount of his prize. The *Daily Mail* printed his letter and telegram and described the gift as a 'splendid and patriotic offer'.

The Baron de Forest was a man who needed such praise. He possessed tremendous wealth but he was not readily accepted in those exalted circles of English society in which he sought to move. Snubs, insults, and slights were his regular portion.

Arnold Maurice de Forest, known as 'Tuty' to his friends, was educated at Eton and Christ Church; but he was born Bischoffsheim and was the adopted son of the great Austrian banker, Baron Hirsch. He was a hereditary baron of the Austrian Empire. In 1900 he received royal authority to use the title of Baron in the United Kingdom. In that year he entered the Prince of Wales Militia as a 2nd Lieutenant. Later, he contested Southport in the Liberal interest, but without success. In 1911 he was elected Liberal Member of Parliament for West Ham North.

One of his childhood friends was Winston Churchill. With that largeness of outlook which always distinguished him, Churchill remained loyal to de Forest throughout the many serious difficulties he encountered in Britain. In 1914 de Forest joined the Royal Naval Volunteer Reserve as a temporary Lieutenant-Commander, and was attached to the Royal Naval Armoured Car Service. At the same time rumours were circulated in England that de Forest's sympathies were 'entirely with the Germans'. This calumny was instantly rejected by Churchill who wrote to the authorities and helped to see to it that the groundless charge was dropped.[5]

De Forest's initiative of 1909 merits a place in the present history because it reflects the attitude of some of his contemporaries in Britain when they learned of the possibility of a successful flight across the English Channel.

Many were frightened that an air invasion of their country was now possible. They were concerned that all the competitors for the *Daily Mail* prize were foreigners. It was felt in these circumstances there was an urgent need to stimulate British energy and enterprise in the field of aeronautics. The British had languished for too long behind their European rivals.

De Forest's offer of so generous a prize in 1909 was designed to win him a measure of the esteem usually withheld from him by his contemporaries. It also reveals the very serious concern felt by the British with respect to the forward march of aeronautical progress in foreign countries when it was compared with their own dismal record in the new sphere.

IV

Blériot's triumph was merely the culminating development in a long series of events which demonstrated Britain's weakness in the air. Shortly after, on 2 August 1909, there was a major debate upon the subject in the House of Commons. On this occasion the House sat as the Committee of Supply. There was not a large attendance but those who did appear revealed a keen interest in the matter. *The Times* on the next day said of these exchanges: 'The House of Commons yesterday had its first debate on aeronautics, a subject which in future will have an assured place in Parliamentary proceedings.'

The historic aeronautic debate of 2 August furnishes us with further insights into the attitudes and reactions of the British people and their leaders. All realized the novelty of their condition. Some scoffed at the significance of the appearance of devices that could enable men to fly through the air. Others believed the nation was now confronted by an entirely new situation, a situation with which it had to deal or else suffer an irreparable disaster in the future. Almost all the speakers in these exchanges emphasized the military importance of aircraft. They were largely concerned with the Air Defence of their country.

In the absence of the Prime Minister who was visiting Cowes, R. B. Haldane took charge of the government's case. He conducted himself with that blandness of manner which frequently marked out his Parliamentary performances. Nevertheless, we must be clear that his outward conduct masked an inner nervousness.

Haldane was a dutiful son. When he was away from home he wrote to his mother every day of his life. On the day after the debate he informed her: 'I am writing from the Imperial Conference. We are dealing with Naval affairs today . . . I had my hands full of Aerial Matters in the House yesterday. . . .' On 4 August he told her: 'Yes the Aviation Debate went off very well, and the public are approving our plans of slow but sure!'[6]

This conclusion was not an entirely accurate one. A son, of course, all will agree, must be permitted a certain leeway in communications of this order. Nevertheless, in the course of the debate several Members offered strong opposition to the government's aeronautical policies. All sought, within limits, to present their arguments in a non-partisan way but important differences of opinion emerged as the debate went on. It was the beginning of air power politics in Great Britain.

Haldane began the proceedings by informing the Committee about 'what progress has been made in aeronautics as applied to war purposes'. He candidly explained that not very much progress had been made. He agreed great strides has been taken in designing machines that could fly but these were not yet reliable enough to be employed in war.[7]

It was at this time that he revealed that the Prime Minister and the First Lord of the Admiralty had handed over to him the task of organizing military aeronautics in Britain. He explained that after reflection he decided there could be no real progress 'unless we proceed scientifically'. He realized this would be a slow process and that some ardent spirits would object but he was certain this was the best way for the government to proceed in the matter.

Haldane informed his audience that Ministers had turned for advice, in the first instance, to the Committee of Imperial Defence. They had recommended a rigid dirigible for the navy and non-rigids for the army. Aeroplanes presented more difficulties. They might be employed by the army in the future but at present they suffered from certain defects. Aeroplanes could not yet rise high enough for employment in war, he argued, and there were problems of control still to be mastered. Even if 'the British Army had 200 aeroplanes of the best present construction', he said, 'we should not be one bit further on than we are at the present moment'.

Since science had to be brought to bear upon the problem the Advisory Committee for Aeronautics, under the Presidency of Lord Rayleigh, had been established. Haldane made clear that the Germans were proceeding in this way and mentioned the chair of aeronautics that had recently been established in 'my old University at Göttingen'.

He next explained that the task of building up an aeronautical service had been divided between the army and navy who would work in co-operation with the Advisory Committee and with the National Physical Laboratory at Teddington. He also mentioned, in his speech, the *Morning Post*'s Airship Fund and the dirigible shed which had been paid for by the *Daily Mail*. He revealed that the War Office hoped to acquire two aeroplanes before long and that experiments would be carried out with them as soon as they were secured.

At this point in the debate Haldane was interrupted by Sir Gilbert Parker, the Member for Gravesend. Parker wanted the Secretary of State to tell the Committee about the 'special difficulties' which prevented flying machines from being employed as effective and accurate instruments of war.

Haldane's reply to this question was of importance because it revealed the attitude of the British War Office to aeroplanes at this stage of their development. British aeronautical historians have speculated frequently upon this subject. We should notice that the Secretary of State never made a secret of his views about aeroplanes. Sometimes, he was not entirely straightforward in his statements but his opinions were made reasonably clear to all who listened to him. He was quite clear when he replied to Sir Gilbert Parker on 2 August 1909.

In the first place, Haldane said, a pilot could not observe accurately if he was flying in misty weather. Furthermore, observation even with the best available machines was not especially good unless one flew at a great height. If the machine was flown at a lower level it was very vulnerable to fire from the ground. The 'position of the unfortunate aeroplanist', said the Secretary of State, 'sitting on a seat like a bicycle seat, and with perhaps a battalion of marksmen below him, is a very disagreeable one'. Moreover, it was extremely difficult to start an aeroplane. Here, Haldane had in mind the Wright brothers' method of launching their machine into the air with the aid of a device they called a 'derrick'. It was also a problem to land flying machines safely so that 'taking one thing with another and considering the difficulties of starting and landing, in the present state and knowledge of construction, the use of these instruments for war is not very great'.

Despite these problems Haldane assured the Committee he realized it was important that Britain should not neglect aeroplanes. The work would cost money but he felt certain the House of Commons would provide the funds whenever the government requested them. He revealed that it was planned to spend the sum of £78 000 for aeronautical development in the year 1909–10.

When Haldane finished Arthur Lee addressed the Committee. He argued that the Secretary of State was too optimistic in his analysis of what the government had achieved thus far. Britain had allowed foreign countries to take the lead in aeronautical development and this might become very serious in the event of a war. Haldane, said Lee, was 'hypnotized by the blessed word "Science" '. He did not object to scientific investigation but men were actually flying in other countries 'and Frenchmen are landing like migratory birds upon our shores'. It was time for the British to 'get on with the matter in a practical way'. Progress had been delayed because the government had not allocated enough money for the proper development of aeronautics. The French and Germans were already spending very large sums. Speaking for the Parliamentary Aerial Defence Committee Lee said: '. . . I think we are entitled to press upon the Government the necessity of spending whatever money is necessary upon this matter . . .'.

Later in his speech Lee referred to the menace of a night air attack upon the British Isles. Haldane had avoided the subject of bombardment from the air when he made his remarks. Lee would not permit this lapse to pass unnoticed. He said:

> I suppose the right hon. Gentleman would not deny that the moral effect of these airships may be very serious indeed. . . . Their power of appearing over such places as the capital of a country, centres of mobilization, bases of operations . . . at the very commencement of hostilities – indeed, almost before war is declared – at a time when these places are considered to be secure against attack, and dropping explosives and bombs quite at random, must have a very demoralizing effect. . . . This applies particularly to the possibilities of their use at night. The right hon. Gentleman did not say anything about that . . . it is just at night that they are able to come closer to us, and do the greatest damage in those particulars to which I have just referred.

Lee ended his speech with a direct challenge to the conclusions Haldane had offered the Committee. The Secretary of State had said not much had been accomplished in the field of aeronautics. Arthur Lee disagreed. Louis Blériot's flight had shown that 'from a practical point of view an immense amount has already been accomplished in aeronautics'.

Cecil Harmsworth spoke next and made it clear that he agreed with Arthur Lee's criticisms of the government's approach to the problem. The practical side of aeronautics had been neglected and too much emphasis placed upon theory and science. This was also the opinion of Sir Gilbert Parker. He spoke in harsher terms, however:

> With regard to aeronautics, the right hon. Gentleman, I am afraid, has not lived up to his reputation as an expert theorist . . . I do not think the right hon. Gentleman has convinced a single Member of this Committee that the Government has done much in the matter of aeronautics to preserve our prestige as a scientific nation . . . I hope the right hon. Gentleman will redeem his character in regard to this question . . . I have no doubt when he makes his statement next year it will be far more satisfactory, so far as practical results are concerned, than the statement which he has been able to make today.

When Arthur Du Cros intervened he made a remarkable proposal. In this debate of August 1909 he suggested nothing less than the creation of a completely independent Air Service, quite separate from the army or the navy. This theme of a separate Air Force was one that would plague those responsible for Britain's military arrangements for decades after because neither the army nor the navy was prepared to give up their control of the nation's air power. Service leaders and Government Ministers in Britain struggled with this problem in the most bitter and savage way for many years. On 2 August 1909 Arthur Du Cros said:

> I am not sure that one Aerial Department would not be the proper and most efficient method of dealing with this matter. I cannot see why two Departments should be dealing with the same subject, when, by making a joint Department, the experience and knowledge of the Executive Officers of two branches of the Service would be combined to the general advantage of both. I hope the right hon. Gentleman will give that point his consideration. I think, as one who has had a good deal to do with the organization of Departments on a large scale, that it possesses some practical merit.

Du Cros proceeded to launch a general attack upon the government's aeronautical policy. He was especially concerned about dirigible airships. He declared that Britain was the only European Power that did not possess an airship and the only one not engaged 'on a definite constructive programme'. There was not a single private firm in the country capable of building an airship. Neither the Admiralty nor the War Office was spending enough to remedy these grave deficiencies.

The Secretary of State, Du Cros continued, had thrown a 'scientific glamour' over the entire subject but there were practical aspects to it as well, and these were being neglected. France, he pointed out, possessed at least seven dirigibles in commission and Germany had even more of them. Factories in Germany and in France were capable of producing at least thirty of these vessels in each year. In view of these facts, Du Cros said, the Secretary of State had no right to declare that 'no practical results have been achieved'.

Arthur Du Cros concluded his speech by warning the Committee that such airships were quite capable of unleashing an air attack upon British targets:

> It is quite apparent that it will be possible in war time, under

certain conditions, for hostile balloons, to find their way over London. I know a great deal has been written, in a somewhat sensational and alarming strain, on these matters, and I do not wish at all to associate myself with that, or to exaggerate the position at the present moment; but I agree that we cannot afford to shut our eyes any longer to the fact that aerial fleets are being created on a large and comprehensive scale by foreign countries. . . . I do not think in this matter we are pursuing the will-o'-the wisp, as has been said. The problem has been solved; flight has been accomplished . . . by years of toil and expenditure and of patient study. . . . I say that no nation can afford to be without the airship as a permanent weapon. . . . Therefore, I do sincerely hope that the Government, in addition to the labours they have undertaken from the scientific point of view . . . will organize our own factories here at home and train men in this advancing science.

Several Conservative Members followed Du Cros with critical remarks similar to those he had made. It was argued there were practical men in Britain who could be put to work in the field of aeronautics and there were foreigners, trained in the art, who would come to England 'if it was made worth their while' to do so. Other countries were progressing rapidly in the aerial sphere. 'What have we?', said Dundas White, 'We have nothing'.

At this point in the debate Harold Cox, the Liberal Member for Preston, intervened in order to defend Haldane from these adverse comments. However, he did so in his own peculiar way.

Cox was an old-fashioned politician and a strict believer in the *laissez-faire*, Free Trade outlook of an earlier generation of Liberals. He objected to those new trends in British Liberalism which held that it was the business of the State to intervene in the working of the national economy and the function of the government to direct and control the lives of its ordinary citizens, in their own interest. He was an opponent of the Lloyd George Budget. It furnished the State with too much power. In this instance he believed the War Office and the Admiralty were already doing too much in the field of aeronautics. He said:

I have the greatest respect for my right hon. Friend and for his versatile talents, but I do not think the efficiency of this country in the matter of airships or aeroplanes can be secured by trusting entirely to him or to any Department, or to any Government. It is not a reflection on him, but on all Government Departments and all Governments. . . . I hope there will be a strict limitation on the amount of money which is spent in experimental work by the War Office or the Admiralty, and that we shall take advantage of all that has been done by foreigners, and challenge our own men of science and our own manufacturers to go one better.

Another Liberal who defended Haldane in the debate was Alfred Mond, later Lord Melchett. Mond was one of the heads of the great firm of Brunner Mond and Company which eventually merged into the gigantic Imperial Chemical Industries. He was a distinguished chemist and also a man of strong opinions. He bluntly told the Committee he hoped the time had come 'when we have heard the last of the so-called practical man whenever a scientific problem had to be solved'.

Mond did not believe airships would be employed to drop bombs on targets of any kind. He advanced two reasons for this conclusion. First, explosives could not be dropped accurately from a machine that was moving through the air. It would be neither useful nor scientific to resort to such a method of attack. Secondly, he was convinced civilized nations would never attempt to destroy property or kill civilians in so random a way. He told the Committee:

> I do not think that nations in the future are going to conduct their battles by scattering explosives over houses. That is very unlikely to take place. It would be the very reversal of the rules of war which have now existed for a long time. Nobody expects an enemy to bombard a seaside place like Brighton. With civilized nations warfare is not conducted by simply destroying property and killing civilians by dropping dynamite about London, Paris, or Berlin. Such a proceeding would have no effect at all on the ending of the war. No nation would make peace because the enemy was killing civilians. . . . It is entirely contrary to all practice to scatter explosives in the way suggested, and that such a brutal and futile proceeding would be resorted to is one which we need not contemplate.

Mond did not think airships could be developed into useful

instruments of war. They were too inefficient for such a purpose. 'No dirigible', he said, 'has yet crossed the German Ocean and come back.' Rather, he argued, the great need for the future was to study the aeroplane. He urged that funds should be supplied for the purpose and also for the purchase of the best types of aeroplanes already in existence.

Furthermore, he explained, men had to be trained in the handling of these machines. By this time Wilbur Wright was instructing pilots for the United States Army. Alfred Mond wanted the British Government to secure his services for the same purpose. He said: 'Mr Wilbur Wright has given instructions to the Americans in the handling of aeroplanes, and I think we should not be too proud to ask him to come here to do the same for us, rather than wait until we have invented an aeroplane for ourselves . . .'

Lord Balcarres, the Unionist Member for Chorley in Lancashire, could not agree with Alfred Mond's ideas about bombardment from the air. He at once declared:

> I do not share that optimism at all. A German airship carried one ton of explosives. These explosives are carried for a definite hostile purpose. The hon. Member says that in modern warfare towns would not be attacked in that way. The enemy might attack Portsmouth by dropping explosives from airships, and they might destroy thousands of lives and smash up the whole town. If the hon. Member is right in saying that no civilized country would ever do such a thing, is it too much to ask the Government to make inquiries of the great Powers of Europe with the view to adding to the Geneva Convention a clause providing that high power explosives will not be used by aeroplanes for the purpose of war? . . . It is, at all events, worth while that the Government should take that aspect of the problem into consideration.

There were a few further exchanges and then Haldane sought to sum up the results of the debate. He argued that the government had investigated aeroplanes and airships with very great care but, he pointed out, the conclusion was that the prospects for employing them as useful engines of war were 'extremely vague at the present time'. He further stated, and here he reflected the views of the Army Council and the War Office, that 'there is only one practical use so far clearly demonstrated, and that is for the purpose of scouting; and even that is attended with very considerable difficulty'. This opinion that aircraft could be exploited most efficiently as scouts, and not for other military purposes, became a factor of significance in later years.

Finally, the Secretary of State sought to gloss over the differences of view which had come to the surface during the course of the debate. The Committee, he said, would have been more pleased if the British Government had 'got on faster and further'. He candidly admitted that this was his opinion also. Nevertheless, a solid foundation had been established and rapid aeronautical progress could now be expected, as a result of it: 'As soon as we have experimented and purchased', he explained, 'I hope we shall be able to construct . . .'. There were a few further comments and then the historic debate of the 2nd August was over. Later, Haldane tried to create the impression that he was entirely content with the result.

V

In the days after the debate the subject of aeronautics dominated the British Press. The reaction to Haldane's exposition and defence of the Government's policy was mixed. All realized the nation was now advancing across a new technological frontier and for that reason criticism was muted. As *The Times* put it in a leading article on 3 August: 'There was evidently an uneasy feeling in the House that something more is wanted'. In the opinion of *The Times* 'Mr Haldane hardly satisfied the House yesterday that the Government are dealing with the matter in a sufficiently practical way.' *The Times* concluded that advances could be secured, not by employing Government committees or departments, but by stimulating private industry and enterprise.

The *Daily Mail* was more favourable in its comment. Haldane took great care to arrange further private communications with the representatives of the newspaper. The first leader in the *Daily Mail* of 4 August reported: 'We are able to publish this morning further information supplementing that which Mr Haldane gave to the House of Commons on Monday, as to what is being done to prevent England from falling behind in the international aeroplane and airship competition . . .'. In the opinion of the *Daily*

Mail: 'The Government, it seems to us, may be heartily congratulated upon their action so far . . .'.

The *Daily News* staunchly supported the Liberal Secretary of State for War. According to its leading article of 3 August '. . . The Minister disposed with a few quiet, but pulverising sentences of those who have regarded the recent French triumphs in the air as symptoms of the hopeless incapacity of our country to keep up with the times.' In yet another article dealing with the aeronautical debate the *Daily News* of 3 August revealed its powerful Radical bias. 'To a friend of peace', the paper said, 'the debate on aeronautics must have been disheartening.' This was so because most of the speakers were concerned only with the military aspects of aeronautics. The debate was, as the *Daily News* put it, 'a sheer rattling into barbarism'. The paper lamented the fact that nearly £80 000 was to be spent upon the 'latest toys in the nursery of slaughter'. Partisan politics were introduced when the article said: 'Indeed, every suggestion that public money is to be lavished upon adding a fresh terror to civilization was received with approval from the quarter of the House which objects to "paying" the Finance Bill.'

On this occasion the *Morning Post* was less critical of the government's aeronautical policy than it had been in the recent past. Haldane had sought to achieve this result by one of those deft touches which gave him so much pleasure when they succeeded. In his speech he had taken care to praise the 'very valuable work which the *Morning Post* is doing in collecting a fund . . . to present an airship to the nation'. The Secretary of State's remark was printed in the *Morning Post* of 4 August. Although it gave some prominence to the speeches of Arthur Lee and Arthur Du Cros, the *Morning Post* did not condemn the government's aeronautical programme. Instead, it urged all its readers to contribute to the National Airship Fund so that it could become 'representative of as many British citizens as possible'.

Arthur Lee, as Chairman of the Parliamentary Aerial Defence Committee, was not impressed by the immediate results of the debate of 2 August. He later wrote:[8]

I . . . had been appointed Chairman of the new 'Parliamentary Aerial Defence Committee', the first body of its kind to concern itself officially with military aviation. It was organized on strictly non-Party lines and its main objects were to impress the importance of air-defence upon the public, whilst at the same time assuring the Government of parliamentary support in any measures that seemed advisable. The first debate on this subject ever held in the House of Commons commenced on August 2 and, in speaking for the Opposition, I took the Government to task for its procrastination in building aircraft and training the necessary personnel, and for confining its efforts to mainly scientific research. . . .

The Government made little response at the time, but looking back at the aeronautical history of the next few years it is clear that the Parliamentary Committee, which was influentially manned and which never relaxed its pressure, was a useful and powerful factor in the development of aerial defence and of the 'Royal Flying Corps' which was the forerunner of our present Air Force. I occupied the Chair of this Committee from its inception . . . until it was dissolved on the outbreak of war in 1914.

Far away in the United States Octave Chanute, a friend of the Wright brothers and a leading aeronautical authority of the day, also observed the British reaction to the aeronautical achievements of the Germans and the French. He was convinced that the British, in the spring and summer of 1909, were the victims of 'panics' and 'scares' because they did not yet understand the technical capacities of contemporary airships and aeroplanes. For example, on 27 April 1909, he told his friend, Colonel Glassford of the United States Army Signal Corps: 'I am sending you today the April clippings so far as now received. They are mostly gush and slush . . . The British papers seem to have worked up a considerable scare but they will get over it when they have the real limitations of Balloons and aeroplanes. . . .'[9]

After Blériot flew across the channel Chanute offered another opinion to Colonel Glassford. On 3 August he explained to him:[10]

I quite agree with you that there is a reaction coming, just as the extravagant hopes excited by the invention of the balloon in 1784 produced a reaction which lasted 100 years until Renard made a round trip in 1885. . . .

. . . Now that Blériot has 'invaded England by his lonesome' the disquisitions will get more and more numerous, but I think that the British newspapers will get over their scares.

Octave Chanute's analysis of the contemporary situation was quite incorrect. No 'reaction' of the kind he predicted set in. Aeronautical development continued and further marvellous results were regularly obtained. Indeed, the British were now to experience new jolts to their national pride and self esteem because of their inability to keep up with the aerial accomplishments and achievements of their foreign rivals.

5

The Forward March of Aeronautics

The Reims Meeting – The Pioneers of a New Epoch – A Landmark in the Progress of the World – Lloyd George's Disappointment – Lord Northcliffe and R. B. Haldane – The Reply of an Underling – Northcliffe's Anger with Haldane – Work of the Advisory Committee – British Interest in German Aeronautics – The Work of Colonel Frederick Trench – Aeronautical Reports – The Secret Service Bureau to be Utilized – The Work of British Air Propagandists in 1909 – Their Warnings – Appointment of Mervyn O'Gorman – First Phases Accomplished.

I

In the last week of August 1909 the first great international flying meeting was held on the plain of Bétheny, three miles north of Reims in France. The event was organized by the champagne industry which provided several large prizes to induce competitors from various parts of the world to come to France to exhibit their skills as pilots and also to demonstrate the technical qualities of their machines. Thirty-eight aircraft were entered for the various competitions.

Contemporaries and aeronautical historians have agreed that the Reims aviation meeting was an event of consequence. C. H. Gibbs-Smith has written: 'Reims marked the true acceptance of the aeroplane as a practical vehicle, and as such was a major milestone in the world's history.'[1]

A leading article published in *The Times* of 25 August 1909 declared:

> We can imagine . . . how centuries hence the students of social, industrial, and probably also military history will look back to these windy days of August as to the turbulent birthday of a new era of locomotion. Now, for the first time, the flying men

have emerged . . . and definitely grouped themselves as the pioneers of a new epoch in the eyes of attentive humanity.

The *Morning Post* in its first leader for 23 August said of the Reims meeting that it 'must be regarded as one of the greatest events of 1909, and may well form a landmark in the progress of the world'.

The French and Americans dominated at Reims. A solitary English competitor, George Bertram Cockburn, also competed but he flew in a French aeroplane, and without much success. Henri Farman won the prize for the flight of longest duration. He remained in the air for more than three hours. Hubert Latham gained the altitude award by flying to a height of more than 500 feet. One of the best speeds in the air was attained by Glenn Curtiss. He rushed around a marked course at 47 miles per hour. In a competition over a shorter distance Louis Blériot flew at a speed of 48 miles per hour.

Great crowds of excited spectators travelled to Reims to observe these miraculous performances in the air. Each day between forty and fifty thousand tickets were sold. People could scarcely contain themselves when they observed the aeroplanes in flight. On 22 August a Special Correspondent for the *Daily Mail* reported:

> The mad enthusiasm of the watchers was a feature of the day. In the open-air restaurant people stood on tables and shouted for joy, completely carried away by the ecstacy of the moment. It is only once in a lifetime that one sees a great crowd so utterly beneath the influence of one common and uncontrollable emotion.

Several British officers of high rank visited the Reims meeting. Lloyd George who was very interested in the subject of aeronautics also took care to attend. When he returned to London he arranged to speak with representatives of the *Daily Mail* and the *Morning Post* so that he could tell them about his impressions.

Lloyd George told the *Daily Mail*'s Parliamentary Lobby Correspondent:

> It was one of the most marvellous sights I ever set eyes on. . . . I thought the experiments at Rheims a great triumph for the Americans. The Wright machine was the most certain of them

all – the most sure and dependable. . . . A feeling one could not resist was this: How hopelessly behind we are in these great and historic experiments. I really felt, as a 'Britisher', rather ashamed that we were so completely out of it.²

On 25 August the Chancellor saw a representative of the *Morning Post* in his room at the House of Commons. An account of Lloyd George's remarks was published in the paper on the next day. He said of his visit to Reims:

I had not realized how far behind we were . . . our backwardness was the subject of general comment and conversation. . . . We were completely out of it. There you had Americans, easy firsts, with their two machines, the Wright and the Curtiss, you had the French, and you had, I believe, the Italians. It is true that the Germans were not there, but there may have been other reasons for that.

Lloyd George, a brilliant and wily politician always sought to charm everyone, even his most bitter political opponents, if something could be gained as a result. He continued:

I am delighted, therefore, to see that the *Morning Post* is taking up the question of aviation in so practical and energetic a fashion. I am sure you are rendering a real service to your country, and I wish you every success in your efforts. . . .

This interview became the subject of the first leading article in the *Morning Post* for 26 August. The paper grasped the opportunity to warn the country that it was vulnerable to a devastating attack from the air:

if Mr Lloyd George's words help to rouse the minds of his followers to the importance of the question, they will do more for the interests of the country than his stupid abuse of landlords. . . . The Chancellor of the Exchequer says that Englishmen would soon grow enthusiastic if they saw airships in flight. In this he is quite right. . . . As Mr Lloyd George no doubt realizes, it is the dirigible balloon that at present is the best adapted for practical uses . . . The terrible damage which such a vessel could work and the tremendous moral effect its operations would produce render it a most formidable engine of war. An attack could only be effectively met by other airships, and the English people cannot realize too soon the urgent necessity of organizing an aerial fleet. Until this is done the position of the country is one of grave and increasing danger.

Lord Northcliffe, as might be expected, was one of the British visitors at the Reims aviation meeting. While he was there the only British competitor, George Cockburn, complained to him that there were not enough flying grounds in England where British pilots could learn how to fly. Cockburn had been trained as a pilot at the Farman School at Mourmelon-le-Grand. Later, when British officers began to train as pilots he assisted in their instruction, without charging for his work. At this time he hoped the War Office would allow him to erect an aeroplane shed at Larkhill, on Salisbury Plain.

When he returned to England after his visit to Reims Northcliffe took up Cockburn's cause. He wrote to R. B. Haldane in order to tell him about Cockburn's complaint and about his difficulties.

The Secretary of State always took care to respond very carefully to any letters Northcliffe sent him dealing with the subject of aviation. On this occasion, however, Haldane followed a different course. He asked an assistant to compose a reply. On 15 September A. E. Widdows wrote to Northcliffe in order to explain the War Office's policy:[3]

> War Office
> Whitehall, S.W.
> 15th September 1909
>
> My Lord,
>
> Mr Haldane received the letter which you wrote on your return from Rheims, but he does not consider that Mr Cockburn's complaint about the English Aviators' difficulty is justified . . .
>
> The Army training grounds are very cramped and crowded but a survey was made for the purpose of ascertaining where aeroplane trials could be carried out, and one or two suitable places were found in the Aldershot and Southern Commands. In these localities responsible aviators are given permission to erect sheds, subject to certain conditions, and to carry out trials,

and among others permission has been given to Mr Cockburn (from whom, at the time of your letter, no application had been received). . . . Mr Haldane can assure you that the War Office is thoroughly alive to the importance of doing everything that lies in their power to assist aviators.

The Rt. Hon. The Lord Northcliffe
  'The Times' Office
    Printing House Square, E.C.

Your Obedient Servant,
A. E. Widdows
Private Secretary

It may be that this response grated. Lord Northcliffe was convinced the British future would be decided by the further development of aviation. He had already made this attitude perfectly clear to the Secretary of State. Ever since his visit to the Wright brothers at Pau early in 1909 he had written to Haldane in order to tell him about the high importance of aeroplanes. He had even arranged a personal interview with Haldane in order to impress him with their importance. Now, when he brought up a further pertinent aspect of the question he found that the only response came from an underling.

Years later, when Haldane was driven from office in May 1915 the Northcliffe Press contributed to the agitation which resulted in his downfall. At that time his experiences of 1909 were very much in Northcliffe's mind.

On 25 January 1915 Northcliffe wrote to Colonel Repington, the Military Correspondent:[4]

> I was, as you know, the person who first brought the aeroplane to the notice of our military authorities – unsuccessfully. It was because I was so well known in the United States that Mr Wilbur Wright, the inventor of the aeroplane, came to me. Had Lord Kitchener been in England I should have taken it to him, and he would have understood it. As it was, the only person who took an intelligent interest in the matter was Lord Roberts. Lord Haldane was suave and obviously bored.

On 5 February 1915 Northcliffe wrote to Admiral Lord Fisher:[5]

> I remember that when I took the aeroplane to Lord Haldane he showed but a sympathetic, scientific interest in the matter – no more. The only person in England who realized the value was

Lord Roberts, who once said that if it were possible to attain an altitude of a thousand feet the whole of warfare would be altered.

Even after Haldane's disgrace in May 1915 Northcliffe continued to be concerned with his experiences of 1909. On 11 July 1915 he wrote to Arthur Balfour:[6]

> Perhaps you will remember your visit to me at Pau when the Wrights demonstrated the first aeroplane for you. Since then I have, as you may know, followed the subject of flight with a good deal of interest, and my newspapers did something to impress the necessity of aviation upon the country.

Sadly, Northcliffe's annoyance with Haldane for his failure to secure the Wright aeroplane in 1909 developed, as the years passed, into a bitter hatred. By the autumn of 1916 Winston Churchill and some of Haldane's other friends sought to restore his reputation and thus undo the grievous wrong which had been inflicted upon him. Lord Northcliffe, a man who knew how to bear a grudge, tried to thwart this movement. At the end of October 1916 Edmund Gosse, the distinguished literary figure, wrote to Lord Haldane:[7]

> I have some information to give you which is very unpleasant. The hounds of Hell are being laid on to your track. A friend of mine who is of importance in the journalistic world tells me that a few days ago Northcliffe gave a luncheon at the Aldwych Club to prominent journalists. He made a speech entirely directed against you. After the bitterest diatribes he adjured all those newspapermen to see to it that you never regained political power. . . . He told them that there was a campaign afoot to reinstate you, but that they must all combine by all means known to them to defeat it. He assured them that you were the greatest enemy to the English state.
> One of these journalists . . . asked him, since he was in possession of all this evidence against you, why he did not rise and expose you from his place in the House of Lords . . . he answered that this would lead to controversy, which would be dangerous and unsatisfactory. 'What you have to do', he went on, 'is perpetually to insinuate into the public mind suspicion

and hatred of Lord Haldane, so that the moment there is any question of his reappearance in public life, public opinion may automatically howl him down'.

I think you should make this odious conspiracy known amongst your friends.

II

After Blériot's brilliant flight across the English channel and after the notable events of the aviation week at Reims the British people, with good cause, were disappointed with the aeronautical condition of their country. The Prime Minister's announcement in the House of Commons in May held out hopes for a brighter future but it could not affect the present situation in any way. Several pioneer aircraft constructors, men who were to win fame and fortune in the industry, were already at work in England. These aspirants were buoyed up when the first British flying meetings were held in October 1909 at Doncaster and at Blackpool. Nevertheless, as one expert said of them: 'All were amateurs compared with France.'[8]

By 1909 three officers in the British Army, all artillerymen, had begun to experiment with aeroplanes. Much of their work was carried out as a result of their own initiatives. Their superiors, for the most part, showed only a tepid interest in what they sought to accomplish.

Captain J. D. B. Fulton, an officer of the Royal Field Artillery, was greatly excited by Louis Blériot's flight in July. He was a capable engineer. He resolved to build an aeroplane for himself but he found the task too difficult. Undeterred, he purchased a Blériot monoplane with his own funds and taught himself to fly it. After the Reims meeting one of his colleagues, Captain Bertram Dickson, bought a Farman biplane and learned to fly at Farman's school in France. The third of these innovators was not a regular army officer. He was Lieutenant Launcelot D. L. Gibbs of the Royal Artillery Militia. Gibbs was a brave young man. He had been Lieutenant Dunne's pilot during the hazardous Blair Atholl experiments of 1908. His disappointing experiences in Scotland did not chill his ardour for the new technology. He went to France and gained his pilot's certificate in the school at Châlons-sur-Marne. Later, he became a test pilot of international renown.

These bold officers, each in his own way, sought to demonstrate to the army the military possibilities of the aeroplane.

Meanwhile, the Advisory Committee for Aeronautics had begun its work under the supervision of Lord Rayleigh and Dr Glazebrook. These high experts lost no time in coming to grips with their new duties.

One of the first matters they took up was the problem of defence against attack from the air. So much had been said and written in Britain about the menace of dirigibles that a paper dealing with the subject was composed for the Advisory Committee as early as 6 July 1909. The author of this document was H. R. A. Mallock, the brilliant consulting engineer who was one of the original members of the Committee.

It is clear from Mallock's introductory remarks that the technical authorities assigned the task of improving Britain's aeronautical position were convinced that their country was in danger because of the physical capabilities of dirigible balloons. In their opinion the nation could be subjected to an attack from the air by these vehicles. Mallock began his paper by pointing out that a method of destroying balloons was a question that was 'likely to become one of some importance'. He explained:[9]

> The question as to what kind of projectile will best serve for the destruction of balloons is likely to become one of some importance. For this purpose the requirements are somewhat different from those desirable for the attack of boats or troops.
>
> The balloons (dirigible or otherwise) are extremely frail structures, requiring hardly any work and very low velocities for their penetration.
>
> They are made of material which can be readily burnt through and filled with inflammable gas, which if once ignited at any place will ensure the complete destruction of the whole fabric.
>
> Such conditions indicate that the projectiles used for attack should be shells which will scatter their contents in a large number of small burning pieces, and in what follows the questions considered are:
> (a) What is the best size for the burning pieces?
> (b) How long should the pieces continue to burn?
> (c) What is the best distribution in space of the pieces after the ignition of the shell?

(d) What horizontal velocity should be left in the pieces after leaving the shell?

The rest of Mallock's paper was technical. However, it was clear he was concerned with the construction of a device designed to destroy 'rapidly moving balloons or airships'.[10] His object was to make it feasible to burn them up, a straightforward solution to a problem that was fast becoming an issue of national consequence.

By 5 August 1909 the Advisory Committee for Aeronautics was able to publish an interim report. This document revealed the Committee had held four meetings. Two of these took place at the War Office, one at the Army's Balloon Factory, and one at the National Physical Laboratory. The Committee's work, it was explained, had been preliminary in character up to this point.

A list of proposed experiments had been drawn up. These included: (1) General questions in Aerodynamics; (2) Questions relating especially to Aeroplanes; (3) Propeller Experiments; (4) Experiments on Motors; (5) Questions relating especially to Airships; (6) Questions relating to Meteorology. It was also revealed that the Committee had communicated with the Aeronautical Society, the Aero Club, and the Aerial League and invited suggestions 'as to the directions in which the assistance of the Committee would be of most value'. These societies had agreed to consider the request and to lay their views before the Committee at an early date.

The members of the Committee were extremely active. When the Admiralty applied to them for advice about the propellers for the airship being built at Barrow Sir George Greenhill, the mathematical expert, drew up a paper designed to assist the naval authorities.[11] Later, in November 1909, it was arranged for the Committee to visit the works of Vickers, Sons and Maxim at Barrow 'to see the Admiralty airship now under construction there'. Lieutenant Trevor Dawson, managing director of Vickers, invited the members to dine with him when they made their visit.[12]

In this way the first steps in a new departure were taken. Although the Liberal Government was severely criticized for its failure to keep up with the French and the Germans in the sphere of aeronautics in 1909, this was the beginning of the fulfilment of R. B. Haldane's great dream. The Advisory Committee was already actively engaged in 'building up the structure of the Air Service on a foundation of science'.

III

As a matter of course certain British authorities began to develop an interest in German aircraft as soon as they became successful. Indeed, as early as December 1905 the British consul in Stuttgart took care to notice what was taking place at the nearby Zeppelin works.[13]

In January 1909 the British Military Attaché in Berlin, Colonel Frederick Trench, began to send a regular series of reports about German dirigibles to the War Office in London. Some of these reports were supplemented, where appropriate, with photographs.[14]

Colonel Trench was a remarkable officer. He was educated at Geneva University, the Royal Military Academy, and at the Staff College. He served with distinction in the Zulu War and in the Boer War. In 1905 and 1906 he was attached to the German headquarters during their operations in German Southwest Africa. He so impressed the Germans during this period of service that they presented him with several decorations and he made many friends in the officer corps of the German Army.

In 1906 Trench was sent to Berlin as Military Attaché. In this post where he was in a position to observe his German friends more closely he became seriously alarmed. He was so agitated as a result of his experiences in Berlin that in 1910 he returned to England and retired from the army. There, he became an advocate of conscription, or National Service as contemporaries called it. He decided it was his duty to warn his countrymen about the 'German Peril'. In 1914 he returned all his German decorations as a 'protest against their barbarities to women and wounded'.[15]

From January 1909 onward Colonel Trench paid careful attention to the development of dirigibles in Germany. He asked his military friends about them; he travelled to different parts of the country in order to see them for himself; he visited the factories in which they were built; and he took notice of accounts about them published in the newspaper Press.

As a result of this activity he was able to send detailed reports to London. For example, on 10 June 1909 Trench drew up a long paper entitled 'Dirigibles' and marked it 'Confidential'. He began this account by referring to a recently published article by General Rohne, 'the great artillery expert'. In Rohne's opinion dirigible balloons had become an 'indispensable' weapon for purposes of strategic reconnaissance. The General was not certain if they could be employed to bombard ground targets from the air because it was not yet known if they would be able to strike with any degree of accuracy.[16]

The firm of Krupp, Trench's account continued, had already constructed four types of guns for use against dirigibles. He reported also that infantry had been employed to practice firing at balloons at Jueterbog, in the previous May. Their fire, the authorities decided, was not effective.

Trench revealed that a 'School for aeronauts' had been formed at Friedrichshafen and that it would begin its work on 1st October 'giving a three years' scientific and practical training to men for the crews of dirigibles'. The first year at this school would be devoted to theoretical instruction. In the second year the men would construct balloons and aeroplanes and in the third year they would be taught to fly them.

The Prussian War Office had recently taken 'an important step', Trench further revealed, by signing a contract with the firm of Fleischer for the erection of plant to produce oxygen and hydrogen. These plants were to be established at Cologne, Metz, Berlin, Hamburg, and at other places. They were to be used for commercial purposes in peacetime but the military authorities would take them over 'on the outbreak of hostilities'. 'No reference to this contract', Trench wrote, 'has been permitted to be published by the press'.

Trench also reported provision was about to be made for an increase in the number of 'balloon troops' in Germany. A '2nd Battalion' would be created and assigned to the 'western frontier and the North Sea Coast'.

Trench knew Colonel Capper well. Occasionally, he wrote directly to Capper to inform him about aeronautical developments in Germany. Early in September 1909 Trench sought to arrange for Capper to visit Germany so that he could observe several types of German dirigibles for himself. He realized great care would have to be exercised whenever this was done. One of his German friends had promised, in confidence, to try to take him, Capper, and Admiral Bacon to see a new balloon at Mannheim. However,

Trench was apprehensive. He explained to Capper: 'I wanted him to take you & not me but he seems to want to take me on account of my German. What do you think of the idea? He seems averse to our being conspicuously foreigners which I don't quite like.' Trench informed Capper in this letter that he had recently had a 'good view' of the latest Zeppelin airship and that he had met Orville Wright who was in Berlin with his sister to demonstrate Wright aeroplanes and to train a flyer for the German Wright company.[17]

The German authorities showed Orville and Katharine Wright every courtesy while they were in Berlin in 1909. Their hotel accommodation was provided for them without charge. They met the Emperor, the Empress, the Crown Prince, and several high military officers. Colonel Trench was concerned with what the Germans did with aeroplanes as well as with their work with dirigibles. He thought Orville's activities in Berlin were so important that he sent a formal report about them to the War Office in London. Orville collected almost $50 000 from the Germans during the course of this visit, tangible evidence of their interest in Wright aeroplanes, but Colonel Trench could not know this. Instead, he reported to the War Office:[18]

> Mr Orville Wright has made some half dozen flights during the past week. One of them lasted 53 minutes. . . . On Saturday there was a big crowd. . . . No enthusiasm is shown by the German crowd who refer to the 'Amerikaner' and condemn his apparatus as a sport-zeug – and of no use for war. Any cheers raised on the Tempelhoferfeld are recognizably American. Wright has been nearly 3 weeks in Berlin.

In 1909 the General Staff in the War Office began to compile reports concerned with military aviation in foreign countries. These reports, marked 'Confidential', were produced each year between 1909 and 1913.

The first report was dated 15 November 1909. It revealed that the possibility of employing airships in an offensive capacity was already being considered in Germany. Reference was made to articles published in the German Press which urged the 'invasion of England in airships'. The German military authorities, for their part, believed airships would prove to be most valuable for the purposes of strategic reconnaissance. The report pointed out that at the end of October and the beginning of November 1909 'important aerial manoeuvres' had been held in the Rhine valley near Cologne. The details of these trials 'were kept rigidly secret' so that not much was known about them. It was said that experiments in wireless telegraphy, bomb dropping, and 'airship chasing' had been carried out. The report offered an interesting conclusion about Great Britain and the new airship station at Cologne:[19]

> With regard to the importance given to Cologne as an airship station, it is to be noted that this town is almost the nearest point to England.

The report for 1910 offered further information on the subject. It declared:[20]

Development of Powers of Offence

Nothing is known definitely about the question of offence from dirigibles in Germany further than that important experiments with regard to dropping explosives from airships are stated to have taken place at the Cologne airship manoeuvres during the past two years. The results have been kept secret.

It remained for R. B. Haldane to suggest that these probings into German airship activity should be carried to a logical conclusion. On 14 July 1910 the Committee of Imperial Defence met in order to discuss 'The Bearing of Recent and Probable Developments in Aerial Navigation on Imperial Defence'. Asquith, the Prime Minister, served as chairman at the meeting. Reginald McKenna, Sir Edward Grey, Winston Churchill, Admiral Sir John Fisher, General Sir W. G. Nicholson and others attended.

After Asquith began the proceedings Haldane said he

> believed that the German Government had recently been conducting experiments with a view to determining the best method of attacking airships. He also understood that experiments had been made in dropping bombs from air-ships, with a view to acquiring sufficient precision to enable these bombs to be dropped down the funnels of war-ships. . . .
> It was important that we should know the results of such experiments.[21]

The Committee of Imperial Defence agreed with the Secretary of State's observation. It was decided at this meeting that:[22]

(*Conclusion*)
The Committee recommend that the Naval and Military Intelligence Services should take special measures to watch foreign trials, and collect and classify all available information on the subject of aerial navigation, utilising the Secret Service Bureau as may be necessary.

IV

During the year 1909 all those who believed that the development of airships and aeroplanes would work serious effects upon the British future laboured to awaken their countrymen to a proper appreciation of the new situation. They were confronted in their propaganda by lack of interest, lethargy, and an inability to comprehend the importance of what was taking place in the aeronautical world.

These 'air enthusiasts' sought to carry their cause forward with articles in the newspaper Press; they addressed learned societies and Service organizations; they spoke on public platforms; and they exerted whatever influence they possessed in Parliamentary circles and in the councils of the Government. Some of them were untiring in their efforts. Some sought to warn while others tried to reassure their audiences about the new technology. A chief theme of all their arguments in 1909 was that Britain could now be attacked in a way that had been impossible before Count Zeppelin developed his airship, in a way that had been unthinkable before the Wright brothers demonstrated the qualities and possibilities of their invention.

In March 1909 Colonel G. F. Stone addressed the Royal United Service Institution on the subject of 'Defence of Harbours against Naval Airships'.[23] Colonel Stone was an artillery expert who had already made a serious study of the problems involved in creating a system to defend the country against aerial attack. His talk and the comments on it offered by the other officers present are illustrative of how men reacted to this novel subject in 1909. They merit attention.

Colonel Stone made clear to his audience that airships could be employed for the purpose of 'Dropping bombs on certain areas which cannot effectively be attacked by ships' guns'. He argued that the time had come for experiments to be made which would provide accurate data about the weight of the bombs to be employed and about the height from which they should be dropped in order to secure the most effective results.

He suggested that airships would carry out their reconnaissance by day but that they would discharge their bombs at night since this method would permit them to approach their targets more closely and thus assure a greater number of hits. Coast defence guns would be of little value in repelling such attacks, Colonel Stone explained, because the airship 'would naturally run inland out of range, and then turn and approach the harbour from the land side'. 'The Germans', he said, 'have tackled the question for land operations, and it is time we did the same for harbour defence.' Colonel Stone believed that artillery defence against airships could be regarded only as an auxiliary method. Aeroplanes and airships would have to be organized in order to form a first line in any proper system of defence against aerial attack.

Colonel Capper was invited to comment on these remarks. He pointed out that airships would not confine their attacks to places on the coast. In the future it would be quite easy for them to bombard a 'vital place inland'.

When Vice-Admiral Sir Charles Campbell was invited to speak he expressed his amazement at the nature of the discussion. Campbell had joined the Royal Navy more than fifty years before and had served in it with brilliant success. He said:

> What I should like to remark about first of all is the extraordinary change that has come over this country with reference to the question of the conquest of the air. In this Institution and everywhere else we had buckets of cold water poured on every suggestion that was made and every idea that was started which had as its object the raising of a body off the earth. I wish to call your attention this afternoon to the extraordinary difference which now exists in the minds of the public when this subject of the conquest of the air is mentioned or discussed. People are beginning to believe in it. When we hear a lecture like this, and the remarks of a man like Colonel Capper, who is a pastmaster of the subject, we know there is something in it and the fact that it is possible is now beginning to dawn on the people. . . .

The question has been raised this afternoon in the paper as to the attack of airships on harbours. . . . There is no doubt that, whether we believe it or not, harbours may be attacked.

At this Major Sir Alexander Bannerman, Bart., rose in his place in order to offer a comment of his own. Bannerman, an officer of Royal Engineers, was an experienced balloonist. He knew little about aeroplanes or airships. However, he explained to the company that he did possess a 'certain amount of practical experience in ordinary gas-bag ballooning'. He knew from this experience that it would be difficult to hit any target from the air and he therefore urged the assembled officers to bridle their enthusiasm. He doubted whether 'one shot in 10,000 will, from a height of 5,000 feet, hit a battleship'.

Even Lord Northcliffe, the powerful champion of the aeronautical movement, was sometimes forced on the defensive in his advocacy. On 28 August 1909 the leading article in his *Daily Mail* was entitled 'A Complaint about "The Daily Mail".' The article revealed that thousands of people had complained the *Daily Mail* was too 'full of aeroplanes' and that it neglected other items of news in its enthusiasm for the new technology. Every post brought letters grumbling about the *Daily Mail*'s aviation reports. The paper responded by pointing out that those who objected to its policy 'have the very human failing of lack of imagination and power to comprehend what is proceeding'. Farman, a French citizen, had just flown a distance of more than 100 miles in France but the *Daily Mail*'s prize of £1000 offered to the first British pilot to fly one mile on a British course 'still awaits a winner'. A 'new age has dawned', the leader concluded, and British citizens had to be made aware of what was taking place.

On 18 November 1909 Colonel Capper delivered a lecture at the Royal Artillery Institution. The subject of his discourse was 'The Military Aspect of Dirigible Balloons and Aeroplanes'. Major-General H. E. Belfield, Commanding the 4th Division, served as chairman at the meeting.[24]

Capper began cautiously enough. He pointed out that the development of airships and aeroplanes was still in its infancy and that no certain grounds of actual experience existed on which opinions could be based. Nevertheless, the progress already made had forced all nations to consider how the new instruments of war might be employed. He begged his audience not to dismiss their military potential too readily. 'Britishers as a rule', he said, 'have all the faults, as well as the virtues, of intense conservatism. We have little as a race of that valuable quality imagination, and in considering the importance of any new invention, we are apt to minimize, rather than exaggerate the purposes to which it may be adapted.'

Having cleared the ground in this prudent way Capper told his audience he believed that the chief use of dirigibles in war would be for the purposes of reconnaissance. He also reckoned that they could be employed for air attack on a variety of targets. He said:

> At night when airships can keep low down, I am of opinion they could be used with deadly effect on transports, and with considerable power of destruction on bridges, dock gates, arsenals, etc., whilst incendiary bombs may play havoc in supply depots and stores, and airships so used would cause both moral and material distress to camps, bivouacs and horses, and undoubtedly would harass, by the need for constant watchfulness, posts on the lines of communication and at places far away from the field army.

Capper did not think the aeroplane could yet be looked upon as a useful instrument of war. In the future, however, it would be able to do everything that a dirigible could: 'every airship of existing pattern will, in my opinion, be driven out of the field, before ten years are passed, by the aeroplane'. No hostile airship would be able to exist in the presence of an aeroplane of the future. Colonel Capper predicted that the British would soon be obliged to maintain a 'very considerable bevy of these vessels, and train large numbers of men to fly and fight them'.

In the discussion that followed Colonel Stone and Colonel von Donop expressed warm agreement with Capper's views and opinions. It remained for Major-General Henry Arbuthnot to introduce a theme that became more and more urgent in the minds of many members of the British public during the course of the year 1909.

Arbuthnot, an artilleryman, had fought in the Crimean War and in the Indian Mutiny. At this time, although he was retired from the army, he was Treasurer of the Aerial League of the British Empire. He warned the gathering that in the field of aeronautics 'England is behind every other nation either in Europe or

America.' He pointed out that the great leader in the new technology was Imperial Germany. 'I dare say you will be astonished to hear', he said, 'that at this present moment the Germans have got sixteen dirigible airships in existence and that in the year 1910 they are going to build six more. . . . It is something enormous compared with what we have done.'

The General argued that in England those who wanted to see a positive aeronautical policy established 'have got very little chance of getting help from the Government'. The Aerial League had been founded 'for the purpose of starting a propaganda' designed to awaken the country to the dangers which menaced it from the air. When General Arbuthnot announced that he would take up subscriptions for the Aerial League at this meeting his remark was greeted with applause.

This note of warning was also sounded at the end of the year by no less a figure than Field-Marshal Lord Roberts. On 8 December he presided at a meeting held at the Royal United Service Institution. At the meeting Major Baden-Powell read a paper entitled 'How Airships are Likely to Affect War'.

In his remarks Lord Roberts said that 'the aerial machine had come to stay'. Unfortunately, Englishmen had not done much with such machines but had been content to wait so that they might 'benefit from the experience of other nations'. Airships, he argued, would be of the greatest value in the next war. Since no one knew when that war would begin it was very important for England to have its own machines and its own trained men ready for action with them. People in Britain, he said, were too apathetic. The 'valour of ignorance' pervaded the entire country: 'They would not be so valorous if they knew what was in store for them. They did not believe in anything happening but were content to remain in ignorance.' When the magazine *The Aero* published these remarks in its issue for 14 December its article was entitled: 'Another Warning: Lord Roberts's Attempt to Awaken the Country.'

V

In the War Office R. B. Haldane was impervious to speeches and declarations of this order. He paid little or no attention to the arguments of the propagandists who sought to hasten him forward in his task of creating an efficient military air organization for the country. When the great aeronautical debate of 2 August revealed that air power politics had begun in England he was almost unmoved by those who criticized the government's cautious policy. He had set his hand to the work of 'building up the structure of the Air Service on a foundation of science'. No one and nothing would force him to tamper with the programme he had fixed upon to achieve this object. The influence of the 'air enthusiasts' complicated his plans and made his work more difficult but he hoped, nevertheless, to be able to advance along the course he had proposed to the Prime Minister in the early months of the year.

The first step was the creation of the Advisory Committee for Aeronautics. Lieutenant Dunne and Samuel Franklin Cody had been removed from their places at Farnborough. Amateurs of their calibre would be allowed no room in the new scientific organization Haldane contemplated. One further preliminary step remained to be accomplished in 1909. In the historic debate of 2 August Haldane had said:[25]

> We have not yet found a head of the construction department of the Army balloon factory. We have been looking about for him very carefully within the last few weeks, and we hope to get the very best man that can be found to put at the head of that work. He will be a practical man, a civilian and an engineer.

Colonel Capper could not fit such a description. The Secretary of State applied to Lord Rayleigh for assistance in choosing his replacement. In response, Rayleigh and Sir Charles Hadden recommended a brilliant young engineer for the post. His name was Mervyn O'Gorman.

In 1909 Mervyn Joseph Pius O'Gorman was already a successful consulting engineer who exercised his talents in the field of automobile engineering. He had been trained at University College, Dublin, and at the City and Guilds Institute, London. His experience included work in the engineering industry in France, as well as in England.

O'Gorman's was a remarkable personality. He possessed wit, courage, and imagination. He has been described as a 'thruster'. It is certain he made warm friends as readily as he made bitter enemies. He wore a gold-rimmed monocle and a fierce moustache.

He dressed immaculately. Above all else he was a capable engineer who was eager to devote all his energies to his new post.

In October 1909 R. B. Haldane selected him to replace Colonel Capper as Superintendent of the Balloon Factory at Farnborough. An 'engineer-scientist', O'Gorman could bring into play in this post qualities and capacities which Colonel Capper lacked. He was more fitted than Capper to co-operate with the scientists of the Advisory Committee for Aeronautics and with the technical experts in the National Physical Laboratory at Teddington. At first his Farnborough appointment was for only three days a week. Eventually, it became a full-time one.

The Secretary of State desired to turn Farnborough into a national institution controlled by a civil servant who would not be dominated by officers like General Sir William Nicholson and those in the War Office who were entirely subservient to him. Years later, O'Gorman wrote: 'Lord Haldane warned me of opposition and obstruction from *everybody* – not least by the military side. If you can get the archives you will find that I was placed direct under the Secretary of State with access to him to avoid this difficulty.'[26] These statements may have slightly exaggerated the nature of the relationship but they do reflect Haldane's attitude in 1909. Percy Walker has stated of this development: 'The single action that inaugurated the revolution in Balloon Factory affairs was the appointment on 19 October 1909, of Mervyn O'Gorman to be Superintendent of the Balloon Factory, and to make him directly responsible to the Master-General of Ordnance at the War Office.'[27]

Haldane's work to create an efficient military aeronautical service for his country was far from complete. With this appointment, however, the first phases of it were accomplished.

6

The Origins of British Air Defence

The Initiative of Sir Arthur Paget – No Response in the War Office – A Meeting at the Admiralty – Danger of Overhead Attack – New Problems for the Royal Navy – Cordite Factories and Naval Magazines Exposed – The Admiralty Approaches the War Office – Negative View of Sir William Nicholson – The Home Ports Defence Committee – Its Recommendations for a Home Air Defence System – Sir Charles Ottley's Note – R. B. Haldane's Comments – The Secret Service Bureau to be Employed – Neglect of the Air Defence of Great Britain – Political Developments in 1910 – Aeronautical Policy of R. B. Haldane – Its Critics – The Aerial League attacks the Liberal Government – The *Daily Mail* Prize – Significance of the Event.

I

While Mervyn O'Gorman settled in at the Balloon Factory, which he was to help develop into one of the great scientific institutions in the world, high authorities in the army and the navy began to turn their attention to the subject of air defence. Generally, the service leaders in Britain did not encourage aeronautical activity but they were now forced to consider this new aspect of their responsibilities as defenders of the nation and Empire.

O'Gorman, described by an expert as the 'great man who founded Britain's scientific approach to aeronautics',[1] knew his exact role in R. B. Haldane's developing plans for the Air Service. Immediately following his appointment to Farnborough he was made a member of the Advisory Committee for Aeronautics so that he could associate, as a matter of routine, with the scientists who served on it and with the expert technicians in the National Physical Laboratory at Teddington.

The service chiefs faced a more difficult task. They had to come to grips with problems that were entirely novel for them.

The agitation of the 'air enthusiasts'; the debates in Parliament;

the articles in the newspaper and periodical Press and the lectures to learned societies and service organizations now produced an effect upon the mind of Lieutenant-General Sir Arthur Paget, the General Officer Commanding-in-Chief, Eastern Command. At the end of October 1909 Paget wrote to the War Office to sound a warning about the dangers to be expected as a result of attack from the air. He was one of the first British Commanders to take such action.

Paget has not been treated kindly by historians. Several of them have reproduced a curious statement about him, published in his obituary notice in *The Times*: 'Had he only devoted to military study a fraction of the time which he gave up to the observation of trees and shrubs, he might have ranked as a learned soldier'.[2] In fairness to his memory his early warning of October 1909 must have a place in this history. He approached the War Office authorities in suitably cautious tones:[3]

SECRET
From, The General Officer Commanding-in-Chief, Eastern Command.
To,   The Secretary, War Office S.W.
                           London, 23rd October 1909
Sir,
The question of the damage which might be caused to magazines by explosives dropped upon them from hostile airships or aeroplanes has recently been occupying my attention.

In this connection, I may point out that magazines, such as those at Chattenden, which cover many acres of ground, offer a comparatively easy target for missiles of this description.

The stage of development to which aeronautics have at present arrived, does not, in my opinion, warrant expenditure upon existing magazines. I would, however, suggest that it is a matter for consideration whether important magazines, should not, in future, be constructed at such a depth below the surface of the ground as to preclude the possibility of their contents being detonated by explosives dropped from air vessels.
            I have the honour to be,
                Sir,
                    Your obedient Servant,
                        A. Paget Lieutenant-General
                        Commanding-in-Chief, Eastern Command.

The Director of Military Operations in the War Office, J. Spencer Ewart, was so impressed by Paget's warning that he prepared a minute dealing with the subject for Sir William Nicholson, Chief of the General Staff. He, too, was restrained in the advice he offered about defence against air attack:[4]

> This is a question which may become important in the future. The Germans and French are both, we hear, making experiments in regard to throwing explosives from airships, but we know little, as yet, as to the possibilities of such a mode of warfare. I certainly think that the question ought to be carefully considered in regard to future types of magazines, but whether we should be justified in spending large sums of money upon reconstruction seems doubtful, in view of our want of knowledge as to the possible development of aeronautics.

Sir William Nicholson offered no response to these reflections. He was still convinced that aeronautics would play practically no role in the warfare of the future. In consequence of his negative attitude Sir Arthur Paget's initiative of October 1909 produced no positive results.

II

However, the Admiralty looked upon the matter in an entirely different way. In order to be able to carry out its duties the Royal Navy depended upon its several naval magazines, cordite factories, and dockyards. At a meeting held at the Admiralty on 4 January 1910 it was recognized that these vital facilities were also static targets now, for the first time, vulnerable to 'overhead attack'.[5]

The naval officers who attended this meeting were the Director of Naval Intelligence, the Director of Naval Ordnance, the Assistant Director of Torpedoes and the Superintendent of Ordnance Stores. They came together to discuss the recent progress made by foreign governments in 'the development of the offensive powers of airships' and to consider what protection was needed at naval magazines, cordite factories, dockyards, and 'other vulnerable points' to guard against overhead attack. The conclusions reached by these high technical authorities were sombre:

> It is well known that Foreign Powers have for a long time been devoting considerable attention to airship construction . . . and that . . . they are at present ahead of us. Germany in particular possesses already a powerful squadron of airships . . . Furthermore the recent tests made with Zeppelins and other German airships of smaller size show that, so far as distance is concerned . . . they are quite capable of reaching this country.
>
> It is unnecessary to emphasize the paralyzing effect which the destruction of our cordite factories, of which there are only a limited number, and magazines would have on the action of the Navy especially, if, as is probable, it took place on or before the outbreak of war. Moreover, the majority of such factories etc. are in the neighbourhood of London or in more or less exposed positions near the Thames estuary.

The officers involved in this discussion at the Admiralty realized they could only arrive at tentative suggestions or recommendations because of the novelty of the subject and the paucity of their own experience. Nevertheless, they had prepared for the meeting with some care since they recognized that the development of aircraft had already created serious problems for the Royal Navy, problems not encountered before in the entire history of the naval service.

They also stressed the urgency of the matter. They made it clear that:

> Under the circumstances, it is considered that the possibility of overhead attack on vulnerable points . . . cannot be neglected and that early steps should be taken to provide some form of defence to meet it. This is all the more important because special forms of guns, projectiles and fuzes will have to be designed for the purpose, experiments abroad having shown that unless the envelope can be destroyed by the explosion of the gas inside it the airship will not be brought rapidly to the ground.

They further decided that attack by airships would probably take one of two forms. High explosive charges or bombs might be dropped on the target or else a light gun carried by the airship might fire high explosive bullets or incendiary projectiles.

In the opinion of these naval authorities an unwieldly airship would not be able to drop bombs on a small target with any degree of accuracy but a large target might be hit effectively. They concluded that 'except with a large target such as a dockyard, this form of attack at present need hardly be considered'.

They were more concerned that a light gun carried by an airship could strike telling blows at their cordite factories and naval magazines:

> it is understood that a .303 bullet striking the nitroglycerine hills, which are very conspicuous objects, would be sufficient to blow up or seriously cripple a cordite factory and that a high explosive cartridge from a small gun (say a 4-bore) fired on the light roof of the present type of magazine would be liable to explode the cordite. Guns of this description could of course easily be carried in and fired from an airship.

After these conclusions were arrived at the subject of 'protection against airship attack' was taken up at the meeting. It was the opinion of those involved that there might be two forms of protection and they classified these as 'Mobile' and 'Fixed'. The mobile defences, it was suggested, might be airships or aeroplanes and 'Balloon guns' mounted on motor-cars. The fixed defences might consist of earth or armour protection placed above the magazines or cordite factories and high angle fire guns 'capable of all round training in the vicinity of the vulnerable points'.

It was thought that the 'best method of meeting airship attacks is by airships'. However, until the British could equip themselves with a fleet of airships it was suggested that all vulnerable points should be defended by Balloon guns mounted on motor-cars or by high angle fire guns, or by both.

The assembled officers believed the situation to be so serious they suggested that the construction of two new naval magazines, those at Bedenham and Crombie, should be delayed and the designs reconsidered so that they could be fitted with some form of overhead protection or else placed below the ground. They recommended also that the entire question of defence against overhead attack should be referred to the Home Ports Defence Committee.

The Home Ports Defence Committee was one of a very small number of permanent and highly important sub-committees of the Committee of Imperial Defence. Established in 1909, it was composed of high-ranking representatives of the Admiralty and the War Office and technical authorities from other departments. This sub-committee carried out for the Committee of Imperial Defence expert examination of all inter-departmental questions relating to the defence of ports in the United Kingdom. The President of the Home Ports Defence Committee was the Secretary of the Committee of Imperial Defence.

When the Board of Admiralty considered this paper of 7 January 1910, drawn up by the Director of Naval Intelligence, Rear-Admiral the Hon. A. E. Bethell, it was decided to accept the suggestions, submissions, and recommendations contained in it. In this way the Board recognized that a new era in the life of the Royal Navy had begun.

The Board was not concerned, at this stage, with the problems of the Air Defence of Great Britain. It was solely occupied with the necessity of dealing with the threat of airship attack against naval magazines, cordite factories, and other naval establishments generally referred to as 'vulnerable points'.

In order to carry the matter further it was now necessary for the Admiralty to approach the War Office for action because the responsibility for protecting these land-based naval establishments lay with that department.

For centuries the key to British defence policy had been the maintenance of supremacy at sea by the navy. If, however, a raiding force evaded the Fleets and landed on British territory it at once became the concern of the army. In accordance with this tradition it was recognized that if enemy aircraft crossed the coast the responsibility for dealing with them would rest with the War Office.

III

On 4 February 1910 W. Graham Greene, assistant secretary to the Admiralty, wrote to the secretary, the War Office, to inform him that the Board of Admiralty had under consideration the question of the protection of naval magazines, cordite factories and other vulnerable parts of naval establishments against overhead attack.

Greene's letter summarized the dangers envisaged from attacks by airships and repeated the recommendations and suggestions for defence which had been brought forward during the discussion at the Admiralty early in the previous month. He requested that the Army Council consider the matter and suggested the question might be referred 'with advantage' to the Home Ports Defence Committee.[6]

This initiative provoked several exchanges inside the War Office and carried the question of home air defence to its next stage. The Director of Military Training, Brigadier-General A. J. Murray, later Chief of the Imperial General Staff, was uncertain if the Home Ports Defence Committee was the body to deal with the Admiralty's request. However, he was certain 'this is a very important matter'. He was also clear that the 'fixed defences' mentioned in the Admiralty letter were not the solution to the problem. 'I strongly recommend', he declared, 'that we concentrate now from the first our attention on mobile forms of aerial defence. If we do not do so we shall have to give up all hopes of establishing a superiority in the air, and we have no idea what this may mean yet'.[7] The advice of J. S. Ewart, the Director of Military Operations, was, if less profound, even more concise: 'I think that the Home Ports Committee should consider this question . . . the Committee is suitably constituted for dealing with the matter'.[8]

Sir William Nicholson, the powerful Chief of the Imperial General Staff, was reluctant to accept the suggestions proferred by his subordinates. On this occasion he managed to leash in his formidable temper but he made clear to his colleagues his hostility to 'recent inventions' in general and to aeronautical developments in particular. Nicholson declared he could not see how the experts of the Home Ports Defence Committee could deal with the technical issues raised in the Admiralty letter. However, he felt the committee might possibly consider, 'with advantage', whether the best form of defence against airship attacks would lie in the employment of aeroplanes or airships against them. For his own part he believed that in peacetime no one would attempt to practice air combat because of the dangers involved in such exercises; and in wartime no one would ever be foolhardy enough to volunteer to fight in the air. He offered his own contemptuous and sardonic opinions about the military potential of aircraft:[9]

It is possible that we might obtain some advantage from the

consideration of this question by the Home Ports Defence Committee. I do not know that with our present knowledge of aeronautics we are likely to obtain much advantage, for we have no experience whatever of one air-ship or aeroplane attacking another, or of an aeroplane attacking an air-ship or vice versa. As experiments in this direction would presumably involve the immediate destruction of the crew of the defeated air-ship or aeroplane, I do not see how they are to be carried out in a practical way in peace time, while even in time of war warfare of this nature would be so deadly that I rather doubt whether men of sufficient nerve and self-sacrifice would be forthcoming to undertake it. In considering the use to be made of recent inventions for war purposes the human element cannot be left out of account.

Eventually, the Army Council requested the Home Ports Defence Committee to consider the entire question raised by the Admiralty in its letter of 4 February. The impressions, submissions and conclusions of the naval officers who met at the Admiralty to discuss the matter in January 1910 were also presented to the committee in order to help it begin upon its novel task.[10]

IV

At this time the President of the Home Ports Defence Committee was Sir Charles Ottley, a former Director of Naval Intelligence and a military administrator and planner of proven quality. Ottley, an officer of considerable and varied experience, had been Secretary of the Committee of Imperial Defence since 1907. He was assisted by a recent recruit to the Secretariat of the Committee of Imperial Defence, Captain M. P. A. Hankey, Royal Marine Artillery, who in 1912 succeeded Ottley as Secretary of the Committee of Imperial Defence and thereafter devoted years of outstanding service to the State in this and similar capacities. Unlike Sir William Nicholson, each of these officers was already very interested in the development of aeronautics for military purposes.

The Home Ports Defence Committee quickly set to work and produced a Memorandum, 'The Defence of Magazines, Cordite Factories and Other Vulnerable points against Airship Attack', by

May 1910.[11] This memorandum may be looked upon as a landmark in the early history of the Air Defence of Great Britain.

It revealed the committee members had invited Colonel J. E. Capper to furnish them with his expert technical opinions. Although R. B. Haldane had already removed him from his place in the Balloon Factory, in order to make room for Mervyn O'Gorman, the 'engineer-scientist', Capper was still looked upon as the leading aeronautical authority in the British Army. At this time he believed airships were more formidable weapons than were aeroplanes. Eventually, he thought, the latter would be able to drive the former from the skies but this could not yet take place in the circumstances of the technology of 1910. Capper told the committee that airships were now capable of dropping heavy weights

> with accuracy on to the earth from heights as great as 4,000 to 5,000 feet. He considered that weights exceeding one ton could be dropped from an airship without the latter suffering any disability. Colonel Capper laid stress on the fact that at night in clear weather . . . it was almost impossible, even with powerful searchlights, to detect its presence or follow its movements if it was in motion.

This information produced an effect. The committee came to the following conclusion which it included in its memorandum:

> In view of the above considerations, the Home Ports Defence Committee are of the opinion that measures should be taken to safeguard ourselves against attacks by airships.

Earlier in the memorandum in a section called 'The probable Nature of Attack' the committee compared aeroplanes and airships as sources of potential danger. With respect to aeroplanes it was decided:

> although they are likely to be rapidly improved, improvements of so radical a nature as would enable them to endanger the safety of our magazines are not at present in sight. The Committee therefore consider that as regards aeroplanes the possibility of an effective attack need not receive immediate consideration.

The conclusion about dirigible airships was entirely different. This conclusion, we should observe, advanced British air defence theory in a substantial way. The reader will recall that when Lord Esher's 'Aerial Navigation' sub-committee reported on the matter in January 1909 it had offered the opinion that air bombardment of warships and dockyards could not be 'regarded as an impossible operation of war'.[12] Now, some eighteen months later, the Home Ports Defence Committee came to a more serious conclusion:

> As regards dirigible airships; there has been great activity both in Germany and France, and from what is known regarding the progress made in these countries, it appears that at the present time, with wind and weather favourable, attacks by dirigible airships against our magazines or stores should be regarded as possible operations of war.

In the remaining portions of its memorandum the Home Ports Defence Committee brought forward several proposals and suggestions that might have formed the nucleus of an effective air defence system, had they been put into effect.

In a section entitled 'The Nature of Defence desirable' the committee pointed out that where local conditions permitted magazines could be safeguarded against the danger of overhead attack by being built below the surface of the ground. However, it was forcibly argued that no form of passive defence by means of fixed armaments or overhead cover or of mobile defence by means of guns mounted on automobiles could be regarded as sufficient in all circumstances.

In order to be effective, the memorandum stressed, defence against dirigible airships 'should be of an offensive nature' and should be undertaken either by other dirigible airships, or by aeroplanes. The memorandum here mentioned Colonel Capper's opinion that aeroplanes, in the next few years, would be sufficiently developed so that they would probably 'in course of time render a successful attack by dirigible airships impossible'.

The committee recognized it would be 'impracticable' to provide sufficient airships or aeroplanes to safeguard all vulnerable points in every part of the country but they did not consider guns mounted on automobiles to be an efficient supplement. Instead, they suggested it might become necessary to adopt 'some kind of fixed armament of a special nature for such points as may be considered vulnerable to aerial attack'. The provision of a fixed defence of this nature, it was explained, 'would set free our aerial fleet for the active operations that may be regarded as their proper role instead of condemning them to a purely local defence'.

The committee further considered that the feasibility of placing important new magazines underground should be studied. When it was impossible or inadvisable to do this the committee recommended the magazines should be dispersed and so constructed as to be inconspicuous from the air.

The Home Ports Defence Committee recognized its recommendations, in the circumstances of May 1910, could only be regarded as provisional. Therefore, they urged that experiments should be conducted 'at an early date' to determine the most effective type of gun and projectile for use against airships; and other experiments to ascertain the effect on structures of heavy charges dropped on them from various heights. When information of this kind was secured an efficient air defence system could then be created.

V

Sir Charles Ottley tried to sustain the momentum of this initiative. He was keenly aware of the negative attitude of Sir William Nicholson and some of his colleagues in the War Office; and of the conservative opinions of certain high-ranking officers at the Admiralty. On 6 July 1910 he therefore prepared a paper of his own entitled 'Aerial Navigation. Note by the Secretary' and laid it before the Committee of Imperial Defence.[13]

In his Note Ottley pointed out that the question of 'aerial navigation' had been considered by a sub-committee of the Committee of Imperial Defence, presided over by Lord Esher, and that its Report, dated 28 January 1909, had been approved by the Committee of Imperial Defence.[14] He then reproduced certain of the Esher committee's conclusions. He continued his Note by reproducing the opinions brought forward by the Home Ports Defence Committee as set out in its memorandum of May 1910. Finally, he contrasted the two reports in such a way as to make clear the increased dangers Britain now faced from a possible air attack:[15]

It will be noted that whereas Lord Esher's Sub-Committee . . . stated that 'attacks upon warships and dockyards by dropping explosives . . . on to them from dirigible balloons, though at present in an experimental stage, cannot be dismissed as an impossible operation of war', the Home Ports Defence Committee, eighteen months later, went a step further and considered that 'at the present time . . . attack by dirigible airships against our magazines or stores should be regarded as possible operations of war'.

Lord Esher's Sub-Committee further recommended that experiments with aeroplanes should be discontinued by the War Office[16] but that advantage should be taken of private enterprise in this form of aviation. The Memorandum of the Home Ports Defence Committee, on the other hand, quotes the opinion of Colonel J. E. Capper, that aeroplanes are likely to develop in such a manner that they will form an efficient means of defence against attack by airships.

Both Committees draw attention to the need for experiments.

On 14 July 1910 the Committee of Imperial Defence considered Ottley's Note. The Prime Minister, H. H. Asquith, presided at the meeting and several Cabinet Ministers and various very high military and naval officers were in attendance. R. B. Haldane, the man charged with creating Britain's new Air Service, took a prominent part in the discussion. In it he offered a significant comment about aeroplanes. Until this time he was, at least in his public statements, at once cautious about their present military capabilities and sceptical about their potential for the future. Now, he said:

> He thought that aeroplanes had developed considerably since last year, when Lord Esher's Sub-Committee issued their report. At that time they had not succeeded in flying more than a few feet from the ground, whereas now they had on several occasions attained a height of more than 4,000 feet.[17]

Haldane further remarked that there had been recent 'great developments' in the construction of airship engines and balloon envelopes for airships and 'these gave a much greater range of action to airships'. He then offered a conclusion which reflected the gravity of the developing situation in the air: 'These and other points render it desirable that we should collect information with a view to seriously discussing the subject of aerial navigation in the autumn'. Reginald McKenna, the first Lord of the Admiralty, supported Haldane's suggestion about the need for further discussions in the autumn. He pointed out that the great naval dirigible, popularly known as the *Mayfly*, was now nearing completion in the huge Cavendish Dock at Barrow-in-Furness. The Admiralty proposed to experiment with this airship in September and October and he hoped the Committee of Imperial Defence would defer any discussions until it could 'have the advantage of hearing the results of these experiments'.

The positive course of the proceedings was now interrupted by Sir George Murray, the Permanent Secretary to the Treasury. Typically, he was concerned about costs. He had been invited to the meeting so that he could give some expression to the Treasury opinion that money could be saved even while an efficient Air Service was being created for the country. He enquired if the Admiralty and the War Office were satisfied with the aeronautical work that was being carried out in the National Physical Laboratory at Teddington. 'A great deal of expenditure was being incurred at this institution', he said, and he suggested that the work done there might 'to some extent overlap that which was being done by the Departments'. This touched R. B. Haldane nearly since he alone was responsible for the elaborate scheme which since 1909 had involved the National Physical Laboratory with the War Office, the Admiralty, and the Advisory Committee for Aeronautics in this novel sphere of scientific enquiry. He

> considered that, as far as the War Office was concerned, there was no overlapping of their work. . . . He thought that the work done by this institution could not be done either better or cheaper in any other way, and by the present arrangement great scientific knowledge was brought to bear on the problems considered.

This exchange brought the discussion to a close. The Committee of Imperial Defence now recommended heightened vigilance. It urged, in its formal '*Conclusion*', that the Naval and Military Intelligence Services should take 'special measures' to watch foreign trials and that they should collect and classify all available information on the subject of aerial navigation, 'utilizing the Secret Service Bureau as may be necessary'.[18] It further recommended that more aeronautical experiments should be undertaken 'as soon as possible' by the War Office and the Admiralty. When these experiments were sufficiently advanced a report about 'the subject as a whole' was then to be presented to the Committee.

It is thus clear that the meeting held in the Admiralty in January 1910 produced significant results. The British Government and its several departments and agencies concerned with the national defence were at last aware, as they had not been before, of the dangers that confronted the country in case of an air attack upon it. Moreover, it had now been established that an attack on British targets by airships was a possible operation of war. In particular, it had been decided that German airships could travel far enough from their home bases to menace naval dockyards, cordite factories, and other vulnerable points in Britain by bombing them from the air with bombs or projectiles exceeding one ton in weight.

At the same time it had also been suggested that several steps might be taken to meet the new situation. Naval magazines, in future, could be built underground. Special guns, projectiles and fuses would have to be designed so that airships could be fired upon effectively by the defenders. The notion that Britain should seek to achieve 'superiority in the air' had been mentioned. Indeed, in a telling passage the memorandum of the Home Ports Defence Committee recommended nothing less than the establishment of an efficient home air defence system. By the terms of this recommendation 'vulnerable points' in Britain were to be protected by fixed gun defences so that 'our aerial fleet' would not be tied to a static role but would be free to attack invading airships wherever they could find them in 'active operations' instead of being confined to 'a purely local defence'. Experiments were to be carried out in order that the responsible authorities could discover the technical difficulties and problems involved in fighting airships and so begin to devise ways of mastering them – with other airships, aeroplanes, and gun defences.

Despite these clear and cogent recommendations we have to notice at this stage of our account that little was done to organize a home air defence system in Great Britain. The military authorities who were given the task of creating the nation's Air Service paid almost no attention to the problem of how their country could be defended against aerial attack. Some of them continued to agree with the opinion of Sir William Nicholson that aircraft would never be able to function effectively in war. Others foresaw a very limited role for aircraft as the reconnaissance arm of the British Expeditionary Force. The Royal Navy, for its part, adopted a somewhat different and more complicated aeronautical policy in the years before 1914. Nevertheless, as we shall see, the requirements of the Air Defence of Great Britain were largely neglected in this period, despite these initiatives of 1910.

VI

While the appropriate authorities in the British Government sought to grapple, however tentatively, with the problems of air defence the British people came to realize they were entering upon a new era of their civilization. It was obvious that a major change was taking place. In a leading article of the time the *Morning Post* declared of aeroplanes and airships:

> If anything approaching the same progress is made in the flying machine as in the locomotive engine the world stands at the opening of a new epoch . . . the aeroplane follows no set route and flies as freely . . . above the frontier as over the hills and valleys of a single country. This peculiarity of the aeroplane and of the airship may disturb and alter the whole life of civilized man.[19]

Two impressions dominated the British outlook. The first was awe at the new developments in the air and their possibilities for the future and the second was fear that their country had already fallen far behind the French, and more especially the Germans, in the technology of aeronautics. This apprehension of a new danger now began to complicate British life even further in a period already distinguished by feelings of tenseness and insecurity.

Politics in Britain in 1910 were not so exciting as they had been in 1909 and there was no 'Phantom Airship Scare'. But the hostility between parties was bitter enough. At the end of November 1909 the House of Lords, controlled by the Conservatives, decided upon an extreme and almost unprecedented step in an attempt to make effective their opposition to the policies of the Liberal Government. They rejected the Lloyd George Budget.

The Liberals struck back at once. Parliament was prorogued in December and a General Election followed, in January 1910. This election decided none of the issues between the two great parties. It merely made their differences more complicated. The result of the election was a stalemate in the House of Commons. The Liberals lost their majority there but the Conservatives, at the same time, failed to secure one. Instead, the Irish Nationalist Party emerged as a powerful new factor in the politics of the country.

These Irish Nationalists existed for one purpose only. They demanded Home Rule for Ireland. If the Liberal Ministers were to retain office in 1910 they needed the votes of these Irishmen in the Commons and for this reason they now had to listen to their demands in a way they had not listened before. This Liberal dependence upon the Irish vote further outraged the Conservatives who were prepared to do almost anything to deny the Irishmen any measure of Home Rule, who insisted that the legislative Union with Ireland had to be maintained for the sake of national and Imperial unity. Indeed, civil war was mentioned as a possibility of the future if a Home Rule Bill was ever forced through the legislature by a combination of Liberal and Irish votes.

At this time sensible leaders in each of the great parties were keenly aware of the problems that faced their country. In their opinion the danger represented by Imperial Germany abroad was matched by an equally critical development at home, the fear that Socialism might attract the working classes to such a degree that the established institutions of the country would be menaced as they had not been menaced before. But the stalemate between parties effectively stultified all positive or co-operative action designed to deal with these national issues.

Throughout the year some politicians sought, almost desperately, to solve the situation imposed upon them by the electors but the deadlock remained unbroken and the country was forced to hold another General Election, in December. For a time it seemed a solution might be found when King Edward VII died, in May 1910. Then, the shock of the monarch's demise produced so serious an effect that the leaders of the Liberal and Conservative parties at last agreed to a political truce. They met together in a 'Constitutional Conference' in an attempt to compose their differences, for the sake of the country. The Constitutional Conference failed but exactly at this juncture Lloyd George launched a bold initiative of his own.

The Origins of British Air Defence 125

Lloyd George, the Radical Chancellor, was a keen partisan and a merciless fighter in politics. He was also a patriot. In 1910 he believed Britain was in great danger. Therefore, he proposed the formation of a National or Coalition Government to be composed of the most capable leaders in each of the great political parties. Their task, according to this proposal, would be to set party bickering aside so that they could solve the real problems confronting the nation and thus safeguard the British future. Arthur Balfour and certain other Tory leaders seriously considered this 'Great Plan', as it was called, but eventually they rejected it.

Lloyd George never forgot this remarkable episode of 1910. Years later, in his *War Memoirs*, he explained:

> I had for some time past been growingly concerned with the precariousness of our position in the event of our naval defence being broken through. . . . Inventions which portended such a menace had already appeared. Whether the peril would come from the air or from under the waters, I knew not. . . . The submarine and the Zeppelin indicated a possible challenge to the invincibility of our defence. . . . It was with these thoughts in my mind that I ventured in 1910 to submit to the leaders of both political parties . . . a series of proposals for national cooperation over a period of years to deal with special matters of urgent importance.[20]

VII

R. B. Haldane, for his part, followed a somewhat different course at this time. In 1910, as we have seen, one of his assigned tasks was to create an Air Service for his country. He still hoped, within limits, to be able to keep this new issue, aeronautics, outside the arena of party politics; and many of his contemporaries agreed with and approved of this attitude. However, his blandness and arrogance now began to irritate his critics who complained that, as the responsible Minister, he was moving much too slowly in this sphere, that the country was in serious danger of an air attack as a consequence of his methods and procedures.

For example, on 7 March 1910 in a debate on the Army Estimates Haldane explained his aeronautical plans in some detail and defended the system he had devised. He said:

These are all our arrangements. . . . Very many people will say: 'You seem to be always making arrangements and never getting any further'. But I am convinced of this, that until we get everything perfectly clear we shall only make very slow progress. . . . The whole subject is . . . very much in its infancy. I am never alarmed when reading of the progress of other nations in this matter. No doubt we are behind. So we were in other matters. In motor cars, for instance. But when we had mastered the thing we went ahead very quickly. And again, so much in this matter that has been undertaken by foreign countries has already turned out to be unsatisfactory. However much they seem to lead the way, I trust I have made clear that when we put our backs into the matter we shall find ourselves all right.[21]

Others, however, were alarmed by the slow aeronautical pace Haldane advocated and followed. Before and after his defence of 7 March they began to speak out, more and more forcibly. Prominent among these critics was Lord Montagu of Beaulieu. His opinions about aeronautics usually attracted a wide notice. On 22 February 1910 he addressed the Aldershot Military Society on the subject. In the chair on this occasion was Lieutenant-General Sir Horace Smith-Dorrien, Commander-in-Chief of the Aldershot Command.[22]

Smith-Dorrien, unlike some of his fellow officers in the British Army, was convinced that 'aerial machines' were certain to play an important role in all warfare of the future. When he introduced Montagu of Beaulieu to his audience on this occasion he explained that 'we soldiers must take every opportunity of studying all there is to be learnt on the subject'.

In his talk Lord Montagu suggested that as soon as aeroplanes were perfected they would force dirigible balloons from the skies. However, dirigibles constituted a grave danger to Great Britain at the present instant. 'It would be possible', he said,

> for dirigibles to leave the frontier stations of at least five Continental powers all within 500 miles distance from where we are now seated, and do much damage at Aldershot, Portsmouth, Dover, Chatham, Sheerness, and other military and naval stations, without taking into account that an attempt would certainly be made to paralyse the heart of the nation by attacking certain nerve centres in London, the destruction of which would

impede or entirely destroy the means of communication by telephone, telegraph, rail and road.

Montagu made it clear that Germany was the country to be feared most in this connection. He next proceeded to link Britain's general anxiety about Imperial Germany to these particular new developments in aeronautics. He said:

> Germany's plans in airships, as in other directions, are not fully known to us; nor would this be the occasion to state how such information can be obtained. But it is beyond doubt . . . that she has at the present moment the best and the only fully-equipped aerial fleet in the world, and one which is more than a match for all the other dirigibles in existence . . . by the end of this year, it is quite possible that Germany could muster a fleet of from twenty-five to thirty dirigibles, the great majority of which would be specially suited for warlike purposes. . . . Today Germany is doubtless mistress of the air. . . . Looking into the future, it is probably correct to estimate that, for some time to come, Germany will be supreme in the air, for no less than 70 German dirigibles will be available by 1912. The need for an awakening on our part is thereby emphasized.

Montagu revealed to his audience that artillery had been employed to fire at aircraft in experiments carried out on Salisbury Plain but he insisted the Germans possessed the best anti-aircraft guns then in existence. He ended his talk on a gloomy note. It was a direct challenge to the policy adopted by R. B. Haldane. He said:

> Before I sit down, I must, without wishing to enter into any political controversy, call the attention of all concerned to the fact that our Government here seems as yet to have hardly realized the importance of the question of aerial warfare which we have been discussing tonight . . . the danger of delay is a very real one. The day is not far distant when England will have to be something besides nominal mistress of the seas. She will have to be at least equal to her neighbours in the matter of aerial defence and offence, and it is our business and the duty of the nation at large to see that the authorities are awakened in time to their responsibilities in this direction.

Lord Montagu's attitude was typical of several of the 'air enthusiasts' of this early period. These men often hoped to avoid political controversy when they discussed aeronautics but they were convinced that Asquith's Liberal Government, guided as it was by R. B. Haldane's outlook and analysis, was neglecting a serious and growing menace.

Montagu, who had been the Conservative Member of Parliament for the New Forest Division of Hampshire until 1905 when he succeeded his father, could not avoid politics when he discussed aeronautics in the House of Lords on 13 April 1910. In order to drive his points home it was necessary for him to speak as a partisan. He pointed out on that occasion that the 'question of aviation as regards our national defence has been for some time a source of anxiety to those who are aware of what foreign Powers are doing'. He continued:

> At present I do not think that the Government fully appreciate the growing importance of the subject, and I hope today to ascertain that Ministers are more alive than the public imagine to the importance of preparing in time. . . . I am not satisfied from the information which I have been able to gather that His Majesty's Government are making adequate preparations or considering seriously enough this new position of affairs . . . a dirigible hovering over the Palace of Westminster could hit either the House of Lords or the House of Commons with unerring accuracy. That shows to what extent progress has been made. . . . I think that the Government must make up their minds that more attention must be devoted to the subject and more money provided in the Army Estimates . . . all other great nations are taking this question up very seriously . . . we are rather at the present moment adopting the attitude of folding our hands and waiting to see what other people can do. . . . The present is the time for taking action, and I hope that His Majesty's Government will give the matter the attention it deserves.[23]

It remained for Lord Lucas, the Liberal Under-Secretary of State for War, to answer for the government. His position was a difficult one because he was unable to challenge successfully the technical opinion of an acknowledged expert like Montagu. He began diffidently enough:

My Lords, I fully realize that there is no person in this country who is more entitled to receive the fullest explanation and assurances that the Government can give than the noble Lord . . . because nobody has identified himself so much with this subject . . . and if I have to differ from him on any point I can assure him that I differ from such an authority with the greatest reluctance.[24]

Lucas offered no arguments of his own but simply repeated Haldane's opinions about aeronautics when he made his reply. The subject, he said, was still in its infancy. 'We do feel', he continued, 'that it is quite possible that, in studying as we are doing all the great and complicated scientific problems . . . and trying to find solutions for them, we may be making just as much progress as the countries which are spending a great deal of money. . . .'[25]

This was not an especially effective answer. The Conservatives, dominated by their fear of the German danger, an immediate menace in their view, could not be expected to accept it; nor would those Liberals who were interested in the subject of aeronautics and air defence be satisfied with such a policy for very long.

At exactly this juncture the Aerial League of the British Empire began to cause trouble for the Liberal Government because of its failure to act more energetically in the field of aeronautics. It will be recalled that this society had been formed early in 1909 as a 'strictly non-party organization'. Its members were patriots who desired to awaken their countrymen, by means of a vigorous propaganda, to the need to 'secure and maintain for the Empire the same supremacy in the Air as it now enjoys on the Sea.'[26] Among those associated with the League were Lord Roberts, H. G. Wells, the Marquis of Salisbury, Lord Montagu of Beaulieu and Lord Esher.

Until this time the League had sought to co-operate with the Liberal Ministry in a non-partisan way. In conformity with this object in July 1909 the Executive Committee of the League invited several departments of the government to nominate representatives to serve as members of the Council of the League. Among other departments the Admiralty and the War Office each agreed to nominate a representative and at the same time it was announced that Lord Esher, a close ally of the Liberal Ministers, had consented to become the League's president. It was the Admiralty's view that the presence on the League's council of several representatives of government departments was 'the best way to ensure it being kept straight and not developing into an agitating body'.[27]

By the end of December 1909, however, the government's failure to adopt a vigorous aeronautical policy began to cause friction and Lord Esher resigned his presidency at that time. Worse followed in March 1910. Then, the Executive Committee of the League published a leaflet which was extremely critical of the British Government and its weak aeronautical programme. The leaflet did not mince words:[28]

> At a time when questions of Imperial Defence are especially occupying the minds of all who recognize the need for a strong and unassailable England . . . the Aerial League has set itself the important and responsible task of watching the progress of aerial activity abroad . . . and of encouraging by every possible means the development of the science and practice of Aviation in Great and Greater Britain, with a view to securing that supremacy in the Air which now lies, to our shame and our danger, in other hands.

The leaflet pointed out that the Germans were far ahead of the British in aeronautic development. The German air fleet, according to the leaflet, would be able to inflict 'incalculable damage' upon British docks, harbours, wireless stations, and stores of ammunition in time of war:

> Compared with the extraordinary energy and presence with which all nations are throwing themselves into the development of aeroplanes and dirigible balloons, backed by boundless public enthusiasm and state encouragement, the backward position of this country, and the extraordinary apathy of its people are deplorable to a degree.

This leaflet was treated seriously in the Admiralty. On 9 March a paper prepared by the Assistant Director of Torpedoes was sent forward for approval to Sir Arthur Knyvet Wilson, the First Sea Lord, and to Reginald McKenna, the Liberal Government's First Lord of the Admiralty. The paper declared with respect to the leaflet and the Aerial League:

As the responsibility for the Aerial Defence of this country lies with the Admiralty and War Office, the foregoing amounts to a direct criticism of the Government, and it is submitted that the Admiralty and other Government departments concerned . . . should at once withdraw their representatives.[29]

When Wilson and McKenna concurred with this suggestion Captain Stuart Nicholson, the Admiralty's representative on the Council of the League, was withdrawn and shortly after the other departments involved also removed their own representatives.

In this way formal co-operation between the government and the 'air enthusiasts' of the Aerial League came to an end. From this time no official of the State would be able to influence or moderate the League's activities as had been attempted earlier. As a result of this development air power politics in Britain were certain to become more and not less strident.

VIII

In November 1906, by Lord Northcliffe's design, the *Daily Mail* had offered a prize in the sum of £10 000 to the first person who could fly by aeroplane from London to Manchester. This princely award seemed unattainable until 1910 when developments in aviation convinced Northcliffe that the feat was at last possible. By April 1910 the magnitude of his prize had attracted two champions. One was a young and charming Englishman named Claude Grahame-White and the other was an experienced French pilot of the day, Louis Paulhan.

Grahame-White was first in the field. Northcliffe saw to it that he received a warm welcome from the *Daily Mail* for his enterprise. The paper's first leading article for 21 April 1910 sought to make clear, in the strongest terms, the high significance of his object and also the pride of his countrymen in the fact that a British pilot could at last come forward to attempt so noteworthy an aeronautical exploit:

> The most remarkable feat in the whole history of aviation and one which will eclipse all the achievements of the brothers Wright, Mr Farman, and M. Blériot is to be essayed on Saturday next. . . . It will indeed be gratifying to our national pride if an

Englishman startles the world this time by proving that he can overcome so many difficulties and accomplish a feat which has hitherto daunted the greatest of foreign aviators.

It was only after Grahame-White had failed in a first attempt that Louis Paulham seized his chance to join in the competition. He and his mechanics rushed from France to a field between Hendon and Edgware, later the site of Hendon aerodrome, and assembled his machine there while Graham-White repaired his own damaged aeroplane at Wormwood Scrubs. The conditions of the prize were that the flight of 186 miles was to be accomplished within 24 hours with no more than two stops for refuelling. Even Lord Northcliffe could not have planned such an outcome as this. The attempt to win his prize had now developed into a contest or race between an Englishman and a Frenchman. As Grahame-White's biographer explained, many years later, the episode became a nation-wide 'sensation'.[30]

Thousands of his fellow countrymen assembled at Wormwood Scrubs to cheer Grahame-White and others continued to cheer him at various places along the route of his flight. Paulhan, however, gained an early and decisive lead. In a desperate attempt to overtake him Grahame-White, on the second leg of his journey, essayed the incredible hazard of a night flight, the first ever attempted in Europe. Despite this brave effort Paulhan won the race. He was given a lunch at the Savoy and presented with his £10 000 in a golden casket. Northcliffe took care to recognize Grahame-White's plucky effort and awarded him a 100-guinea cup.

The effect of this episode upon the contemporary mind was profound. It was in response to it that the *Morning Post*, as we have noticed earlier in this chapter, declared that the 'world stands at the opening of a new epoch'. In a leading article of 29 April the *Morning Post* also said: 'The imagination is simply unable to keep pace with the development of the new art. The ability to be astonished is exhausted; these amazing exploits follow one another too fast. . . . As a test of the aeroplane in what may be called practical work this flight is final'.

This was exactly the opinion of the *Daily Mail*. Its first leading article for 29 April was entitled 'The Triumph of the Aeroplane':

the age of experiment in aviation has reached its close . . . and the age of achievement has arrived. . . . Man, the all-conquering, who shall overcome all things, save only death, has achieved his greatest triumph. He has conquered the air. For ourselves we believe that our purpose in offering this prize has now been attained. We trust that our countrymen will recognize the lesson of this portent.

In its leader for 1 May 1910 *The Observer* explained to its readers the tremendous implications of what had taken place; it paid tribute to Lord Northcliffe's role; and like many other contemporary journals it warned that England's future turned upon the nation's ability to seize the first place in the air.

These things are no mere sensation of the hour. They may mean, and in the end must mean, one of the most wonderful revolutions in the conditions of human life, one of the strangest expansions of the powers of man, that has been known from the beginning of time. It is not enough to compare this change with the introduction of the steam engine. That mechanism, mighty as it was, only increased our command over elements that were traversable before. . . . Look at it how we will, the new power of man to fly . . . is one of the very miracles of time, and posterity will regard it all as something more romantic and thrilling – perhaps in all ways more significant – than the voyage of Columbus. Unless we wish to leave our children an England that will hold a much humbler status than it has occupied for centuries, we must realize that we are more vitally concerned to strive for national mastery of aviation than is any other people in the world. Where all other efforts had failed to seize the popular imagination, one energising newspaper has succeeded.

In this way the competition between Grahame-White and Paulhan contributed to and reinforced the awareness of many in England that their country was now entering upon a new phase of its national life. This phase, full of marvels for the future, also contained within it a grave danger. By the spring of 1910 it could be and was argued that the British people were more imperilled than any other by the technical advances that had at last been made in aeroplane development. Lord Northcliffe's years-long campaign in the *Daily Mail* to make the nation 'air-minded' had now achieved at least a part of its goal.

7
The Paris Conference and its Consequences

The First Air Conference – Initiative of the French Government – Surprise of the British Delegation – 'Sovereignty Over the Air' – Analysis by the Secretariat of The Committee of Imperial Defence – Captain Hankey's Fear of Air Bombardment – Memorandum by the General Staff – An Interim Report – The Foreign Office Intervenes – Opinions of the Admiralty – Sir E. Grey's Policy – Reaction of Lord Esher – He Calls for a 'New Arm' – Initiative of Sir Francis Bertie – The Attitude of Louis Renault – The Conference Adjourned – The Aerial Navigation Bill – Effects in the Foreign Office, War Office and Admiralty – The Reaction of Captain Hankey.

I

Shortly after Louis Paulhan's famous victory the first International Conference on Aerial Navigation was held in Paris, in May 1910. The British delegation to this conference expected to deal with certain routine matters the new technology had thrown up and which now required the attention of the international community. On the agenda of the conference were such subjects as papers of identity for aeronauts, permits, identification marks for aircraft, international certificates of proficiency, police and customs precautions, and those 'rules of the air' civilized nations might agree upon in the novel circumstances of the day.

However, before any convention or treaty was signed at the Paris Conference the actions of the German delegates caused serious alarm. They raised strategic, political, and diplomatic issues which aroused British suspicions and caused the British Government to look again, in the most concentrated way, at all those problems connected with the air defence of their country.[1]

The original proposal for holding such a conference was made by the French Government at the end of April 1909. A remarkable number of German balloons had recently descended on French territory and the authorities in Paris decided something should be done to deal with the situation. In August 1909 the French furnished the British Government with a detailed *questionnaire* they proposed to submit to the conference; and they also asked for some expression of Britain's views upon the subjects and issues contained in this document. It was hoped in Paris that the deliberations of the various European powers there would result in a treaty or convention upon which all might agree.[2]

In London, in response, the Foreign Office suggested the appointment of an interdepartmental committee composed of representatives of the Home Office, Admiralty, War Office, Customs and Excise Department and the Board of Trade. The Foreign Office did not participate in the work of this committee because the subjects mentioned in the French *questionnaire* seemed to be routine and not of the highest significance. The Inter-Departmental Committee issued a Report in October 1909. Its recommendations were straightforward. The committee members suggested that Britain's centuries' old experience of sea shipping should be exploited in order to enable her to deal with this most recent advance of technology:[3]

> The Committee feel that, so far as is compatible with the safeguarding of national interests . . . it is highly undesirable to introduce restrictions which may tend to retard the development of aerial navigation. It must, however, not be forgotten that it will be difficult to impose, at a future date, any restrictions affecting foreign aeronauts which may be waived at the present Conference. Caution is, therefore, necessary not to depart radically from our present tried practice in analogous undertakings. The similarity between sea shipping and aerial locomotion is sufficiently close to furnish a useful and safe basis in discussing the questions relating to our inquiry. Our present regulations as regards shipping are the outcome of several centuries of experience. We therefore consider it desirable that, where restrictions are called for, the precedents of sea shipping should be followed as nearly as possible in aerial navigation.

The French Government next circulated the replies to its *questionnaire* received from all the countries involved. The replies furnished by Germany were considered to open a wide field of discussion and to go somewhat beyond the limits of the original programme. In these circumstances the Foreign Office at once applied to Paul Cambon, the French Ambassador in London, in order to ask if a new agenda for the conference was to be issued or if the original programme was to be adhered to when the various powers met in Paris. In due course Cambon replied that the conference itself would determine the extent of its deliberations.

The British Government then nominated Rear-Admiral Sir Douglas Gamble, recently naval adviser to the Turkish Government, as senior British delegate to the conference. He and his colleagues received no written instructions other than the report of the Inter-Departmental Committee.[4]

The International Conference opened on 18 May in the French Foreign Office and a distinguished French jurist, M. Louis Renault, was elected its President. At once the British delegates discovered they had been inadequately prepared for the role they were required to play. In their Interim Report they declared:[5]

> It was not our original expectation that we should have to take any prominent part. We considered that aerial navigation was still in an embryonic stage, and that many of the problems created by it affected continental countries with land frontiers more nearly than Great Britain. As the Conference progressed, however, we became convinced of its importance, and what we learnt from other delegates led us to think that aerial navigation was a more urgent matter, especially from a military point of view, than it has been considered hitherto in this country. There were divergent views as to the likelihood of air-craft being largely utilized in the immediate future for the transport of passengers and cargo. There was practical unanimity among the foreign expert delegates as to their utility in time of war.

At an early stage of the conference 'the question of State-sovereignty over the air' was brought up even though it had not been mentioned in the *questionnaire* previously circulated by the French. Here was an issue of the very highest consequence. The British delegation had not expected questions of such an order to be raised; and they realized they were unprepared to deal with the problem now confronting them.

The lead in this matter was taken by Dr Johannes Kriege, the chief German delegate. Oddly enough, Louis Renault, the French President of the Conference, agreed with him:

> Both advocated that navigation above foreign countries should be declared free in principle, but that the State should be permitted to disregard this principle in the interests of national defence or of the security of its inhabitants and their property. To this the German delegation added a further clause stipulating that foreign air-ships should not be treated less favourably than those of nationals.[6]

Sir Douglas Gamble and his colleagues understood that too much was at stake for them to accept such proposals without further consultation with London. As a first step they drew up a memorandum of their own which criticized the German arguments. Then they noticed something else. It became obvious that Dr Kriege regarded the matter very seriously; and that he meant to have his way in it. In their Interim Report the British delegates remarked of him: 'It was evident that he attached the utmost importance to gaining his point . . .'; and '. . . Dr Kriege offered the most strenuous opposition . . .'; and '. . . the pertinacity of the German delegation . . . was certainly remarkable . . .'[7]

In these curious circumstances the British delegation on 10 June sent a letter and a telegram to London, asking for instructions. Sir Douglas Gamble's letter was concise:[8]

> The main question of principle which is being discussed at the Conference is whether airships shall be permitted to circulate above, or to land in, foreign States, and under what conditions. The French and German delegations proposed that aerial navigation should be declared free in principle, but that each State might impose restrictions in the interests of its security. . . . The British delegates objected to laying down any abstract principle.

In London, the full Committee of Imperial Defence was hastily called together by R. B. Haldane, on 14 June. He and Winston Churchill were alarmed at the course pursued by the Germans at the conference. Further despatches from Sir Douglas Gamble 'affecting the strategical aspects of the question' were dealt with by the Standing Sub-Committee of the Committee of Imperial Defence which held meetings on 20 and 25 June.[9]

After the meeting on 20 June, presided over by Churchill, Captain Hankey was sent to Paris with a letter and with further oral instructions for the British delegation. This was the first time Maurice Hankey acted as an official emissary of the British Government. Eventually, after the meeting on 25 June the matter was looked upon so seriously in London that the British delegates were directed to ask for an adjournment; and the conference was adjourned until 29 November 1910, as a result of their request.

II

For a time the British authorities were puzzled by the attitude adopted by Dr Kriege and his colleagues in Paris. 'The more closely we looked into the matter', Lord Hankey has explained, 'the less we liked the look of it'.[10]

Then, the Secretariat of the Committee of Imperial Defence produced a strategical analysis of the motives behind the stance taken up by the German delegation to the conference. This paper clearly reflected that fear of Imperial Germany which dominated British minds in this period. Dated 23 June 1910, the paper was entitled: 'The Strategical Aspect of Certain Proposals before the International Conference on Aerial Navigation':[11]

> During the course of the International Conference on Aerial Navigation the German delegates are reported to have laid exceptional stress on the importance of the principle of equal treatment of national and alien airships in the matter of the right to fly over and land in every country. . . . It is unreasonable to suppose that so much trouble would have been taken for the mere assertion of an abstract principle, and the question as to whether there is any underlying military reason merits examination.
>
> 2. From a glance at the map of the North Sea it would appear extremely probable that the real object is to tie the hands of the minor Powers adjacent to the German Empire, and to deprive them of the right of making regulations prejudicial to the free passage of German airships.
>
> 3. Consider, for example, the use of German airships during a war between Germany and Great Britain. It seems very probable that Germany may hope to compensate for her relative

inferiority in cruisers suitable for scouting by the use of her great superiority in dirigible balloons. . . . The strategical areas which Germany would most desire to watch would be the Straits of Dover, the mouth of the Thames, the entrance to the Baltic, and the approaches to the German enclave in the North Sea. With the exception of the latter the direct routes to all these areas lie across the territory of minor Powers, viz. – Belgium, Holland, and Denmark.

4. If these Powers can be induced by International Convention to admit the principle of equal treatment to foreign and home airships, thereby surrendering the right to make regulations directed against foreign airships, it will be possible for the German airships (other than military) to move freely across them during a war with Great Britain without the fear that, in the event of an unpremeditated descent, they may be liable to detention, confiscation, or other . . . regulations.

5. It is true that the proposed rules regarding the freedom of movement of airships is not to apply to military airships. At present however privateering is not forbidden in aerial warfare. . . . The position of a weak Power, with little to gain and much to lose by provoking a strong Power like Germany, in restricting the movements of her privateer airships would be particularly difficult if her hands were tied by International Convention. . . .

9. It would appear that the proper policy of this country is to reserve not only for herself, but also for minor Powers, the right to make whatever . . . laws and regulations they think fit regarding the land(ing) and navigation of all airships.

This secret paper was most probably drawn up by Captain Hankey himself. Years later, in his Memoirs, he explained his contemporary attitude with very great care. Lord Hankey wrote:

Germany, at that time, was far ahead of all other countries with her Zeppelins, which had become a serious menace. . . . If the International Convention were signed in the form in which it was presented, we saw that in an Anglo-German war German Zeppelins would be able to take the shortest route to England which lay above Holland and Belgium (which was assumed to be neutral)[12] to attack London, and the Governments of these countries would have no right to interfere or protest. After delivering their attack the Zeppelins might return by the same route, and once they were over neutral territory it would be an awkward business for our aircraft to attack them. It seemed to us wrong that a neutral country, placed in such a position, should surrender the right to impose restrictions and to make them effective. We felt that, until the whole question had been explored . . . it would be inadvisable to accept the draft convention.[13]

III

The British Government now at last embarked upon that exploration of the subject it should have taken up in the very first place. In this period some of the service chiefs in Britain were so conservative in their outlook they were not in any way impressed by the possibility of employing aircraft for military purposes. We have already noticed, in this connection, the harsh attitude of the Chief of the Imperial General Staff, Sir William Nicholson. For his own part Admiral of the Fleet Sir Arthur Knyvett Wilson, the First Sea Lord at the Admiralty at this time, was reported to have said that in his opinion 'the aerial needs of the Royal Navy could be met with two aeroplanes'.[14] Now, however, R. B. Haldane and Winston Churchill, aided by Captain Hankey in his more limited sphere, were alerted to the new and serious danger confronting the country. The two politicians demanded the Asquith Government at last adopt a firm and vigilant attitude with respect to the deliberations in Paris.

A first step in this new departure was taken by the General Staff at the War Office, in Whitehall. On 11 July 1910 the General Staff submitted a Memorandum to the Committee of Imperial Defence. The Memorandum dealt with the issues raised at the conference in Paris. This document reflected the conservative outlook about aircraft held by several of the highest soldiers but it revealed also that some of the military were now aware of the gigantic changes that were certain to take place with the further development of aeronautics. This Memorandum of July 1910 contained elements and ideas from the two separate but contemporary points of view.

The Memorandum, entitled 'The International Conference on Aerial Navigation' began by stating that in the opinion of the General Staff the questions 'unexpectedly' raised at the Conference

The Paris Conference and its Consequences 141

in Paris 'appear to be of considerable strategical importance to the safety of the United Kingdom'. These questions, in the opinion of the General Staff, required careful examination by the Committee of Imperial Defence before any final instructions could be sent to the British delegates in Paris.[15]

The General Staff further advised:

> To admit the principle that foreign air-ships were not to be treated less favourably than national ones was tantamount to saying that foreign air-ships were to have the same privileges of passage over, say, the Thames or Portsmouth defences as our own Naval and military air-ships. It seemed imprudent to accept such a proposition and . . . the General Staff urged that as an absolute minimum we should insist upon our right, in the interests of national defence, to forbid aerial navigation by foreign airships over certain zones of reasonable extent.

The next passage in the Memorandum revealed a peculiar blend of the attitudes of those military authorities who regretted the recent and rapid development of aircraft and also of those soldiers who already realized that basic changes in the strategic situation of the country were taking place even as they sought in London to deal with the situation that had developed at the conference in Paris:

> It may no doubt be argued that we cannot arrest or retard the perhaps unwelcome progress of aerial navigation by standing out of an agreement which may commend itself to a majority of the other European Powers. . . . It will be said that our attitude displays timidity, and that it will encourage our rivals to renewed efforts to perfect a weapon which causes us so much anxiety. On the other hand, we must realize that, in according absolute freedom of approach to our coasts and passage over the United Kingdom, we are, under existing aeronautic conditions, relinquishing the advantages of our insular position and giving facilities in time of peace for over-sea reconnaissance and practice from dirigibles in wireless telegraphy which might be turned to useful account against us in time of war. In this connection it must be borne in mind that the distinction made in the Convention between military and private air-ships and aeroplanes is a distinction without a difference. They will all be

employed against us in war, and German writers make no secret of the hope that the command of the air which they are striving to obtain will also give them command of the sea.

The conclusion of the General Staff Memorandum was sober. Despite its tentative and provisional tone, this conclusion carried Great Britain to the threshold of a new era:

> The comparative security of the United Kingdom as contrasted with Continental States depends no doubt on our insularity, but only so long as that insularity is coupled with superiority at sea . . . it is submitted that, if aerial navigation becomes a practicable art, if air-ships can be designed able to move with reasonable certainty and to drop explosives with reasonable accuracy on war vessels, docks, magazines and factories of warlike stores, neither our insular position nor the restrictions which we may now seek to impose will avail us. We shall, in such circumstances, have to rely on acquiring and maintaining, if not air command, at least air equality, by constructing air-ships capable of defeating the air-ships of an enemy, and by developing guns and other appliances capable of destroying those airships. . . .
>
> Some time, however, must yet elapse before aerial navigation reaches a state of development which will enable us to determine its precise value. . . . In the meantime we are confronted unexpectedly, and it is thought, prematurely, by a request that we will forgo the right to impose regulations and restrictions which might prove a safeguard to this country during the experimental stages of a novel method of warfare.

A second development took place on the day after the General Staff submitted its Memorandum to the Committee of Imperial Defence. On 12 July 1910 Rear-Admiral Sir Douglas Gamble and his colleagues of the British delegation in Paris completed their 'Interim Report'.[16]

This paper, addressed to Sir Edward Grey at the Foreign Office, set out a detailed account of what had occurred in the Paris negotiations; it stressed the importance accorded to the subject of military aeronautics by the European powers who were much more concerned with it than were the British; it described the problems encountered by the delegation during the course of the conference; and it asked for further instructions from the authorities in

London. The delegates, in order to impress Sir Edward Grey with the high significance of their work, reported further that: 'On the 18th May the Conferece met (for the first time) at the Foreign Office. The importance attached to it may be judged by the fact that all the principal Powers were represented by diplomatists, lawyers, or military men of the highest standing . . .'.

The delegates emphasized that the most serious problem encountered in the course of their negotiations was, as they put it, 'The Admission of Foreign Airships'. In this connection their Report stated: 'the problem of creating a right of international circulation without infringing rights of State-sovereignty was difficult and delicate. . . .' Indeed, as the discussions proceeded it became evident to the British delegates that this was the most serious issue that would confront them during their stay in Paris. They were also surprised, given the diplomatic climate of the day, that 'throughout the discussions and negotiations . . . the Presidents of the French and German delegations acted in complete unison' in opposition to the arguments of their British colleagues.

The conclusion of the Report was clear. It required further action by the British Government:

> We have now reported on the draft Convention which was drawn up by the Conference, and we submit it for consideration by the various departments of His Majesty's Government with a view to its adoption, if found acceptable. We have also explained fully our action in regard to the difficult question of the admission of foreign air-ships, up to the stage which the negotiations had reached when the Conference was adjourned on the 29th June at our request in order that the Governments concerned might have an opportunity to give the whole question their mature consideration. The concessions made by the German and French delegates to our views as regards the admission of foreign air-ships appeared at the time of the adjournment of the Conference to be the maximum they were prepared to give . . . We think that if we refuse to accept the final text . . . it would involve our standing out of the Convention altogether – a course which, it appears to us, would be most regrettable, as it would both retard the progress of aviation in this country and deprive British aeronauts of facilities for flying abroad. . . . The Conference is due to reassemble on the 29th November

next; and we should be glad to receive your further instructions before that date.

IV

In these circumstances the Prime Minister, H. H. Asquith, took decisive action. On 18 July 1910 he requested that the Standing Sub-Committee of the Committee of Imperial Defence should meet to consider and report upon the various problems thrown up by the conference at Paris. For the purposes of this particular enquiry the following were asked to serve as members of the Sub-Committee: R. B. Haldane was selected as Chairman; the others included Winston Churchill; Lord Esher, who often furnished Liberal Ministers with strategic advice; Sir Charles Hardinge, Permanent Under-Secretary of State at the Foreign Office; Admiral Sir Arthur Knyvett Wilson, the First Sea Lord; Rear-Admiral the Hon. A. E. Bethell, Director of Naval Intelligence; General Sir William Nicholson; Brigadier General A. J. Murray, the Director of Military Training; and Sir Edward Troup, the Permanent Under-Secretary of State at the Home Office. Asquith's instructions, or terms of reference as British official jargon had it, were concise. He asked this distinguished group:

(a) to consider the strategical and political aspects of the Draft Convention drawn up at the International Conference on Aerial Navigation, in relation both to the international interests of this Country and to those of such States as France, Denmark, Belgium and the Netherlands, not only as regards their own security, but also as affecting their duties as neutrals in the event of war; (b) and to make recommendations regarding the action to be taken previous to the proposed re-assembly of the Conference, fixed for the 29th November next.[17]

V

Meanwhile, the Foreign Office now began to intervene, and with decisive effect. It will be recalled that when the French authorities first approached the British Government in order to propose an International Conference on Aerial Navigation, in April and in

August 1909, officials at the Foreign Office showed almost no interest in the proposal since they believed the subjects mentioned by the French were routine and not especially significant or momentous. The experiences of the British delegates at the Paris Conference changed this attitude.

On 8 July the Foreign Office addressed a 'secret' letter to the Admiralty in order to secure its technical views and W. Graham Greene, at this time the assistant secretary of the Admiralty, replied to it on 21 July 1910. Greene's letter, a remarkable document, at once set out the technical advice offered by the Admiralty in response to the enquiry from the Foreign Office and it also revealed the strange attitude toward aeronautics shared by some of Britain's highest naval officers of this period. 'Their Lordships' explained for the benefit of the Foreign Office that the

> only practical use that has been hitherto proposed for air-ships is for purposes of war. . . . Aeroplanes have been made for the purposes of sport, especially that form of sport which consists in providing a spectacle to draw a crowd, but neither air-ships nor aeroplanes have yet been put to any practical use which tends to the advantage of the community at large.

Greene's letter offered the further opinions of the Admiralty that

> In all probability the most important use of both air-ships and aeroplanes in war will be obtaining information as to the enemy's position. . . . Direct attack by dropping explosives etc. though possible, is not likely to have so much effect on the war as a whole, though it is more calculated to create panic in the popular mind.

The remaining parts of the letter were somewhat more forthright than these excursions:[18]

> Up to the present aeroplanes that have visited this country have come obviously only to take part in competitions *etc.*, and to advertise their machines, and consequently no legislation has been necessary; but if it appeared that they were being used to obtain information on matters that we desired to keep secret, stringent regulations would have to be made, and Great Britain

must reserve her right to make these whenever the occasion may arise. . . .

It would seem advisable that the question of the right of each nation to grant or refuse permission to foreign air-craft to fly over its territory should be withdrawn from the consideration of the Conference, and left for the present at all events to be settled by each nation independently, or in concert with its immediate neighbours.

VI

After the Admiralty analysis was received in the Foreign Office Sir Edward Grey took charge. In this period the great object of Grey's foreign policy was to maintain the balance of power in Europe in order to deter aggression and thus preserve the peace.[19] Grey's policy has also been associated with the idea of *Einkreisung* – the 'encirclement' of Germany by Britain and her allies and associates of this era. Expression was given to this idea by the German Kaiser who said in a speech at Döberitz in 1908. 'Yes, it now appears as though they wanted to encircle us. We will know how to bear that. . . . Just let them come on. We are ready'.[20] A standard account declares: 'The opinion grew and became widespread in the German nation that British politicians, especially King Edward, were engaged in an attempt to encircle Germany'.[21]

Most historians reject this notion of an *Einkreisungspolitik* waged by Britain against Germany. R. C. K. Ensor in his classic account referred to it as a 'baseless myth'. A later British writer argued there was 'conclusive evidence' that 'Grey's policy was not one of encirclement'. A standard American account offered a slightly different view. It suggested British foreign policy

> produced a conviction that Germany was being 'encircled' . . . that this encirclement . . . aimed at strangling German commercial and colonial expansion, and even at crushing Germany's political and military position. There is no substantial evidence that there was any deliberate encirclement with such aims. . . . But there was nevertheless something of a diplomatic encirclement.[22]

Fritz Fischer, the German expert, has offered an entirely

different analysis. He argued that the idea of *Einkreisung* was given life years earlier when Germany first decided to become a 'world power'. The fear of 'encirclement' by hostile enemies was employed by German propagandists as a pretext for the building of a great new German navy, beginning in the year 1896–7. Fischer wrote:

> The year 1896–97 . . . marks the end of the nineteenth century for Germany. . . . The idea of building a fleet – an idea designed to combat political stagnation at home – could be made acceptable . . . only by creating the impression . . . that Germany was in grave danger. This would have to be accomplished by a propaganda campaign . . . the nation as a whole would have to accept both an image of itself as a world power and the consequent necessity of naval power. . . . In short, the idea of 'encirclement' was born along with the naval policy and before there was any sign of the Entente. This argument for building a fleet had to be exploited and amplified more and more . . . until it became a staple of Admiralty propaganda, and in the course of these . . . years, the danger of 'encirclement' which had at first been used as a mere pretext for expanding the navy, gradually became a bitter reality.[23]

In July 1910 when Sir Edward Grey decided to intervene in the situation created by the impasse at the International Conference on Aerial Navigation he did so by proceeding to put Fritz Fischer's 'bitter reality' into effect. He prepared a letter for the British Ambassador in Paris, Sir Francis Bertie, later Lord Bertie of Thame, instructing him to approach the French Government at once in order to win its support for the position taken up by the British delegation at the conference. At the same time exact copies of this letter were also separately addressed to 'His Majesty's representatives at Christiana; The Hague; Stockholm; Brussels; St Petersburgh; Lisbon; Madrid; Berne; Rome; Copenhagen'. The letter to Bertie and these other British diplomats in the various capitals of Europe explained:[24]

> With reference to previous correspondence respecting the International Conference on Aerial Navigation, I transmit herewith, for your most confidential information, a memorandum setting forth the views of His Majesty's Government respecting

the draft convention drawn up at the recent meeting of the conference. This document is sent to you as an *aide-mémoire* in order that you may explain the views of His Majesty's Government in strict confidence to the Government to which you are accredited. . . .

You should inform the Minister for Foreign Affairs that His Majesty's Government will be glad to learn that his Government, on a mature consideration of the whole matter, share the views of His Majesty's Government, and are prepared to support the British delegates in urging them upon the conference when it meets again at the end of November.

The memorandum mentioned by Grey in his circular letter was marked 'Secret'. It set out the formal opinion of the British Government with respect to the draft convention already prepared by the delegates of the various nations at the Paris Conference. It began by explaining that Britain had asked for an adjournment of the conference in order to afford each of the powers ample time to consider those proposed regulations which would 'have the effect of fettering the freedom of action of the various Powers . . . in a manner which might prove detrimental to their own interests in time of war or strained relations, or might further hamper them in fulfilling their obligations as neutrals'. The memorandum pointed out, as the Admiralty letter of 21 July had advised, that airships and aeroplanes could only be used for 'purposes of war'. It followed that it was 'supremely important' for each country to study the draft convention 'from the aspect of its bearing on the rights and duties of belligerents and of neutrals'.[25]

The British Government believed, the memorandum declared, that the 'tendency' of the proposed regulations would deprive the various powers of the 'control of air-ships within or passing over their own territories. . . .' The proposed convention proceeded on the principle that

> the air is free to all and that every reservation is grudgingly admitted as a concession made by Europe to an individual State. It ignores the elementary right of a State to take each and every measure which it considers necessary for self-preservation, and only concedes this right as if it were a privilege. It thus exposes a State, and particularly a weak State, to constant pressure to withdraw its reservation.

The Paris Conference and its Consequences 149

It will be recalled that when the Secretariat of the Committee of Imperial Defence prepared its strategical analysis of the German proposals, in the previous month, emphasis was placed upon the fact that weak states like Holland, Belgium, and Denmark might hesitate, in time of crisis, to provoke Germany, their powerful neighbour, by restricting the movements of German 'privateer airships'.[26] This conclusion was reflected in the heart of the Foreign Office memorandum of 29 July. It declared:

> The special restrictions in the Convention regarding the movements of military aircraft . . . do not appear sufficiently to safeguard national interests as, in the absence of well-defined rules respecting privateering and the rights of neutrals in aerial warfare, private air-ships would be used for reconnoitring purposes, and during the important period of strained relations their use in this manner would be extremely difficult to prevent.
>
> For the above reasons His Majesty's Government adhere to the principle . . . that 'no regulations should be framed which in any way . . . exclude or limit the right of any State to prescribe the conditions under which the air above its territory should be navigated', and they are of opinion that in the present experimental stage of aerial navigation, more particularly in its application to war, it would be premature for any nation to agree to regulations and restrictions which in any way fetter its sovereign right to make such regulations as may be necessary for its own security or for the proper discharge of its duties as a neutral.

Asquith's Standing Sub-Committee of the Committee of Imperial Defence, charged by him earlier in the month to consider and report upon the issues thrown up by the Paris Conference, met at exactly this time and studied, among other papers, this Foreign Office memorandum. The Committee's conclusion was: 'The Memorandum prepared for the use of His Majesty's representatives in certain countries which had taken part in the International Conference on Aerial Navigation, was agreed to as correctly stating the views of the Sub-Committee'.[27]

Subsequently, Sir Edward Grey decided it would be desirable to arrange a preliminary meeting between the British and French delegates at the conference. This meeting, Grey believed, might enable the two groups to discuss those controversial issues which would arise when the conference reassembled. It would also provide the delegations with the opportunity to 'arrive at a general agreement between the two countries on these points'.[28]

VII

The International Conference on Aerial Navigation produced other consequences of substance. When H. H. Asquith decided to secure the advice and assistance of the Standing Sub-Committee of the Committee of Imperial Defence, on 18 July, one of the members of this body was Lord Esher, the influential technical authority who often co-operated with the Liberals when they worked out their strategic and defence policies. It will be recalled that when the British Government undertook its first serious examination of the entire question of 'Aerial Navigation', in October 1908, Asquith had appointed a powerful Sub-Committee of the Committee of Imperial Defence for the purpose, and he had selected Lord Esher as chairman of this body. When this Sub-Committee issued its report and conclusions, at the end of January 1909, it recommended that the experiments with aeroplanes then being carried out by the War Office at its technical establishment in Farnborough 'should be discontinued'.[29] Now, almost two years later, Lord Esher changed his mind. Early in October 1910 he submitted a 'Note' to the Committee of Imperial Defence urging the value of aeroplanes for the use of the British Army and, indeed, proposing the establishment of a British Air Corps. One British aeronautical expert has called Lord Esher's 'Note' nothing less than an 'instrument of recantation, since it withdrew without reservation much that had been said before'.[30]

By October 1910 Lord Esher, ever vigilant in the service of his country, had become impressed by the 'immense advance in the construction and handling of aeroplanes' since his Sub-Committee's work had been concluded in the previous year. In his 'Note' he called the attention of the Committee of Imperial Defence to an article published in *The Times* of 3 October 1910. The article dealt with some recent activities of the French Army when 'Interest centred upon certain special features of the manoeuvres . . . the use of dirigibles and aeroplanes . . .'. The author of this article, which Esher reproduced as an accompaniment to his 'Note', was the famous Military Correspondent of *The Times*, Colonel Charles À Court Repington.[31]

By 1910 Colonel Repington, a brilliant if untrustworthy military thinker, had become convinced that aircraft were certain to play a significant role in any wars of the future. In July of that year he published the second of two articles in *Blackwood's Magazine* which began by stating that 'When M. Blériot landed at Dover he opened a new chapter in the military history of the British Isles'. Years later, in his memoirs, Repington, who possessed a good opinion of himself, wrote:

> there was one subject which I took up in Blackwood when I wrote two articles under the title of 'New Wars for Old!' This subject was the submarine and aircraft . . . in my second article, which dealt with the future of aircraft, I stated that superiority in the air had become an imperative obligation to the State . . . and I laid stress on the advantages which Germany would derive from her aircraft and the damage they were sure to do to us, both in scouting and bombing. . . . These ideas were extremely revolutionary when I propounded them, and my articles were bitterly attacked.[32]

As a result of his new interest in aircraft Repington travelled to France to interview General Jean Brun, a former Chief of the French General Staff, and later Minister for War. Since 1908 Brun had been experimenting with aircraft for purposes of tactical and strategic reconnaissance and, as a result, French experience in this sphere was more extensive than anything yet attained by the British. Repington's conversations with Brun formed the basis of his article for *The Times* of 3 October. Lord Esher was so impressed by this article he saw to it that it was printed for the information of the Committee of Imperial Defence; and it was attached to his 'Note' of 6 October.

The French authorities, convinced by their recent experiences in the field, told Repington that in their opinion huge rigid dirigibles of the Zeppelin type were too cumbersome to be employed for the purposes of land warfare. His article of 3 October revealed that in the French view aeroplanes would prove to be superior for 'tactical reconnaissance in the air on land frontiers'. Repington paid particular attention to the opinions of General Brun who believed aeroplanes possessed 'an incontestable superiority over dirigibles in speed' and that for land warfare they must form the bulk of the air fleet of the future. Dirigibles were still of value because at that period of aeronautical development they could still travel greater distances than aeroplanes and thus might be employed for strategic purposes. General Brun's object was to secure the 'largest possible number of dirigibles and aeroplanes, and the largest possible supply of trained personnel'. In embarking upon this new phase of military activity the General did not propose to confine himself merely to the military resources of France: 'He is determined to profit by all the resources of private industry and to procure, above everything else, a select personnel, well trained and organized, which will receive a permanent impulsion from the general staff'.

In his article Repington offered certain conclusions of his own for the information of his readers. He believed that while the aeroplane would 'revolutionize the whole service of reconnaissance' it would never replace cavalry because it could not perform by night or on misty days or in very strong winds. He pointed out that the French experiments dealt especially with the issue of tactical reconnaissances. He believed that the aeroplane capable of extended strategic reconnaissance was 'sure to be invented' but it had not yet appeared. The large dirigible still led the field in the distances it could travel and in its lifting powers, subjects of moment to a 'State which desires to destroy hostile ships, docks, and maritime establishments by the discharge of projectiles and explosives'. These are matters, Repington concluded, 'of greater concern perhaps to us than to the French. . . .'

A second news item, published in the Press on 4 October 1910, also contributed to Lord Esher's desire to prepare his 'Note' for the attention of the Committee of Imperial Defence at this time. This was an official statement issued by the War Office. The statement declared:

> With a view to meeting army requirements it has been decided to enlarge the scope of the work hitherto carried on at the balloon school at Farnborough by affording opportunities for aeroplaning as well as by developing the training in employment of dirigibles more fully than has hitherto been the case. The object to be kept in view will be to create a body of expert airmen, both officers and other ranks, from which units capable of acting with troops operating in the field can be drawn.[33]

The official statement went on to reveal that Major Sir Alexander

Bannerman, Bart., RE, had been selected to replace Colonel Capper as Commandant of the Balloon School and that he would be 'at the head of the new organization'. Although these changes were a part of R. B. Haldane's continuing scheme to establish a capable and competent British Air Service it should be mentioned that they also reflected the negative and hostile attitude of those conservative elements in the War Office led still by General Sir William Nicholson, Chief of the Imperial General Staff. One aeronautical expert commented later:

> The War Office attitude was demonstrated still further in its treatment of the purely military Balloon School. Colonel Capper retained his post until he became due to promotion from brevet to substantive rank on 7 October 1910. The War Office refused to let him continue as Commandant at this higher rank, but instead downgraded the post to that of Major, and appointed Major Sir Alexander Bannerman as his successor.[34]

Nor should it be supposed that Sir Alexander Bannerman possessed any experience of aeroplanes or that he was one of those 'air enthusiasts' of the day who believed in the military possibilities of the new technology. As we have seen in an earlier chapter, he was an amateur 'gas-bag balloonist' who was convinced that it would not be possible to hit a target from the air with any degree of accuracy and who doubted whether 'one shot in 10,000 will, from a height of 5,000 feet, hit a battleship'.[35]

In his 'Note' Lord Esher was at pains to explain that conditions had changed since his 'Aerial Navigation' Sub-Committee had reported so negatively about aeroplanes, in January 1909. He now suggested it was 'evidently desirable' to purchase thirty or forty machines for the use of the army but he raised further pertinent questions: 'Is this new arm', he asked,

> for that in reality is what it is, to be handed over like a mechanical contrivance of war to the navy on the one hand and the army on the other, to be used as an adjunct to naval and military forces. . . . Or should a new corps be specially raised, not necessarily sailors or soldiers, trained to handle dirigible balloons, and especially aeroplanes. . . . For these reasons I suggest that the time has arrived when the Committee of Imperial Defence should consider the following proposals:[36]

1. Whether a corps of aviators should be formed.
2. Whether a school of practical aviation should be established, and if so, where.
3. From what class the personnel should be selected.
4. By what Department of State such a school should be organized and controlled.
5. Whether the personnel of officers and men should be handed over in time of peace respectively to the navy and army authorities, and left under their control.
6. Or, whether they should be interchangeable, and should be lent to the navy and army authorities for the purpose of manoeuvres.
7. Under what head of the estimates the cost of the purchase of aeroplanes should be charged.

These far-ranging suggestions were followed, in Lord Esher's 'Note', by a conclusion that was nothing less than a warning to his colleagues of the Committee of Imperial Defence:

I am concerned to make these remarks and suggestions, because in view of the German fleet of airships, and of the use of aeroplanes in the recent French manoeuvres, it must be obvious to those interested in forecasting the probabilities of the next great war, that unless some immediate and rapid development takes place in aerostation and aviation by the British naval and military authorities, Great Britain may find herself labouring under grave disadvantages, for which the Committee of Imperial Defence would not unnaturally be blamed.

VIII

Meanwhile, in Paris the British delegates were preparing for the preliminary meeting with their French colleagues so that they could work out, in concert, the policies they should adopt when the International Conference reassembled. On 18 October they sent a letter to the Foreign Office asking for further instructions. As a result, the Standing Sub-Committee of the Committee of Imperial Defence reassembled, and, after deliberation this body advised the Foreign Office about the instructions to be sent to the delegation.[37]

The initiative, however, now lay in Paris. There, Sir Francis Bertie replaced Sir Douglas Gamble as the chief of the British delegation. In order to assist him in what was now looked upon as a delicate task the authorities in London sent Bertie three technical experts of high quality. These were: Captain M. P. A. Hankey of the Secretariat of the Committee of Imperial Defence; Admiral Henry Jackson, later a First Sea Lord; and Colonel George Macdonogh of the General Staff, later the Director of Military Intelligence in the War Office.

This was Hankey's first diplomatic misson. He realized from the start that the problem confronting him and his colleagues was an exceptionally difficult one. They were not in Paris to argue against the proposals advocated at the conference by the delegates of Imperial Germany. They were there to try and win the French to the British side so that the two nations could co-operate, as partners, when the business of the conference resumed. Years later Lord Hankey wrote of his experience: 'I had taken a very active part in bringing to light the dangers of the draft convention'; and he also explained that 'above all, we were not in line with the French Government on this subject'.[38]

As a preliminary Hankey and his service colleagues called at the French Ministry of Marine and then at the War Office where they received a 'sympathetic if non-commital hearing'.[39] However, the chief and very formidable obstacle in their path was the attitude adopted by the French President of the International Conference, M. Louis Renault, a great authority on international law. He agreed with the stance taken up by the German delegates to the conference and he was not prepared to change his mind for any reason.

Renault had good cause to believe in the validity of his own views. He had served as a delegate of the French Government at the Peace Conference at the Hague in 1899 and he was a member of the permanent Court of Arbitration. In 1907 he was awarded the Nobel Peace Prize. He was Professor of International Law in the University of Paris, a Commander of the Legion of Honour, and Jurisconsult of the Ministry of Foreign Affairs. He had received an honorary degree from Oxford and he was a Fellow of the British Academy. His legal opinions usually counted in the deliberations of the French Government and he hoped they would count still in any decisions taken at the International Conference on Aerial Navigation.

The meeting between the British and French delegations was postponed a number of times because of the French attitude but they came together at last on 7 November 1910. In these discussions Sir Francis Bertie believed he possessed an almost impregnable position. As Lord Hankey put it years later: '. . . the dice were loaded in our favour, for Bertie knew that, whatever happened, the French Minister for Foreign Affairs would not adopt the German point of view'.[40]

Nevertheless, Renault refused to change his mind. Although Bertie teased him unmercifully and the French service delegates whispered messages of sympathy to Hankey the chief French delegate held firm to the opinions he had expressed from the beginning of the conference. In these circumstances Bertie decided upon a diplomatic initiative of his own.

Bertie's first idea was to call upon Stephen Pichon, the French Foreign Minister, but Pichon was unable to see him. Instead, Bertie visited the Minister of War, General Brun and told him of the unpleasant meeting with Louis Renault. At this interview Bertie gave the General a copy of the Foreign Office memorandum of 29 July 1910, which, as we have seen, had been circulated to the various British ambassadors in several of the capitals of Europe. He informed Brun of 'how uncompromising M. Renault's attitude had been' and 'expressed the hope that the French War Department would not adopt his views, for His Majesty's Government could never accept what M. Renault proposed. They considered that the proposals as they stood would be dangerous to national defence; and the security of England had become an important French interest'. The General replied that he was 'not at all in favour of entering into engagements in a hurry on a matter about which so little was known'. He also declared that he 'quite appreciated our apprehension that we might have unwelcome visitors coming from a distance and making use of Holland and Belgium as bases for their flights'.[41]

Later on that day, 9 November, Bertie visited Admiral Boué de Lapeyrère, the Minister of Marine, and complained to him also about Renault's uncompromising attitude. In response the Admiral stated pithily 'that the important military and naval considerations involved in the questions before the conference must not be settled in a hurry by jurisconsults'.

On 12 November Bertie managed to see the Minister for Foreign Affairs. In this instance his reception was somewhat different from the two earlier ones with General Brun, and Admiral de Lapeyrère. After Bertie criticized Louis Renault, Pichon said to him: '. . . the position of France as regarded Germany was rather delicate, for the French Government had convoked the conference . . .'. However, Pichon at once took care to reassure his British visitor: '. . . he assured me', Bertie wrote to Sir Edward Grey, 'that the French Government would proceed in accord with His Majesty's Government'.

Pichon then asked the ambassador to provide him with a document setting out those portions of the draft convention to which the British Government objected. This Bertie undertook to do and later he asked Admiral Jackson to draw up such a paper, for communication to the French Minister for Foreign Affairs. Pichon went on to mention that a convention might be signed by France, Germany, Italy and Switzerland. He also told Bertie the Italian Ambassador had suggested that in view of the opposition of England and Switzerland 'to some of the most important provisions of the draft convention, the best course would be to adjourn the conference'.

Bertie's reply was an exact reflection of the attitude the British authorities had come to by this stage of the proceedings. He told Pichon:

> His Majesty's Government have no objection to the conclusion of a convention to regulate matters ripe for decision. What they desire is to exclude those which require much fuller consideration than had been given to them by the several Governments represented at the conference.

Despite Pichon's assurances no accord or agreement between the British and French could be secured and the International Conference on Aerial Navigation was adjourned *sine die*, at the end of November 1910. It did not meet again.

IX

Although the international conference proved a failure the British Government attempted to secure at least a part of its objects by resorting to domestic legislation. A Bill entitled 'the Aerial Navigation Bill 1910' was drawn up and in its preamble it was asserted that 'the air above all parts of His Majesty's Dominions and the territorial waters adjacent thereto was inviolable'.[42] Owing to a crowded session this Bill did not become law but in 1911 a shorter Bill was produced which, in addition to giving effect to this preamble, authorized the appropriate authorities to suppress dangerous flying and also scheduled certain areas, ports and wireless telegraphic stations as places over which flying by foreign and by British civilian pilots was prohibited. This Bill became law early in 1913.

There were other results. In the Foreign Office Sir Edward Grey and his colleagues in that important department of the State were more alert about the air danger to the country than they had been before. In the War Office and in the Admiralty the attitude and the proposals of the German delegates in Paris had caused alarm even if some of the older service leaders continued in their belief that aircraft would not prove to be of very great military value in time of war.

A further significant development was the effect upon the outlook of Captain Hankey. His experiences in Paris, which he remembered for the remainder of his life, took place almost at the start of his long career in the Secretariat of the Committee of Imperial Defence. As his power there, and in similar posts, increased his influence in the nation's councils became more and more important and from this time he kept in his mind the German air threat to his country. The Paris Conference, at which he had seen 'the whole of Europe . . . divided into two camps . . .'[43] marked the beginning of his interest in air power as a factor of consequence in Britain's strategic arrangements. Moreover, many of his contemporaries in the service departments continued to look upon aircraft as instruments that could be employed for purposes of reconnaissance only but, as we have seen, he was already convinced that they could be used by the Germans to bomb targets in Britain. In 1912 an Air Committee was set up as a permanent sub-committee of the Committee of Imperial Defence and Hankey saw to it that he became its first secretary. When he analysed the results of the International Conference on Aerial Navigation of 1910, Lord Hankey's official biographer took care to point out that Hankey managed to involve himself '. . . in the military and naval implications of aviation from the very beginning'.[44]

8
The Agitation for a National Air Force

Holt Thomas writes to the *Daily Mail* – Brooklands – Hendon – Eastchurch – Larkhill – Northcliffe's campaign – the War Office and the Air Corps – Air Battalion of the Royal Engineers – Sir Alexander Bannerman's Opinion – Haldane defends his accomplishments – The Royal Aero Club and the Admiralty – C. R. Samson's History – Claude Grahame-White and the Parliamentary Aerial Defence Committee – The Hendon Air Display – A Deputation sees the Ministers – Bitter Party Politics in 1911 – The Agadir Crisis – A National Scandal – Mutually Exclusive Plans for War – Winston Churchill becomes First Lord of The Admiralty – His Interest in Aviation – The Mayfly Disaster – An Attack in Parliament – Asquith's Response.

I

Colonel Repington and Lord Esher were not the only ones in Britain to be impressed by the French employment of aircraft in the military manoeuvres of the autumn of 1910. Lord Northcliffe's friend, George Holt Thomas, a British pioneer in the field of aircraft manufacture and an 'air enthusiast' of quite exceptional vigour, also attended the manoeuvres. He was so affected by what he saw that he wrote a long letter to the *Daily Mail* upon his return to England in order to implore the paper to publicise the novel importance of aeroplanes in the conduct of modern land-warfare.

In his letter published in the *Daily Mail* of 17 September 1910 Holt Thomas explained that 'The chief point . . . to which I would like to see you give great publicity is the fact that the generals in command of the opposing forces were compelled entirely to change their plans of campaign owing to aerial scouts and the information they were able to give . . .'. He took care to contrast the French interest in aeroplanes with the neglect of them shown by the British military leaders up to that time: 'I ask you to give publicity to the success of aviation as adapted to military purposes because it is high time that our authorities took some steps to follow the example of the French'.

Lord Northcliffe's response was to launch a new campaign in the *Daily Mail*. Its object was to force Asquith's Liberal Government into a more active course with respect to military aeronautics. Eventually, Northcliffe offered another huge prize in the sum of £10 000 to the winner of an air race called the Circuit of Britain. Thousands flocked to the newly developing aerodromes in various parts of the country in order to watch the competitors in flight. To the great disappointment of the English people, the prize was won by a French naval officer and aeroplane pilot named Jean Conneau. This result served yet again to make clear to many in Britain that their country was falling behind in the new technology, and in a most lamentable way.

George Holt Thomas was a formidable man and one to be reckoned with in this period of British aeronautical history. He was a member of a wealthy family, owners of the *Graphic* newspaper. In 1906 when the Brazilian pioneer Alberto Santos-Dumont was making his first tentative aeroplane hops at Bagatelle near Paris Holt Thomas offered, through the medium of the *Graphic*, a prize of £1000 for a flight of one mile. It was Holt Thomas who helped to inspire Northcliffe's interest in flying at that time.

It has been said of Holt Thomas that he had an instinct for the shape of things to come. He wanted to make certain his country was prepared to fight a war in the air if it should be necessary and at the same time he realized that a new industry, in which money could be made, was bound to develop as the technology of aircraft construction advanced. He was an active and energetic advocate who never hesitated to make public his various criticisms of the Liberal Ministry in the field of aeronautics.

He played a major role in the development of some of Britain's earliest aerodromes. In 1909 Holt Thomas saw Louis Paulhan flying at the famous Reims aviation meeting and was so impressed with him he decided to bring the Frenchman to England there to demonstrate his skill and to attract popular, and also official, interest to the subject of aeroplanes. He persuaded the proprietors of Brooklands, the motor race-track, to prepare an aerodrome and Paulhan flew many demonstration flights at this ground. The owners of Brooklands, Mr and Mrs Locke-King, now decided they would turn part of their establishment into a permanent airfield. Aeroplane hangars and sheds were built as regular additions to the place and in this way the great Brooklands aerodrome, where flying first attracted popular attention in England, came into being.

Shortly after, Holt Thomas persuaded Paulhan to travel to England again, in April 1910, in order to attempt to win the £10 000 prize offered by the *Daily Mail* for the first aeroplane flight from London to Manchester. One of his initial tasks was to find a suitable starting place for his champion. He selected a field at Hendon in the northern suburbs of London. After Paulhan's success Claude Grahame-White, the loser in their competition, visited Hendon and chose it as the place where he would open a flying-school in Britain. Thus began one of the early centres of flying in the country. Shortly after, it came to be known as the London aerodrome. At about the same time Eastchurch aerodrome on the Isle of Sheppey was developed as a centre for the training of a small number of naval aviators while Larkhill, on Salisbury Plain, was employed for the instruction of military airmen.

Lord Northcliffe did more than print Holt Thomas's letter in the *Daily Mail* of 17 September 1910. He used the letter to begin a new phase of his campaign designed, in his own phrase, 'to make the nation air-minded'. The first leading article in the *Daily Mail* of that day was devoted to the subject of military aeronautics. Entitled 'The New Arm', the leader argued that the aeroplane was now a weapon of the 'utmost importance in war' because of its ability to enable a pilot and his observer to discover the military dispositions and arrangements of an enemy force. It stated also that land battles of the future would be preceded by a 'battle in the firmament above' so that one side or another would dominate in the air and thus decide the eventual issue of the fight on the ground. This article concluded: 'More than ever is the possession of a force capable of holding its own in the air a matter of necessity and of national security to the British people'. It was obvious to all concerned with the subject that the British Army did not possess such a force and that the French and Germans did.

A few days later the British Army began its own autumn manoeuvres. The first leader in the *Daily Mail* for 20 September pointed out that, unlike the French, the British soldiers would employ one small dirigible and only two borrowed aeroplanes for the purposes of reconnaissance. The conclusion was blunt: 'Our Army needs not two flying machines, borrowed for the purpose, but a whole fleet of its own'. The article then attacked the cautious, leisurely, and deliberate aeronautical policy adopted by R. B.

Haldane at the War Office. It also stressed the value of an intelligent and aroused public opinion if the official policy was to be altered into a more positive course:

> We in this country should see to it that action is taken here and now. Let it be remembered that it was only by the pressure of public opinion in France that the French War Office was induced boldly to purchase a large number of machines. . . . No one on this side of the Channel can rest content till the British War Office has followed suit. . . . The public should not allow its attention to be diverted from that aim by the hackneyed argument that it will be well to wait till the aeroplane is further perfected. The nation has waited long enough.

These issues were to become the themes of Lord Northcliffe's campaign.

As we have seen in an earlier chapter, it has sometimes been argued that Northcliffe engaged in campaigns of this type merely in order to sell more newspapers. He used, it has been said, sensational articles and 'scare exploitations' in order to enhance his circulation figures. This was not always the case, however. Asquith's Liberal Government would not readily alter the aeronautical programme and policy devised for it by R. B. Haldane. As a result this particular Northcliffe campaign or agitation continued for many months. In June 1911 a 'sensational' article appeared in the *Daily Mail*. It dealt with the effects to be experienced as a result of bombardment from the air by airships. Contemporaries, because they had no experience of it, could only guess at the results of air bombing. Some argued there would be devastation on a serious scale while others believed there would be but little effect in consequence of such attacks. Lord Northcliffe, in this period, accepted the latter point of view. When his *Daily Mail* published an article that disagreed with this opinion he at once wrote to Thomas Marlowe, Editor of the paper, in order to dismiss as 'rubbish' the idea that bombing from airships could be an effective form of attack at that stage of their development. He ordered Marlowe to publish more sober articles on the subject:[1]

> I read that very good letter . . . about bombs dropped from airships being of no effect. This I know to be the view of all high authorities on the subject, and I wish you would get another . . .

article, which you might back up with a leader calling attention to this ridiculous bomb theory. I discussed the whole thing with German, French, and American authorities. . . .

The Paper has been very good lately, but I was sorry to see the airship rubbish. And it is something of an anticlimax after all my experience and study in connection with this airship business to find that my own newspaper should behave so foolishly.

II

While the *Daily Mail* and other newspapers, aided by enthusiasts, advocates, and partisans, continued to agitate for the creation of a national air force, the responsible authorities in the War Office approached the problem in their own way. On 12 July 1910 a conference had been held under the presidency of Sir Charles Hadden, the Master-General of the Ordnance. This War Office conference decided 'that it was desirable' to organize a new unit to be known as the 'Air Corps' which, in time, would replace the old Balloon School at Farnborough. This new unit would be 'instructional and experimental in its nature'. Shortly after, on 30 July 1910, the Army Council, with R. B. Haldane present, decided that such a unit should be organized.[2]

This was merely the first step in a complicated progression. In October 1910, as we have seen in the previous chapter, the War Office issued an official statement declaring that the scope of the work carried on at the Balloon School would be enlarged and that Major Sir Alexander Bannerman would replace Colonel J. E. Capper as Commandant. Training in the employment of aeroplanes would now be carried out at Farnborough while practice with dirigibles would be continued and extended.

In the War Office the authorities continued to work, at their own pace, upon the problem of the 'Air Corps'. Eventually, a new organization designed to replace the old Balloon School or Balloon Companies, as they were sometimes called, was created. This new unit was the Air Battalion of the Royal Engineers.

On 28 February 1911 an Army Order was issued. It explained that it had been decided to organize an Air Battalion whose duty it would be to train a 'body of expert airmen'. Furthermore, the training and instruction of men in handling kites, balloons, aeroplanes and 'other forms of aircraft' would also devolve upon this new unit, the Air Battalion of the Royal Engineers.

The Army Order revealed that the battalion would be organized with a headquarters and two companies. Under the new scheme airships were assigned to No. 1 Company and aeroplanes were the concern of No. 2 Company. The aeroplane company was stationed at Larkhill while the balloon unit and the headquarters were based at South Farnborough. The new scheme came into effect from 1 April 1911 and the Balloon School then ceased to exist. The old Balloon Factory where all kinds of aeronautical devices had been built for years past now became a separate but co-related unit. Mervyn O'Gorman continued as Superintendent but the title of the establishment was changed as from 26 April 1911 to the Army Aircraft Factory.

The publication of this Army Order did not succeed in reducing the criticism levelled at the Liberal Ministry because of its aeronautical policy. Even the radical *Daily News*, an organ staunchly loyal to the government, expressed its reservations in clear terms. In an article by 'Our Aviation Expert' the *Daily News* of 2 March congratulated Haldane upon the issue of his special Army Order creating the Air Battalion but it added: 'It is to be hoped, however, that this will be only the initial step, for this country stands far behind other countries in the development of dynamic flight for war purposes'.

III

The Air Battalion was merely a tentative advance in the development of a national air force. By the time it was created aeronautics had begun to pass from the experimental to the practical stage. But the higher authorities in Whitehall were uncertain about which branch of the technology to endorse. Would aeroplanes or airships develop into the most efficient engines of war? The old debate on this subject had reopened and those in charge were uncertain of the answer and unable, therefore, to fix upon the proper course to follow. The very organization of the Air Battalion, with one company concerned with aeroplanes and another concerned with dirigibles, was a reflection of the issue. In the winter of 1911–12 two important sub-committees of the Committee of Imperial Defence attacked the problem of the air force and, as we shall see, made positive recommendations fo the next forward step. The Air Battalion, in these circumstances, was looked upon only as a transitional unit that would soon be replaced by some more permanent organization. As a consequence, it received little support. In particular, as the official history put it, 'The number of aeroplanes supplied to the flying camp at Larkhill was almost ludicrously small. . . .'.[3]

The selection of Sir Alexander Bannerman as the unit's commander was not an especially happy choice. A technical expert once dismissed him as a 'stout, moustached, dyed-in-the-wool R. E. officer . . .'.[4]

Sir Alexander came from an ancient Scottish family. Indeed, he was the eleventh baronet of his line. He fought with distinction in the South African War and succeeded his father in 1901. He was subsequently attached to the Japanese Army during the Russo-Japanese War and was present at the fall of Port Arthur. Later, he was appointed a General Staff officer at headquarters and remained there until he was picked to replace Colonel Capper, at Farnborough.

This connection with the General Staff played a large part in his appointment. Bannerman was a balloon expert. In April 1912 he qualified as an aviator but before that time he made it clear to his subordinate officers in the Air Battalion that he did not believe aeroplanes would prove to be of much value in war. This attitude caused annoyance. Bannerman was not popular with the pilot officers attached to the Battalion's No. 2 Company. Nor was he a dedicated soldier. He retired from the army in 1912 at age 41.[5]

Early in February 1911 Sir Alexander spoke at a dinner meeting of the Royal Aero Club. His remarks shocked the assembled company. He said: 'For military purposes I doubt whether the aeroplane is very far ahead of what it was at the time of Wright's first flight'. The experts in his audience at once responded with loud cries of 'Oh!' and 'No, no'.[6] They were astounded by his ignorance of the technical advances made since Wilbur Wright's pioneering achievement. Bannerman's remarks produced a storm of hostile criticism in the Press. It lasted for weeks.

The *Daily Mail* of 4 February was first in the field and the technical journals like *Flight* and *The Aero* made haste to condemn the Major's 'appalling' ignorance. However, the most significant attack appeared in the *Westminster Gazette* of 7 February. This journal, controlled by one of the most powerful Liberal Editors of the time, J. A. Spender, was usually unswerving in its loyalty to Asquith's Government. It now condemned Bannerman's remarks as poisonous and ruinous. Referring to him, the *Westminster Gazette* declared:

> Needless to say no one agreed with him; but coming as it did from a man holding such an important position the effect of his words can hardly fail to be baneful. His was evidently a studied pronouncement; and if it reflects the official view, as most people thought and still think it does, it is high time that some change should be brought into the mind of the War Office.

A further blow was struck in *The Observer* for 12 February. Although it did not mention him by name *The Observer* attacked Sir William Nicholson, Chief of the Imperial General Staff, as the real culprit. The paper explained:

> The trouble at the War Office with regard to aviation is not Major Sir Alexander Bannerman . . . it is essentially due to the presence at the War Office of an engineer officer of exalted rank, who is not merely frigid towards aviation, but is even violently opposed to the idea that it has any military value. It is unfortunate that a man with such ideas should happen to be in a position in which he is able to place obstacles in the way of every suggested forward step.

It remained for the *Daily Mail* to carry the attack to a higher level. The *Daily Mail* of 13 February placed the blame for the situation upon R. B. Haldane himself. The first leading article for that day explained:

> When Mr Haldane went to the War Office five years ago he told us that he would introduce 'clear thinking'. But the result appears to be rather the reverse. The Secretary of State has absorbed the traditional atmosphere of the War Office. . . . An institution which can produce Sir Alexander Bannerman must be capable of a great deal of intellectual resistance, and it has evidently had a good deal of influence upon the Minister for War.

R. B. Haldane was not visibly moved by this agitation in the

Press. He continued to demonstrate an unruffled calmness in his public conduct. Later in the month he prepared a memorandum, for publication, dealing with the Army Estimates for the year 1911–12. In this memorandum he coolly pointed out, with pride, that there would be a decrease in the estimates of £70 000 when compared with the total required for 1910–11. When he turned, in this paper, to the development of military aeronautics for the coming year he was almost serene. His memorandum was published in full in the *Morning Post* for 25 February 1911. In the aeronautical section of it the Secretary of State offered no defence of Sir Alexander Bannerman but he declared that 'a good deal had been accomplished' in the previous year. He explained, suavely, that the War Office had received help in the acquisition of dirigibles from the proprietors of the *Daily Mail* and the *Morning Post*; and he allowed that it had also benefited from the advice offered by the Parliamentary Aerial Defence Committee. He remarked that the Balloon Factory had been reorganized and enlarged at Mervyn O'Gorman's direction while he described O'Gorman as 'an expert throughly in touch with all the developments of modern science in this field'. In the new estimates £85 000 would be provided for the acquisition of aeroplanes and dirigibles. He mentioned the Balloon School would be reorganized and transformed into an Air Battalion. Trials, he revealed, would be conducted with aeroplanes and when these were completed particular types of machines would be chosen for further development. He refused to admit the existence of any serious problems in the matter of British military aeronautics. He was content with what he had accomplished for it up to this point.

IV

In the Admiralty not much attention was paid to the subject of aviation. The chief aeronautical project of the Royal Navy in this period lay in the construction of the great dirigible known in the service as *No. 1 Rigid Naval Airship*, and more popularly as the *Mayfly*. There were high hopes for this vessel which was looked upon as an instrument of war of quite remarkable potential.

It has been well said that 'The Board of Admiralty (1910–11) were not impressed with the naval possibilities of the aeroplane . . .'.⁷ Nevertheless, pressure from outside the service forced the Admiralty into action, however limited in its nature.

In November 1910 the Royal Aero Club informed the Admiralty that one of its members, at once generous and patriotic, was prepared to lend two aeroplanes, free of all charges, so that naval officers could learn to fly. Moreover, the club itself proposed to allow these officers free use of its aerodrome at Eastchurch, Isle of Sheppey. In addition, members of the club who were certified aviators were willing to act as instructors so that the proposed project could be completed successfully.

On 6 December 1910 a General Fleet Order was issued by the Commander-in-Chief at the Nore setting out the details of the Royal Aero Club's initiative. The number of officers who would be allowed to accept this offer was, by Admiralty direction, limited to four. They were to use any experience they gained during their training for the good of the naval service; and they would be required to make periodic reports to the Commander-in-Chief at the Nore. These reports would be sent on to the Admiralty. It was decreed that the officers selected would have to be unmarried and that they would have to take up membership in the Royal Aero Club at their own expense. The Admiralty stipulated there was to be no flying on Sundays. The Admiralty, after an interval, issued a call for volunteers and about two hundred officers came forward. The naval contingent of four arrived at Eastchurch on 1 March 1911. Each had been given six months' leave of absence on full pay. This was the origin of what later developed into the famous Eastchurch naval air station.

Among the officers selected for training at Eastchurch was Lieutenant C. R. Samson, RN. He developed into one of Britain's most capable and dashing pilots. Later, one of his superiors, Admiral Sir Murray Sueter, said of Samson: 'He seemed to be in his element taking risks. . . . He always reminded me of what one reads of Francis Drake'. A subordinate officer later wrote of him: 'he was the best leader I ever served under'.⁸ In 1918, when the war ended, Samson was asked to prepare a brief history of British naval aeroplanes and seaplanes. In his paper Samson revealed concisely and succinctly exactly what he and his fellow pioneers achieved for British naval aviation, even though they carried out their several tasks without the enthusiastic support of their superiors. Samson's document, or history, merits careful attention.

The Agitation for a National Air Force

It began with the offer made by Frank McClean of the Royal Aero Club:[9]

> Mr Frank McClean offered 2 Aeroplanes and his sheds free of cost to the Admiralty for the use of the Navy. Mr Cockburn volunteered to teach Naval Officers to fly free of cost.
>
> The Admiralty called for volunteers. About 200 names were sent in and 4 were picked. Myself, then a Lieutenant R. N. R. Gregory, Lieut. R. N. A. Longmore, Lieut. R. N. and Wildman-Lushington, Lieut. R. M. A; the latter was sick, so he was replaced by Lieut. Gerrard, R. M. L. I. We were sent to Eastchurch to undergo a course in Technical work at Messrs. Short Bros., and in theory under Mr Short, in flying under Mr Cockburn, using Mr McClean's two Aeroplanes.
>
> I was the first Naval Officer to take my Certificate, which I did in April 1911. . . .
>
> We went on flying McClean's Aeroplanes until October 1911, and then I went up to the Admiralty with proposals. They bought the two Aeroplanes . . . and sent me 12 Naval ratings to start a school and experimental work. I was put on the Committee of Imperial Defence to discuss and propound schemes for Naval and Military Aviation. . . .
>
> Experiments we carried out in 1911 were:– Lieut. Longmore and Mr Oswald Short designed and tested Air bags for a Land Aeroplane, a machine was alighted successfully in the water. I designed and had built in Chatham dockyard a platform for flying off the decks of ships. . . .
>
> Other experiments we *were the first to make* were:–
>
> (1) Bomb dropping. I dropped a 100 lb. Bomb (dummy) in March 1912.
>
> (2) A machine gun was fitted and fired (the first time this was ever done) to the Short Seaplane . . . The gun was fired from the air with success.
>
> (3) Signalling with Klaxon horns and flashing lamps . . . .
>
> In 1913 Naval Stations began to develop all round the coast, and officers and men we had taught were sent to them.

V

While Sir Alexander Bannerman struggled with the problems of his new command and Lieutenant Samson and his friends settled in at Eastchurch, the agitation for an adequate national air force continued. It was at this time that the *Daily Mail*, in a leading article published on 21 March 1911, announced the details of its second £10 000 prize. This Circuit of Britain, as it was called, would require competitors to race for more than 1000 miles in the air. The article explained that one object of the competition was to reveal to the people of Britain, in every part of the country, the great recent advances that had been made in aviation. There was a second object connected with the prize, also: 'This great race of 1,000 miles should also awaken our War Office to the full importance of the new art, the value of which in war cannot any longer be denied'.

On 4 April the Parliamentary Aerial Defence Committee met in the House of Commons in order to listen to a lecture by Claude Grahame-White who, by this time, had become an active partisan for a strong military air force. As a result the Committee decided to ask Lord Haldane,[10] the Secretary of State for War, and Reginald McKenna, the First Lord of the Admiralty, to receive deputations so that its views could be laid before them.

At this meeting Arthur Du Cros, the secretary of the Committee, explained to the Press that by 1911 the aeroplane had become 'an indispensable weapon of war'. France, he declared, possessed almost 100 aeroplanes while the British Army had but ten, and some of these were obsolete. While Du Cros emphasized that aeroplanes would be needed to serve as aerial scouts in war-time he also placed stress upon the fact that the 'aeroplane's efficiency now gives it destructive as well as reconnoitring possibilities'. In his opinion aeroplanes could now be employed, easily enough, to carry out bombing attacks against selected targets on the ground.[11]

By this time Claude Grahame-White had established himself at Hendon aerodrome where he built aeroplanes and also taught people to fly in his school there. Lord Northcliffe had already assured him that he would support his activities in the columns of his 'numerous influential papers' while Arthur Du Cros, a very wealthy man, had lent him the sum of £10 000 so that he would have enough capital to carry his various ventures forward.[12]

With this kind of support behind him, Grahame-White spoke

boldly to the assembled members of the Parliamentary Aerial Defence Committee. He explained he could foresee three possible military uses for the aeroplane; reconnaissance; dispatch-carrying, and the bombing of targets from the air. He even predicted that Britain's vaunted new battleships of the *Dreadnought* class would be vulnerable to aerial attack. This statement at once aroused the warm attention of Admiral Lord Fisher who, years earlier, had introduced the *Dreadnought* type into the Royal Navy. Grahame-White offered to carry out a demonstration of air bombing for Fisher and others at Hendon aerodrome. This was the origin of the first of the famous 'air displays' at Hendon that so thrilled British contemporaries.[13]

This first display was held under the auspices of the Parliamentary Aerial Defence Committee. In order to attract national attention to the event invitations were sent to the Prime Minister and to other high dignitaries. When Lord Haldane agreed to attend Northcliffe saw an opportunity to further the cause. He published a special article in the *Daily Mail* of 4 May entitled 'When Lord Haldane goes to Hendon'. The article was nothing less than a plea for action by the State before Britain found herself vulnerable to an air attack or an air invasion from the Continent:

> When an invention of the first importance is so far upon the road to final success, when private enterprise has done its utmost . . . when it is at a stage of development at which it affects problems of war even more than of peace, then it is surely time that the State took it most seriously in hand. Suppose that all the hopes of the enthusiasts are baseless and that the aeroplane is never destined to become a commercial 'proposition', that will not excuse its neglect. Submarines bring us no commercial gain, nor does the Maxim . . . yet we dare not neglect them. We have ventured to neglect the aeroplane for two years . . . Armies are kept from our borders by the silver streak and the fleet. We have shut our eyes to the weapon which recks nothing either of waves or warships. Faster and faster goes its development upon the Continent. We mark time while others move forward at the double.

It was almost a national holiday at Hendon on 12 May 1911. The Duke and Duchess of Connaught and Prince Arthur of Connaught were among the thronged spectators. Asquith, the

Prime Minister, and many members of his Cabinet were there while other visitors were Lord Lansdowne, Lord Fisher, Lord Northcliffe, Lord Roberts, and Lord Rothschild. Arthur Balfour, the leader of the Opposition, and more than 200 members of Parliament were present, including Ramsay MacDonald, the Socialist; and representatives of the Army Council and the Board of Admiralty also attended.

Grahame-White began the military display by dropping, from a height of 2000 feet, a 100 lb. sandbag squarely on the outline of a battleship whitewashed on the grass. He repeated this feat several times and while some were overjoyed by the spectacle Admiral Fisher showed his concern. When the set-piece demonstrations by Grahame-White and other fliers were over Grahame-White invited his distinguished guests to fly with him as passengers. Arthur Balfour accepted the invitation at once but Asquith excused himself rather blandly by pointing out that he had greater responsibilities than a leader of the Opposition. Reginald McKenna, the stern First Lord, removed his tall silk hat and also flew with Grahame-White.

On the next day the Press expressed delight at these developments. The *Daily Mail* declared: 'It was a great day for airmanship – the greatest in the history of flying in this country'. According to the *Daily Mail* a 'stream of enthusiasts' flocked from London to Hendon. 'The flying' attracted 'at least as many spectators as a popular football match'.

Contemporaries were excited by the demonstration. Admiral of the Fleet Lord John Hay wrote to his First Lord, McKenna, to congratulate him upon his exploit in the air at Hendon. Lord John had entered the navy in 1839 and had served in the China War of 1842 and before Sebastopol in the Crimean War. He had retired from the naval service as long ago as 1897 but now, in his eighty-fifth year, he wrote to McKenna: '. . . I envied your trip aloft immensely. It is the one thing worth doing in my life that I have not had a try at . . .'.[14]

VI

About a fortnight after the Hendon display, on 23 May, a deputation of some sixty members of the Parliamentary Aerial Defence Committee was received by Lord Haldane, and later by Reginald McKenna. Despite the high enthusiasm generated by the Hendon meeting, the members of the deputation were disappointed by the attitude of the two service Ministers on this occasion.

Arthur Lee, Chairman of the Committee, began the proceedings by pointing out to Haldane that Britain was far behind other countries in aeronautical development. In reply the Secretary of State declared that 'although considerable progress had been made in the development of heavier-than-air machines, the subject of aerial science was yet in a state of evolution'. His conclusion was firm: 'it was desirable to move cautiously and not commit themselves to an idea which might become obsolete in a few months . . .'.[15]

Having delivered these judgements Haldane retired and his place was taken by Colonel J. E. B. Seely, the Under-Secretary of State for War. He candidly admitted 'there was no disguising the fact that we had fewer flying men than either France or Germany'. However, he assured the deputation that the War Office was interested in aeroplanes and he emphasized that Haldane had promised that a War Office prize would be offered 'in connection with the effort to obtain the best type of aeroplane for military purposes'.

Later, Reginald McKenna received the deputation. He asked them 'to remember that the Admiralty had devoted itself to experiments with the dirigible balloon not because the possibilities of the aeroplane had been overlooked but because by arrangement the Army had undertaken experiments with the aeroplane while the Navy had dealt with the dirigible'. The First Lord's conclusion, similar to Haldane's, was: 'At present . . . matters were rather in the experimental stage'.

The 'aeronautical enthusiasts' were not satisfied by these exchanges. Their attitude was well stated in *Aeronautics* for June 1911. *Aeronautics* was perhaps the most sober and capable of the technical air journals of this period but it expressed itself in very strong terms in its account of this deputation of 23 May. In its first opportunity to comment on the deputation *Aeronautics* declared:

> The up-shot of it is that the War Office actually admit our hopelessly backward position. To remedy this, they are giving their careful consideration to the possible advisability of offering a prize. . . . The Admiralty's statement has, at all events, the merit of frankness, for, in reply to further questions, the First

Lord definitely stated that the Admiralty had no present intention of giving any attention whatsoever to aviation.

Lord Haldane still hoped, within limits, to keep the subject of aeronautics outside the realm of partisan party politics; but as the campaign of the 'air enthusiasts' continued exasperation began to set in, on each side.

At this time, it should be recalled, politics in Britain were especially bitter, and for good reason. In April 1910 the famous Lloyd George Budget of 1909, after a year of fierce struggle, at last received the Royal Assent. The Conservatives in the House of Lords had been forced to give way in the matter of the Budget. At the same time, however, resolutions curtailing or restricting the ancient powers of the Lords had been introduced in the Commons by the Liberals and these were subsequently carried, by varying majorities. On 14 April 1910 a Parliament Bill, based on these resolutions had been read for the first time in the Commons. The stage was thus set for violent political action.

If the members of the House of Lords refused to surrender their traditional powers by rejecting this Parliament Bill, the Liberal Government might force the King to create enough new Liberal Peers to ensure its passage. If this happened no one could foresee the consequences for the monarchy, for the country, for the Empire, for the Parliament, or for the two great political parties. This gigantic political and constitutional battle was raging toward a climax exactly at this time. Liberals and Tories despised each other and within the Conservative ranks terrible disagreements had arisen. Some Tories wanted the Lords to give in and pass the Bill while others demanded 'No Surrender'. The entire future of the Union with Ireland, it was thought, would turn on the action of the Lords in 1911. Eventually, several of Arthur Balfour's outraged followers sought to drive him from the leadership of their party because of their strong disapproval of his conduct during the crisis. This general condition of affairs was reflected in the Liberal *Manchester Guardian*'s comment on the deputation of the Parliamentary Aerial Defence Committee of 23 May. In a leader published on the next day the *Manchester Guardian* permitted itself some especially harsh remarks:

> The deputation of that somewhat unnecessary body the Parliamentary Aerial Defence Committee which met Lord

Haldane and Mr McKenna yesterday to urge them not to overlook the aeroplane as a fighting weapon got little but sympathetic generalities for their pains. They deserved no more. . . .

When they talk of 'aerial defence' they mean offence, and their dreams are those of the sensational press and sensational fiction – bomb-dropping on ships and armies and arsenals, – dreams which are, happily, nothing near realizable at present, and which it is our duty to keep in the realm of hectic and heated imagining.

Ministers now began to respond to the pressures of the agitation. It was announced that Colonel Seely would make a major policy statement on the subject of aeronautics in the House of Commons on 18 July. The announcement aroused a wide interest.

In his statement Seely revealed there would be War Office prizes for military aircraft designs. He also explained that the Army Council had come to the conclusion that from 80 to 100 officers would be required to be trained as aeroplane pilots.

In its first leading article on the next day the *Daily Mail* condemned the statement as inadequate, marked out by 'the timidity' which had characterized War Office policy for years past. The air agitation was now certain to continue. However, exactly at this juncture a major international crisis arose. Liberal Ministers suddenly found themselves facing the possibility of war with Imperial Germany. This Agadir crisis, as it was called, produced significant consequences for the further development of British naval and military aeronautics.

VII

Since 1909, in conformity with the policy of imperialism it pursued in these years, the French Government sought to extend its authority and power in Morocco and to establish a protectorate there. In the spring of 1911 disturbances broke out in Morocco and the French used this development as a pretext to send troops into the interior to occupy Fez, the chief city. The Germans at once claimed they possessed certain interests in Morocco. As a result conversations began between the French and German Governments to explore the possibility of a French occupation of Morocco in return for adequate compensation for the Germans. While these diplomatic exchanges were still in progress the German Government, on 1 July, suddenly despatched a small gunboat, the *Panther*, to the port of Agadir on the Atlantic coast of Morocco. The despatch of the *Panther* transformed the Moroccan problem into an international crisis.

The chief inspirer of German policy at this time was the recently appointed Foreign Minister, Alfred von Kiderlen-Wächter. He was seen by the nationalist Right in Germany as a 'new Bismarck'. He now decided to embark upon what he thought was a Bismarckian forward move in the diplomacy of his day. One reason for Kiderlen's brusque action at Agadir was to convey the impression of a return to a policy of strength in the conduct of Germany's foreign affairs and thus win popular support for the government. Secondly, Kiderlen hoped the other countries of Europe, France in particular, would feel the German pressure and so be more amenable in the matter of the compensation now demanded by Imperial Germany.

Large sections of the German public were at once overjoyed by the despatch of the gunboat; but the diplomatic results, especially in London, were not so positive for Kiderlen's design.

There, the highest elements in the British Government were unclear about the course they should follow. The Ministers who conducted British foreign policy at this time came from the Liberal-Imperialist wing of their party. They were convinced the German object was to humiliate France and thus wreck the Anglo-French *entente*. Sir Edward Grey was determined to thwart the Germans by supporting France but the members of the Radical wing of his party were suspicious of his policy and meant to rein him in whenever he seemed about to move too boldly in behalf of the French, and thus involve Britain too deeply in the affairs of the European Continent.

In the Foreign Office some of Grey's professional advisers argued that the Germans wanted Agadir because it was on the flank of Britain's highly important south-Atlantic trade routes and astride the line of her route to the Cape. In the Admiralty, however, it was seen that a base at Agadir, some 1500 miles from the homeland, could only be a source of weakness to the German High Seas Fleet. Despite the alarm in the Foreign Office the Admiralty was not especially concerned by the prospect of Germany acquiring a port on the Atlantic coast of Morocco.

Above all else, Sir Edward Grey could not discover exactly what the Germans wanted. In these circumstances tensions mounted, dangerously. On 21 July Lloyd George, a leader of the Radicals, delivered his famous Mansion House speech, after consultation with Grey and Asquith. In this speech, which was a warning to all the powers, Lloyd George said that peace at any price 'would be a humiliation intolerable for a great country like ours to endure'.[16]

The speech produced a shock in Germany. There, many people were convinced the British were now determined to prevent them from attaining their 'place in the sun'. Eventually, Britain became the main adversary in the eyes of the German public. In these circumstances as the atmosphere became more strained Grey feared the British Fleet might be attacked at any moment. Bridges in England were put under guard and key railway lines patrolled by troops. At the same time a series of bitter syndicalist strikes shook the country and rendered the situation even more complex. On 18 August at the request of the General Staff the House of Commons passed without debate, and contrary to every parliamentary practice, the second and third readings of an Official Secrets Act which furnished the government with drastic and unprecedented powers.

By September, however, Kiderlen decided to cut his losses. He remained resolute but the Mansion House speech had undermined the Kaiser's resolve. Early in the month when it seemed that war or peace was the issue there was a run on the German banks and then on 9 September, 'black Saturday', the bottom dropped out of the Stock Market. Negotiations with the French were resumed and eventually, in November, a Morocco treaty was signed by the two powers. For many in Germany this outcome of the Agadir crisis was looked upon as an intolerable insult. For these ambitious but disappointed citizens Great Britain was now a hated enemy.

VIII

Before the crisis subsided in this way Lord Haldane believed the situation was so threatening he asked the Prime Minister to convene a meeting of the Committee of Imperial Defence to discuss what should be done in case of a war with Imperial Germany. The result was a special all-day session of the Committee, held on 23 August 1911.

At this meeting the civilian Cabinet Ministers were shocked to discover that the Admiralty and the War Office had worked out their plans for war with Germany in isolation. The war strategies advocated by each of the services were mutually exclusive. Historians have blamed H. H. Asquith for this almost incredible situation which was nothing less than a national scandal.

Both departments agreed that the protection of the United Kingdom against invasion depended upon the Royal Navy. However, they were unable to concur about the role to be assigned to the army when war came.

The General Staff in the War Office were convinced that the issue of the war would be settled very early in the fighting after a series of gigantic clashes on the French frontiers. Their representatives at the meeting argued that the speed of arrival of the British forces would be highly important in checking the great German attack that was certain to come. They proposed to rush the bulk of Britain's expeditionary force to the left flank of the French Armies and thus provide them with assistance that might be decisive in its effect.

The Admiralty representatives revealed they had worked out an entirely different strategy for war. Although Admiral Fisher had retired his strategic views continued to dominate the minds of his successors. He believed it ridiculous and even grotesque to thrust the small British Expeditionary Force into a Continental war where it would be dwarfed and eclipsed by the huge European armies of the day. Instead, the so-called Fisher–Wilson school of thought insisted that the army should be employed in amphibious operations where it could exploit the tremendous power of the navy. Their idea was to hurl all or part of the expeditionary force ashore at some strategic point on the German coast and thus create a diversion that would draw significant numbers of German troops away from the main theatre of operations.

The army and navy officers of this period were men with powerful personalities. Each side in the dispute was convinced of the validity of its own view. General Henry Wilson, the Director of Military Operations, who spoke for the War Office at the meeting of 23 August had only contempt for the Admiralty proposals while he despised the civilian Cabinet Ministers as mere 'frocks', men lacking in the ability to grasp the strategic principles involved in his arguments, and concerned only with narrow political advantage for themselves and for their party. Admiral Sir

Arthur Wilson who explained the navy's strategy was equally firm in his belief that the strategic plan of the General Staff would destroy the army in a Continental campaign for which it was entirely unprepared. After a day of bitter disagreement it became clear that the War Office case, in the circumstances of August 1911, was the more valid one.

The effect of this meeting on Lord Haldane was profound. In September he wrote to Asquith to tell him that the 'grave divergence of policy' between the admirals and generals might have involved the country in a 'disaster'. He demanded that a properly trained and organized War Staff should be established in the Admiralty, a War Staff prepared to work in the closest cooperation with the military General Staff. He bluntly warned his friend: 'I have after mature consideration come to the conclusion that this is, in the existing state of Europe, the gravest problem which confronts the Government today and that unless it is tackled resolutely I cannot remain in office'.[17]

Haldane had reorganized the British Army and made it ready for war. He now proposed that he should be appointed First Lord of the Admiralty in order to carry out the same tasks for the Royal Navy.

H. H. Asquith, the Prime Minister, was not to be rushed in such matters. He retired to Scotland for a change of scene. Then, he invited Haldane and Winston Churchill to join him there in order to talk over the problem. Eventually, to Haldane's distress, he asked Churchill to take over at the Admiralty. This appointment has been called the 'most significant result'[18] of the meeting of 23 August 1911. It also had profound consequences in the sphere of British naval and military aeronautics.

IX

In 1911 Winston Churchill was 37 years old. He had begun his professional life as an officer of cavalry. In his early youth he worked also as a military correspondent and as a military historian. His account of the campaign in the Sudan, *The River War*, published first in 1899, has been called, properly, 'one of the best military books in the language'.[19] However, a soldier's life could not satisfy his ambition. He meant to play a part in the history of his country and only a career in politics could provide him with the opportunities for which he craved. In 1899 he resigned from the army.

By the autumn of 1900 Churchill was elected to Parliament, as a Tory. When Joseph Chamberlain introduced his plan for Tariff Reform and Imperial Preference Churchill, a convinced Free Trader, would not accept it. In 1904 he crossed the floor of the House of Commons and took his seat on one of the Liberal benches. The Conservatives were outraged. They condemned him as a traitor to his class, as a scoundrel who had abandoned his party because he calculated he could secure political office for himself more quickly as a Liberal than as a Conservative.

When the Liberals won the General Election of 1906 Churchill was given a place in Sir Henry Campbell-Bannerman's Government. In 1908 Asquith became Prime Minister and he chose Churchill as his President of the Board of Trade. In this period Churchill was already recognized, despite his age, as a politician and statesman of very great promise. He was marked out at this time by his boyish energy and by his enthusiasm for work. For a time he abandoned his military interests and threw all his strength into campaigns for social reform. He became friendly with Lloyd George and ranked second only to him as a Radical champion. After his appointment as First Lord of the Admiralty, in 1911, Churchill was thrilled by the scope and challenge of his duties. He devoted all his time to mastering the technical requirements of his new post. 'Winston', Lloyd George eventually complained, 'was taking less and less part in home politics, and getting more and more absorbed in boilers . . . .'[20] He was also fascinated by the subject of military aviation.

In February 1909 the Committee of Imperial Defence had met to consider its aeronautical policy. On that occasion the Committee had before it a letter from the Hon. C. S. Rolls. In this letter Rolls informed Lord Esher that he had purchased an aeroplane from the Wright brothers. Rolls hoped the authorities would give him facilities for experimenting with this machine on 'Government ground' in exchange for the technical information he might acquire during his experiments. All the members of the Committee of Imperial Defence voted to accept this curious proposal. There was one exception. Winston Churchill objected in the strongest tones:

> MR CHURCHILL thought that there was a danger of these proposals being considered too amateurish. The problem of the

use of aeroplanes was a most important one, and we should place ourselves in communication with Mr Wright himself, and avail ourselves of his knowledge.[21]

As First Lord Churchill did everything he could to advance the cause of naval aviation. He even took flying lessons and spent many hours in the air in the primitive aeroplanes of that day. Years later his son Randolph proudly reproduced the opinion of one contemporary naval aviator, in the official biography of his father:

> Those who were sceptical of heavier-than-air machines always deprecated WSC's interest in them. . . . Then the war came. . . . They began to clamour for greater activity in the very sphere in which they had shown such hostility. (Richard) Davies (Lieutenant RN, later Admiral, VC) said . . . 'They have pissed on Churchill's plant for three years – now they expect blooms in a month'.[22]

C. R. Samson, the Eastchurch pioneer, expressed a similar opinion in his *History of Naval Aeroplanes and Seaplanes*, which we have noticed in an earlier section of this chapter. Referring to Churchill and Admiral Sir Murray Sueter, Samson wrote in his history.[23]

> Commander Murray F. Sueter was an outstanding figure in helping the Naval Wing to develop, and . . . Mr Winston Churchill pushed us along very fast when he became First Lord of the Admiralty. These two are certainly . . . the two people responsible for anything the Navy did to help Naval Aviation.

Shortly after Churchill's appointment as First Lord, the pressure of public opinion forced H. H. Asquith to take decisive action in the matter of the nation's air force, as we shall see. As a preliminary, several Ministers were consulted. For his part, Winston Churchill prepared a minute about the new Corps which revealed his keen interest in the subject while it reflected that largeness of outlook so characteristic of him throughout his life. He saw to it that a copy of his minute was sent eventually to Colonel J. E. B. Seely at the War Office. Churchill's minute, written in his own hand declared:[24]

1. Whatever happens the R(oyal) E(ngineers) must have nothing to do with H.M.'s Corps of Airmen, wh.shd. be a new & separate Organization drawing from civilian as well as military & naval sources & providing air-scouts, skilled pilots, & tuition, both for the fleet & the army.

2. Terms and conditions must be decreed to make Aviation for War purposes the most honourable, as it is the most dangerous profession a young Englishman can adopt.

3. No regard to military or naval seniority shd. prevent the real young & capable men who have already done so much for the new arm, from being placed effectually at the head of the Corps of Airmen.

WSC   9.11.11.

X

Before the Agadir crisis ended British aeronautical development suffered another heavy blow. It will be recalled that in 1909 the Committee of Imperial Defence had recommended that the Royal Navy should be provided with a large rigid dirigible. This work was undertaken, in great secrecy, by the firm of Vickers Sons and Maxim in their Cavendish Dock at Barrow-in-Furness. At the same time a special branch was organized at the Admiralty to deal with all matters relating to the construction of the vessel. Sir John Fisher's original intention was to put his trusted aide Captain Reginald Bacon in charge of this new branch but Bacon resigned from the naval service. The post was given to Captain Murray F. Sueter who was designated Inspecting Captain of Airships. From time to time in the months that followed information about this *No. 1 Rigid Airship*, known popularly as the *Mayfly*, was released to the public. It was hinted that novel designs and new technology would result in the production of a British dirigible even more efficient than the German Zeppelin.

Unfortunately for these hopes, the *Mayfly* never flew. When she was being drawn out of her shed on 24 September 1911 the airship was struck by a sudden gust of wind and smashed against the shed door, breaking her back. The *Mayfly* was reduced to wreckage and became a total loss. The shock to the British public was profound.

For his part, Captain Bacon was greatly distressed. He wrote to

Reginald McKenna, the First Lord of the Admiralty, in abject terms: 'I . . . must write you . . . to say what a beast I now feel myself for having deserted you over the airship . . . I left it and you in the lurch . . . I owe you an apology for not having stayed & seen her through . . .'.[25]

There were further results. The Admiralty ordered a Court of Inquiry to determine the cause of the accident. The President of this court was Rear-Admiral Sir F. C. Doveton Sturdee who later won a great victory at the Battle of the Falkland Islands, in December 1914. He was astonished when he looked for the first time at the mass of twisted girders and wrecked gondolas. His reaction was recorded in a document entitled *Rigid Airships of Zeppelin Type*, prepared during the First World War:[26]

> A Court of Enquiry was held under Admiral Sturdee.
> Admiral Sturdee, on seeing the airship and never having seen one before, said:
> 'It is the work of a lunatic'.

Sturdee also recommended that no more rigid airships should be built for the Royal Navy. By Admiralty decision Captain Sueter's special branch was broken up. He pleaded, in vain, to be allowed to continue his work but he and his subordinate officers, despite their protests, were returned to general service.

As a matter of course the disaster to the *Mayfly* increased the general dissatisfaction of the public with the aeronautical policies of the Liberal Government. Ministers were spared from parliamentary attack on this occasion because on 22 August the Parliament had been prorogued until October. However, on 30 October Captain George Sandys, the Conservative Member of Parliament for the Wells Division of Somerset, raised an aspect of the question in the House of Commons. Captain Sandys concentrated on the subject of 'military aeroplanes' and began by stating that a 'large number of persons in this country who speak with authority feel a certain amount of anxiety . . . in connection with this particular branch of military science'.[27]

Sandys pointed out that a few days earlier Colonel Seely had declared that while the French Army possessed no less than 200 aeroplanes the total provision of the British Army in these machines was 16. Sandys declared that this was a 'very unsatisfactory statement'. Colonel Seely, in his reply, admitted that the

British Army had too few aeroplanes but he sought to assure the House that 'the Government as a whole realize the vital importance of this service and will press forward all necessary matters to maintain the defensive forces of this country at a high level of efficiency'. Seely's parliamentary performance on this occasion was ineffective. It was obvious to all concerned that the Under-Secretary of State for War could not dominate the House of Commons in the way Haldane had done when he sat there. As an immediate consequence, the aeronautical position of the government began to deteriorate even further.

By November 1911 the campaign for a more efficient national air service produced a significant effect in the Liberal Cabinet. This agitation of outsiders, as so often the case in British history, forced Ministers into action. The situation at that time was explained by C. F. Snowden Gamble, in his classic work, *The Air Weapon*:

> During 1911 a feeling of dissatisfaction . . . had been growing concerning the inferiority in aerial strength of the country as compared with our continental neighbours. . . . The criticism grew in volume and weight until the Cabinet felt that some steps should be taken to see whether or not the existing system could be improved.[28]

It was in these circumstances that the Prime Minister, H. H. Asquith, requested the Standing Sub-Committee of the Committee of Imperial Defence to carry out a wide-ranging enquiry into the subject of Aerial Navigation so that the country could at last secure for itself an 'efficient Aerial Service'. His instructions, or Terms of Reference, dated 18 November 1911, were concise:[29]

> The Prime Minister desires that the Standing Sub-Committee of the Committee of Imperial Defence . . . shall meet to consider –
> (1.) The future development of Aerial Navigation for naval and military purposes.
> (2.) The measures which might be taken to secure to this country an efficient Aerial Service in war both as regards matériel and personnel.
> (3.) Whether steps should be taken to form a corps of aviators for naval and military purposes, or otherwise to co-ordinate the study of aviation in the Navy and Army.

The Agitation for a National Air Force

By the Prime Minister's instruction, the membership of this Standing Sub-Committee was especially strong. The Chairman was Viscount Haldane, the Secretary of State for War. Other members included Winston Churchill, First Lord of the Admiralty; Lord Esher, a permanent member of the Committee of Imperial Defence; and Colonel Seely, the Under-Secretary of State. Among the service members were Vice-Admiral Prince Louis of Battenberg, the second Sea Lord; Lieutenant C. R. Samson, the technical expert; Major-General C. F. Hadden, Master-General of the Ordnance; and Brigadier-General David Henderson of the General Staff. Mervyn O'Gorman represented the Army Aircraft Factory while Sir R. Chalmers, Permanent Secretary to the Treasury completed the membership. As a result of this initiative of the Prime Minister, a new and important phase in Britain's aeronautical development was at last begun.

9

The New Arrangement and its Lapses

Captain Burke's Lecture – Bombing and Air Defence Neglected in It – The Opinions of Colonel Aylmer Hunter-Weston – Haldane in The House of Lords – Emphasis on Air Reconnaissance to the Neglect of Everything Else – Colonel Seely's Technical Sub-Committee – The Royal Flying Corps – A Real Air Service at Last – Seely's Success in the House of Commons – The Intervention of the Marquess of Tullibardine – Arthur Lee's Opinion – Weaknesses in the New Scheme – A Unified Air Service – The Question of Leo Amery – An Air Office – Other Lapses – Neglect of Home Air Defences – David Henderson and Frederick Sykes – The Attitude of Colonel Seely – Sykes Neglects Home Air Defence – Winston Churchill in *The World Crisis* – His Ideas About Aeroplanes – Churchill's 'Aerial Defence' and Its Revelations.

I

While Lord Haldane and his colleagues settled down to the tasks assigned them by the Prime Minister, the government itself was exposed to further attacks owing to the position it had taken up with respect to military aviation. On 15 November 1911 Captain C. J. Burke, the Royal Irish Regiment, and a member of the Air Battalion, delivered a lecture at the Royal United Service Institution. This lecture attracted a wide notice in military and civilian circles.

Captain Burke was not an especially capable pilot nor was he a profound military thinker; but he was convinced that aeroplanes would be weapons of significance in the next great war. As he saw it, the ability of aeroplanes to reconnoitre and to gather information quickly would be their most valuable function in wartime. In his opinion the war would begin with a series of battles in the air. When one side attained command in the air after victory in these early battles it would be in a position to dominate the fighting on the ground because it would then be able to carry out its aerial reconnaissance missions unchallenged. We should notice that in this lecture Captain Burke made no mention of bombing from the air and the related issue of air defence by aeroplanes of military and civilian targets.

In the discussion that followed the lecture these serious lapses in his analysis of the value of aeroplanes in war were not commented upon because most British officers of the day neglected the matter of air bombing and air defence when they thought about the wars of the future. What did attact attention were the opinions of Colonel Aylmer Hunter-Weston, RE, a General Staff officer then serving in the War Office. Hunter-Weston, who was destined to have a brilliant career in the First World War, announced on this occasion that he had attended the lecture in order to give some expression to the views of the General Staff. His remarks shocked the technical audience in attendance that evening and they disturbed the public at large when they were eventually reported in the newspaper Press.[1]

Colonel Hunter-Weston, known to his colleagues as 'Hunter-Bunter', had several things to say. Speaking on behalf of the General Staff he declared that the military authorities were fully alive to the possibilities of employing aircraft in war but he deprecated the notion that any invention could 'revolutionize' warfare. He thought aerial warfare required 'special heroism' on the part of pilots but that no 'reasonable officer' would engage in air fighting since a combat between two aeroplanes involved 'certain death'. This attitude, it will be seen, was an exact reflection of Sir William Nicholson's ideas. Hunter-Weston particularly warned the public not to be 'carried away with wild ideas'. In his opinion civilian airmen would be 'perfectly useless for reconnoitring work' since this could only be carried out by trained officers who knew the kind of information that was required and who could 'identify . . . the various military formations, and what they portended'.

The 'air enthusiasts' were annoyed by these statements and angry that they were presented as the considered and official opinions of the British War Office. The comments about them which appeared in the periodical *The Aeroplane* were typical. At this time the Editor of *The Aeroplane* was C. G. Grey, a truculent but capable journalist who devoted his life to the subject of aeronautics. In time, Grey became a leader in the field of aeronautical journalism. He was outraged by what Hunter-Weston had said at the meeting of the Royal United Service Institution. In *The Aeroplane* for 23 November he dismissed him as one of a school in the War Office who were '. . . so utterly ignorant on the subject that they are even more harmful than if they were actively in opposition'. He pointed out that Hunter-Weston had brought to light 'the narrow and biassed views of the War Office' and compared his attitude to that of Sir Alexander Bannerman. He proceeded to link the opinions of these two officers to those of Sir William Nicholson, Chief of the Imperial General Staff, 'the General Officer who controls our military aviation in so disastrous a fashion'.

The general feeling of uneasiness about the condition of British military aviation at the end of 1911 was also expressed in Parliament. On 6 December the Earl of Hardwicke spoke on the subject in the House of Lords. He pointed out that the usefulness of aeroplanes had been clearly demonstrated in the recent French and German manoeuvres, and in the fighting in Tripoli.[2] He stated that the French already possessed 'between 200 or 300 war-planes' while Britain 'has only six officer aviators and only two serviceable aeroplanes'. In these circumstances he asked if the government would not make a statement 'of its policy with regard to the training of the necessary mechanics, pilots, and military observers'. He further asked if the government proposed to give 'any practical encouragement to the manufacturers of this country with a view to enable them to supply the Government with British built machines'.[3]

Lord Haldane replied for the Liberal Government. He explained with that blend of arrogance and smoothness that infuriated his political opponents that he felt a great responsibility whenever it became necessary to spend public money. He was always ready to spend it, even lavishly, when he was convinced that good results would be obtained but he would never do so 'in the hope that something will come of it'. He admitted that Britain was behind other countries in the sphere of military aviation but he felt this situation could be remedied, easily enough. In his opinion, many technical aeronautical problems had not yet been solved and before the 'requisite scientific knowledge' had been obtained it would be foolish to invest heavily in aeroplanes that might become obsolete in a short space of time. 'We have been buying a variety of different types', he said, '. . . in order to see which is the best'.

This line of argument earned Haldane a rebuke from an elder-statesman who had won an almost universal respect in the

The New Arrangement and its Lapses 189

country. This was the Viscount St. Aldwyn, formerly Sir Michael Hicks-Beach, 'Black Michael' as he had been called. Hicks-Beach had served as Lord Salisbury's Chancellor of the Exchequer at the end of the last century. He now brought this experience to bear on the contemporary problem of military aviation. His speech on this occasion was an effective touch in the debate. He said:

> My Lords, all my official experience . . . as ex-Chancellor of the Exchequer would induce me to refrain from pressing the War Department in regard to anything that would involve unnecessary expenditure. But I do think that in this matter there has been very slow progress indeed. . . . The noble Viscount appeared to me to be waiting for some design of perfection in this matter. Now that has never been the policy of the Navy, and quite rightly too. I remember ships built on designs which were considered reasonably good . . . in order that we might be provided with what at any rate was the best that could be obtained at that time . . . I would venture to tell the noble Viscount opposite that I am afraid under present circumstances if we should unfortunately be involved in war we should be quite unprepared with regard to appliances of the kind which have been the subject of discussion. I do hope, therefore, that the War Office will take care that there is some provision in respect of aviation without waiting for some possible design in the future which may not be attained until it is too late.

In this debate Haldane made statements that merit the reader's careful attention. In attempting to defend the government he pointed out that the French and Germans possessed 'enormous' armies when compared with the British. It followed, in the argument he now developed, that the British needed a smaller air service than did the Germans or the French. 'Your air service', he said, 'is required as part of that Army . . . organization'. This curious statement completely neglected the idea of home air defence by aeroplanes. Instead, it concentrated solely on the air reconnaissance requirements of an army in the field engaged in a war of movement, which was the kind of war most contemporary experts thought about when they contemplated the future. This remarkable attitude which was widely held in the War Office and in the army was to have serious effects upon Britain's defensive arrangements, as we shall see in the further course of this history.

With good reason C. G. Grey attacked Haldane's ideas in *The Aeroplane* for 14 December 1911:

> Another statement by the noble lord seems to call for comment, namely, that we require an air service for a comparatively small army. The writer begs . . . to suggest that we require an air service not for a very small army, but for an extremely rich, an extremely helpless, and an extremely thickly populated country. . . .
>
> One most important point must be considered by all those concerned with our national safety, namely, that military aviation is not concerned merely with a small aeroplane contingent intended to operate in conjunction with our overseas striking force, but is concerned with the aerial defence of every foot of our enormous coast line . . . we must have a regular military air station at every one of our important ports, for it must be recollected that coast defence is a military and not a naval job.

II

Lord Haldane's Standing Sub-Committee acted quickly and within a short time the 'broad principles' of a scheme for a national air service were worked out. At a meeting held on 18 December 1911 a Technical Sub-Committee, under the chairmanship of Colonel Seely, was assigned the 'task of elaborating all the details necessary to give effect to the policy proposed'.[4]

By the terms of the new policy a British Aeronautical Service called 'The Flying Corps' was to be created. This Corps would consist of a Naval Wing, a Military Wing, and a Central Flying School for the training of pilots. By design, the Flying Corps would work as closely as possible with the Advisory Committee for Aeronautics and with the Army Aircraft Factory. Further, a consultative committee named The Air Committee was to be established as a permanent sub-committee of the Committee of Imperial Defence. It was to be a solely advisory body without executive or administrative powers. In due course the Cabinet approved these various proposals and many of them were published in a White Paper, *Memorandum on Naval and Military Aviation*, 11 April 1912 (Cd. 6067).

The New Arrangement and its Lapses

The memorandum explained that with the permission of the King the new service would be called The Royal Flying Corps and the existing Army Factory would henceforward be known as The Royal Aircraft Factory. The proposals contained in the memorandum became effective by Royal Warrant on 13 April and in the following month the Air Battalion of the Royal Engineers was absorbed into the new organization.[5]

The Liberal Ministers did not wait for these important developments to occur. By design, they began to reveal some of the details of their new course as early as February 1912. The Agadir crisis had startled some contemporaries into an acute awareness of Britain's weakness in the air in case a war should break out; and Haldane and Colonel Seely were now anxious to show they were taking steps to remedy the situation.

For example, a special article in the *Daily Mail* of 9 February complained that the British Government was not spending enough on aeroplanes. As a result

> London, for the first time in her history lies immediately open to attack without means of defence. If we should ever quarrel with our neighbours and find an aerial fleet over our city, the War Office . . . could only throw stones . . . is it not time . . . that we ceased to rely upon the mere good will of others for safety from attack of our capital?

However, on 27 February 1912[6] Haldane explained in public that the new Army Estimates for the year would provide the 'substantial' sum of £308 000 for the air service. He also revealed that a joint navy and army flying school was to be established. On the next day a leading article in the *Daily Mail* entitled 'A Real Air Service at Last', declared: 'The Government and Lord Haldane are warmly to be congratulated upon the substantial vote contained in the new Army Estimates for the nation's air service'. The article called attention to the paper's own contribution to this result:

> Thus after several years of effort on the part of *The Daily Mail* the importance of 'the new arm' has been recognized by the British military authorities. The great competitions which we have organized have produced their effect. The crossing of the Channel by M. Blériot in 1909, an event which opened new horizons before mankind; the first £10 000 race, won so brilli-

antly by M. Paulhan in 1910, and the yet more wonderful 1,000 miles circuit of Britain, the £10 000 prize for which was carried off by Lieutenant de Conneau last year, have opened the eyes of the nation to the promise and potency of the flying machine.... The first steps have now been taken; it only remains for the Navy and Army to go firmly forward in the path which Lord Haldane has traced.

Colonel Seely enjoyed an even greater success in the House of Commons. There, on 4 March he explained the entire scheme and also revealed it had been 'this morning approved by the Prime Minister, and will now be carried into effect'.[7]

Members were impressed by Seely's declarations and explanations. Several Conservative opponents of the government rose in order to congratulate him and his colleagues for what they had done. Captain Sandys, who was often critical of the government's aeronautical policies, said:

The Government have now at least abandoned what I may call the Micawber-like attitude of waiting for something to turn up.... So far as the scheme of organization which the right hon. Gentleman outlined in his speech this evening ... I will ... say it seems to me an admirably thought-out scheme, and one which certainly is going to get an aviation service for our Army and Navy far and away beyond what we had reason to anticipate a year or two ago.

Another Tory champion, the Marquess of Tullibardine, declared: 'I should like to congratulate the Under-Secretary on the success of his endeavours so far as aviation is concerned, and also on keeping the Debate out of politics so far as it was possible for him to do so'. Tullibardine also said: 'So long as he takes that line and does his best to keep the Army up to the mark he will certainly find no better friend, talking from the Army point of view, on either side of the House than myself'.

The Marquess of Tullibardine was the eldest son of the seventh Duke of Atholl. At this time he represented West Perthshire in the House of Commons, as a Unionist. In politics he was a fierce partisan. Two days before this debate took place he wrote to the Conservative leader, Andrew Bonar Law, to ask him to 'let me know the chief points you are going for as then I can rub it in afterwards . . . I am quite confident I can flatten Seely on any point he chooses to raise including Aeronautics . . .'.[8]

However, Tullibardine was more closely involved with the developing aeronautical situation than these rather commonplace political remarks suggest. In 1907 he had helped to secure one of his father's remote grouse moors near Blair Atholl in Perthshire for use as an airfield where secret trials were held to test an aeroplane designed for the War Office by Lieutenant John William Dunne, the Royal Wiltshire Regiment. Unfortunately, the Dunne machine could not fly in 1907 nor could it do so in subsequent military trials held at Blair Atholl in the following year. Nevertheless, Tullibardine and some of his wealthy friends were convinced of the validity of Dunne's ideas. Together they formed the Blair Atholl Aeroplane Syndicate and Tullibardine became a director of this commercial firm. Their object was to make it possible for Dunne to continue his work as a designer of aeroplanes. Early in 1912 after Seely was put in charge of the Technical Sub-Committee to work out the details of Britain's new aeronautical programme, Tullibardine wrote to him in order to urge that the War Office purchase the latest version of the Dunne aeroplane. The Marquess, a gallant and much decorated soldier, expressed himself in the most forceful tones:[9]

<u>Private</u>   22nd January 1912

Dear Jack

Just a line to let you know privately that Dunne's aeroplane is all right, and has a long way surpassed anything that we expected. . . .

Now quite honestly I have put in a lot more money than I could afford, and so have several other officers, simply to save the invention for the War Office, who were stupid enough to throw it away when they had it for nothing, after having spent a lot of money over it. We could have sold the patent abroad easily, in fact we had several offers. . . . It is impossible to go on unless one gets an order, and we shall have to consider accepting offers that have been made to us for purchase both at home and abroad. . . . We have not advertised, and we have kept very quiet, as we all along felt that what we have got must and will be the military machine of the immediate future. . . .[10]

This is the first time I have ever asked the War Office practically to give us an order because it is only now that I am perfectly satisfied that we have got a bigger thing than even we anticipated. I have never bothered or touted any of them, except to come and observe the progress of events. O'gorman [sic] I don't know and Bannerman only slightly. . . .

Communications had better be sent to me not to Dunne.

Yrs.

Tullibardine

This debate of 4 March 1912 was something of a triumph for Haldane's policy. He hoped, within limits, to keep the subject of military aeronautics outside the realm of party politics. His success in this object was confirmed, however temporarily, when Arthur Lee, chairman of the Parliamentary Aerial Defence Committee, spoke in the House of Commons on 6 March.

Lee, the Conservative Member for the Fareham division of Hampshire, was a savage enemy of the Liberal Government. He was usually as abrasive as he could be in speeches he made in the House or on public platforms. However, on 6 March he spoke in tones that, for him, were reasonably moderate. His remarks reflected the tensions of the Agadir crisis of the previous summer:[11]

> We on this side, and I am sure hon. Gentleman in every quarter of the House, must welcome the substantial though lamentably tardy action which the Government have taken. . . . The Government blindness and indifference in the past with regard to military aviation, with regard to this new arm which may quite conceivably revolutionize modern warfare, have gone so far as to become a national peril, and but for the action which has been taken by the Government in the present Estimates there would undoubtedly have been a concentrated and determined attack by all who are interested in this question. Particularly so is that the case when we realize what the situation would have been last summer if our Expeditionary Force had to go to war.

Arthur Lee had been one of the chief organizers of the air display held at Hendon in the previous May. Years later, in his memoirs, he emphasized the importance of that event when he wrote of the Hendon meeting: '. . . Immense publicity followed in the Cinema as well as the Press and Aerial Defence as a Parliamentary and live issue was from that moment definitely on the map'.[12] On 6 March Lee referred to the Hendon meeting in order to pay Colonel Seely a compliment. He said:

> I know the right hon. Gentleman himself is not to blame in this matter, and we have been particularly lenient towards him about it. I know he has done his best, at least since he was inoculated at Hendon . . . at any rate, I may be permitted, on behalf of the Parliamentary Aerial Defence Committee, heartily to approve of the Government's action. . . . I venture very respectfully to congratulate the right hon. Gentleman . . . as chairman of the committee which produced this scheme, to beg him not to rest on his oars in regard to this great matter, and to assure him that if he requires or is willing to accept any assistance from this side of the House . . . he will receive it in no small measure.

III

Despite this high praise of the government's new plan, contemporaries and later writers recognized several weaknesses in the scheme as it developed. According to its terms the country's aeronautical service was 'to be regarded as one'.[13]

However, rivalry between the two great service departments destroyed this idea almost at once. An official document entitled 'The Unified Air Force' baldly states: 'The Royal Flying Corps was . . . at the outset conceived and formed as a unified air service . . .'. This document further points out that the Admiralty worked independently from the first and it concludes: 'Thus from the earliest days there were two separate air services'.[14]

It will be recalled that the idea of a unified air service had been mentioned years earlier in the great aeronautical debate of 2 August 1909 when Arthur Du Cros of the Parliamentary Aerial Defence Committee suggested the creation of 'one Aerial Department' for the nation's air force.[15] Now, Leo Amery touched upon this matter again, in a debate in the House of Commons on 12 March 1912. Amery, a Tariff Reform champion, was also looked upon by contemporaries as an expert military critic and as one of the ablest opponents of the Asquith Government. In his remarks he referred to the 'peculiar position that the Air Corps is occupying as an independent body meeting the requirements of both the Army and Navy . . .'. He pointed out that these separate requirements were completely different for each of the two services. The Admiralty, he argued, might decide it wanted to employ dirigibles and 'hydro-aeroplanes,[16] capable of going very long distances and of carrying a large supply of bombs'. On the other hand, he said, the War Office might require a 'large number of swift monoplanes' for the purposes of reconnaissance. He pointedly asked: 'How is that to work?'[17]

The idea of a separate department of State for the air was certainly thought of by contemporaries at this stage. When the Air Committee held its first meeting on 31 July 1912 Colonel Seely was the chairman. He said at that time:[18]

> unless the Hague Conference decided on forbidding aerial warfare, there would probably be an Air Office in a few years time having the same status as the War Office and the Admiralty have at the present moment. Until this came about, the Air Committee would be the only direction in which the Cabinet could look for advice in aerial matters.

The creation of such an 'Air Office' was delayed for years. Inter-service jealousies and rivalries were so intense that they were only overcome after the bitter failures and experiences of war; and it was not until 1918 that a central controlling body, the Air Ministry, and a unified air service, the Royal Air Force, were at last established in Britain.

IV

There were other gross lapses in the new arrangement. Even though there was a real fear that Britain, in 1912, lay open to air attack or air invasion the home air defence requirements of the country were almost entirely neglected. In its report Colonel Seely's Technical Sub-Committee set out the functions of the two Wings of the Royal Flying Corps as follows:[19]

> The Naval Wing of the Flying Corps, entry to which should ultimately only be obtainable by qualifying at the Central Flying School, should for the present have its headquarters at the Naval Flying School at Eastchurch. It is impossible to forecast what its ultimate organization and development will be, as this depends

to a great extent upon the results of experiments, which are about to be commenced with hydro-aeroplanes.

The Military Wing of the Flying Corps should consist at first of eight squadrons. . . . The whole of these squadrons are required for use in connection with the Expeditionary Force.

This neglect of the Air Defence of Great Britain was not the result of any accident. The plans adopted by Colonel Seely's Technical Sub-Committee had been prepared in private by a small informal body of three officers. They worked very closely with Seely and influenced his outlook from the start. Each of these officers believed that reconnaissance would be the most important function of aeroplanes in war. These men played a tremendous role in the early development of Britain's air service. They were: Brigadier-General David Henderson, shortly to become Director of Military Training at the War Office; Captain Frederick Hugh Sykes, a General Staff officer of much experience; and Major Duncan MacInness, RE.

Of this trio the first two will always be associated with that pre-war school of thought in Britain that emphasized the reconnaissance aspect of air warfare, almost to the exclusion of the other functions of aircraft.

V

David Henderson was born in 1862 to a well-known family of Glasgow shipbuilders. He originally planned to become an engineer but after study at the University of Glasgow he entered Sandhurst and eventually, in 1883, he joined the Argyll and Sutherland Highlanders. In 1896 he graduated from Camberley and later served at the battle of Omdurman. Before the South African War broke out he was sent there on an intelligence mission. After the hostilities began he served as an Intelligence Officer on Sir George White's staff and was wounded at Ladysmith. Eventually, he became Director of Intelligence under Lord Kitchener at Pretoria. He impressed his colleagues there as an officer of high capability and great devotion to the service.

In 1903 Henderson returned to the War Office in order to set out the lessons of his intelligence experiences in South Africa in a book, *Field Intelligence*, published officially in 1904. In 1907 he produced another work, *The Art of Reconnaissance*, which increased his military reputation, significantly.

In 1911 the wound he received at Ladysmith began to bother him and an operation was performed to correct the problem. In order to recuperate from this ordeal he went to Harrogate in Yorkshire and there he had an experience that changed the course of his life. One day an aeroplane competing in Lord Northcliffe's round-Britain air race was forced down in a field near the hotel in which Henderson was staying. He walked up to the aircraft and had a long talk with the pilot. It was at this time he decided he must learn to fly.[20] His chief professional concerns were field intelligence and reconnaissance and he realized the aeroplane had profoundly affected these branches of the military art. He learned to fly at Brooklands in 1911 when he was almost fifty years of age.

From this time he devoted himself to the subject of military aeronautics. He became the 'moving spirit' in the group of men who organized the Royal Flying Corps. In 1913, as the War Office came to recognize the significance of flight, Henderson was appointed Director-General of Military Aeronautics. Years later, his friend John Buchan wrote of Henderson and the Royal Air Force that: 'To David Henderson this service owes more than to any single man, and his name must for ever be linked with it'.[21] This was also the opinion of Lord Trenchard, the great leader of the RAF. He once said of Henderson: 'he was the founder & father of the Air Force'.[22]

For his part, Captain Frederick Sykes was an officer who had been interested in aeronautics for a considerable period of time. In the South African War Sykes served with the Imperial Yeomanry and was severely wounded. After further service in Africa and in India he became a General Staff officer in the Directorate of Military Training. As early as 1904 he was attached, for temporary duty, to the balloon units practising on Salisbury Plain.

Sykes was a studious soldier who read all he could about contemporary military developments. After reflection, he decided a major change was taking place in the world at the beginning of the present century. In the Victorian era Britain's military leaders had to watch Russia's measured advances towards the Indian frontier but this situation was changed completely as a result of the Russian defeat in the war with Japan: 'I realized more and more strongly that the political centre of gravity was swinging back from east to west'.[23] After witnessing the German army manoeuvres in 1907 Sykes further decided that the Germans were determined to 'usurp' Britain's position as a world power. When Blériot crossed the channel in 1909 Sykes was convinced the aeroplane would soon take the place of the horse for the purposes of strategic reconnaissance and raiding. As a result of this conviction he enrolled in a course in aerodynamics at London University and learned to fly, in June 1911. He was by now certain the aeroplane would play a very important role in the 'coming struggle'.[24]

Sykes, an aeronautical enthusiast, was not liked by all his colleagues in the army. In his memoirs he complained that Sir William Nicholson 'was of the opinion that aviation was a useless and expensive fad' and that Sir Douglas Haig once said to a friend of his: 'Tell Sykes he is wasting his time; flying can never be of any use to the Army'.[25]

VI

Nevertheless in October 1911 Brigadier-General Henry Wilson, the Director of Military Operations in the War Office, ordered Sykes to visit aerodromes in France and then prepare a report upon the French Air Service. When he returned to London Sykes produced, in November, a detailed paper entitled 'Notes on Aviation in France'.[26] Sykes began this paper with an introductory section he called 'General principles of the use of Aeroplanes in war'. These ideas or principles determined, in good part, the way in which the Royal Flying Corps was developed in the period between 1912 and 1914 because in the former year Sykes was appointed to recruit, train and command the Military Wing of the Corps.

Sykes's paper was an exposition designed to convince a hostile reader of the value of aeroplanes in war. His first sentences stated: 'The French manoeuvres of 1911 and the reconnaissance work effected in Tripoli have demonstrated the use of the aeroplane in war. There can no longer be any doubt as to the value of aeroplanes in locating an enemy on land and obtaining information . . .'.

He then set out the various 'functions' of the aeroplane in war. The first of these, in his opinion, was for 'Strategical Reconnaissances'. An aeroplane, he argued, could carry out such a reconnaissance in four hours while it would take an officers' patrol of cavalry at least three days to complete the same mission.

The second function of the aeroplane in war would be 'to make raids against vital points of the hostile strategical deployment such as bridges, junctions etc., and destroy dirigible sheds near the frontier'.

A third function lay in the realm of 'Tactical Reconnaissance'. Aeroplanes would be able to ascertain, for a commander, the nature of the ground to the front, to the rear, and on the flanks of his position. They would also seek out and find suitable targets for their own artillery. This latter function involved aeroplanes in the fourth category, which was the control of artillery fire to be effected by means of signals from the aeroplane. Finally, the aeroplane would be of great use in 'the inter communication between Forces'. It would perform as a supplement to the telephone and telegraph in obtaining news of what was happening during a battle, and 'drop bombs against enemy in stockades or fortified villages'.

In a later part of this paper Sykes argued that the advantages of an aeroplane scout able to transmit information would be very great. At present, he explained the question was receiving 'serious attention' from various manufacturers of wireless sets and he had no doubt 'great improvements will shortly be effected in this direction'.

VII

The ideas of Henderson, especially, and also those of Sykes about the employment of aeroplanes in war were accepted, completely, by Colonel Seely. In June 1912 Asquith selected his close friend, Haldane, to succeed Lord Loreburn as Lord Chancellor and Seely was picked to take Haldane's place as Secretary of State for War. Seely was an enthusiast. He had fought gallantly in the South African War. Asquith and his friend, Venetia Stanley, liked to call him the 'Arch-Colonel'. He plunged into his new duties with vigour. In intellect he could not compare with his predecessor but he was eager to serve his country and the British Army to the best of his abilities. This attitude extended to the field of military aeronautics. Years later, Lord Hankey remarked of Seely: 'I can testify to the splendid work he did in developing the new arm and in making a start with all kinds of technical devices including anti-aircraft guns and searchlights'.[27]

The New Arrangement and its Lapses

Nevertheless, Seely, together with Henderson and Sykes, musst bear part of the blame for the inadequatte home air defence system Britain possessed upon the outbreak of war in 1914. In a debate in the House of Commons in October 1911 Seely revealed his views about the value of aeroplanes in war. He neglected, in thee remarks, everything about the several capabilities of aeroplanes in order to concentrate on one aspect only, that of reconnaissance:[28]

> I may say . . . that we fully realise the immense importance of aerial scouting in war. An aeroplane . . . enables one to see the other side of the hill . . . it is vital for any country which has an army to have an efficient aeroplane survey. That is realised by both the War Office and the Admiralty, and the Secretary of State for War and the First Lord of the Admiralty are giving their earnest attention not only with a view to getting a good aerial scouting service for each Department but are cooperating with the idea of working together in this as in other matters in order to provide a really efficient scouting service for both the naval and military services.

This remarkable attitude with respect to aeroplanes produced a curious result in 1913. It must be realized that in this period the War Office claimed 'complete and sole responsibility for the aerial defence of Great Britain'.[29] As a matter of policy the military leaders guarded this responsibility jealously and they usually opposed any attempts by the Admiralty, their great rival, to have any share in it. Occasionally, when some agreement between the two services was reached with respect to home air defence the War Office sought to overthrow it and to reassert its sole control in this area.

In December 1913 Sykes, as Commander of the Royal Flying Corps, Military Wing, was working at the details of its organization. In order to formulate his proposals he 'found it necessary to make an arbitrary assumption as to what is likely to be the ultimate future of the Military Wing . . .'. Given the attitude of the War Office with respect to home air defence, Sykes wrote, in terms almost impossible to believe, as follows:[30]

> Eventually, I assume, the duties of the Wing will include the provision of an Air Service for Home Defence, and for the defence of fortresses and defended ports at home and abroad.

At present, however, its functions are limited to the provision of an air service for the Expeditionary Force, and as long as its horizon is thus confined, it would appear necessary to design its war and peace organisations mainly with the idea of these functions.

We should note here that when war came in August 1914 the War Office, despite its high claims about its responsibility for home air defence, at once despatched almost all its aeroplanes to France for reconnaissance service with the Expeditionary Force. The Admiralty, shortly after, had to take over much of the burden of the Air Defence of Great Britain by itself.

Winston Churchill, in his *The World Crisis*, later tried to explain and account for these singular developments. His ideas about the employment of aeroplanes in war were not like the ideas of those who controlled the Military Wing of the Royal Flying Corps in the pre-war period. Churchill encouraged C. R. Samson and his colleagues at Eastchurch to carry out their various trials and experiments with aircraft. He sought to acquire for the Royal Navy machines that were more powerful than those used by the army because he foresaw that aeroplanes would perform as bombers, air fighters, and also as scouts. Shortly after the Naval Wing of the Royal Flying Corps was created Churchill, as First Lord, established a new department to supervise it, and placed Captain Murray Sueter in charge, with the title of Director of the Air Department. When, in his *The World Crisis*, he turned to the confusion over Britain's home air defences in August 1914 he wrote with typical generosity that the War Office

> owing to the difficulties of getting money . . . were unable to make any provision for this responsibility (home air defence), every aeroplane they had being earmarked for the Expeditionary Force. Seeing this and finding myself able to procure funds by various shifts and devices, I began in 1912 and 1913 to form . . . flights of aeroplanes as well as of seaplanes for the aerial protection of our naval harbours, oil tanks, and vulnerable points, and also for a general strengthening of our exiguous and inadequate aviation. . . . The War Office viewed this development with disfavour, and claimed that they alone should be charged with the responsibility for home defence. When asked how they proposed to discharge this duty, they admitted

sorrowfully that they had not got the machines. . . . They adhered however to the principle.[31]

This published account failed to reflect the bitterness which marked out the rivalry between the two services in this sphere. In October 1914 Churchill prepared a paper for the use of the Cabinet. Entitled 'Aerial Defence', the paper was a more accurate analysis than the one presented in *The World Crisis*. Churchill's 'Aerial Defence', although it was the work of a partisan, made clear the absolute failure of the War Office and the Admiralty to co-operate in the matter of home air defence in the period before the war:[32]

> Until more than a month after the war began, the sole responsibility for the defence of all vulnerable points in England by gunfire, sea-planes, or any other method, against aerial attack, rested with the War Office. The only exception to this was that at a conference between the heads of the War Office and Admiralty on the 19th November 1913, it was agreed that, where a vulnerable point was in close proximity to a naval air station, the naval aeroplanes would be available. Even this position was, however, challenged by the General Staff on the 21st July, 1914, when . . . General Robertston, representing the War Office, claimed for the War Office the sole responsibility, not only in regard to everything inland, but in regard to naval ports and vulnerable points of all kinds, even those exclusively of naval interest.
>
> Notwithstanding these views, the War Office had not up to the time of the declaration of war provided any aeroplanes for home defence; they had limited themselves exclusively, and as a matter of prior urgency, to the development of the expeditionary squadrons, and, on the outbreak of war, practically all the army aeroplanes were sent abroad. Not only were none available for guarding the vulnerable points, but none could be found even for the temporary purpose of coast-watching during the passage of the army to the Continent.

10

Home Air Defence and Lamp-posts

Failure to Create an Efficient Home Air Defence System – Vulnerability of Woolwich Area – Experimental Units – Gun Defences – Winston Churchill's Complaints – Aeroplane Defences – The War Office Attitude – Naval Air Stations – Sir David Henderson's Plan – Sefton Brancker and Lord Kitchener – The Airship Problem – The Mission of O'Gorman and Sueter – Their Visit to Germany – They Report to Colonel Seely – The Government's Aeronautical Policy – Coastal Air Stations Established – The Opinion of Sir Arthur Wilson – An Incident at Sheerness – Public Agitation – The Tories Attack Winston Churchill – Social Reform or National Defence – The Air Defence Problem – Lamp-Posts in 1913 – A Lamp-post in 1917 – A Lamp-post in 1940.

I

It will be recalled that in January 1910 the Admiralty, alarmed by the novel prospect of aerial attacks against its naval magazines, cordite factories, dockyards, and other vulnerable points, had caused the whole question to be submitted to the Home Ports Defence Committee for consideration. After deliberation this subcommittee of the Committee of Imperial Defence made recommendations which, had they been put into effect, might have resulted in the establishment of an efficient home air defence system in Britain before 1914.[1]

The Home Ports Defence Committee, as an immediate response to the problem, had urged the Admiralty and the War Office to carry out trials to determine the proper use of aircraft in aerial defence and experiments to discover the most effective guns and projectiles for employment against airships and aeroplanes. When the Royal Flying Corps was established both the Military Wing and the Naval Wing created experimental units and some experiments were carried out. However, the situation with respect to home air defence in August 1914 was quite unsatisfactory.

Although the Committee of Imperial Defence had approved the air defence recommendations of its sub-committee it was unable to force the two great service departments to carry them out.[2]

The military authorities responded to the recommendations of the Home Ports Defence Committee in due course. In January 1911 General Launcelot Kiggell, the Director of Staff Duties in the War Office, drew up a paper in which he recommended that experiments should be undertaken 'as soon as possible . . . with a view to elucidating the problems proposed by the Committee'.[3]

In his paper Kiggell suggested a 'programme of trials'. The first of these would attempt to ascertain the vulnerability of dirigibles to the fire of 'field, howitzer, heavy etc., Artillery, both by day and by night, when travelling at different heights and speeds . . . in different meteorological conditions'. Kiggell suggested further that the value of tracer for the purpose of checking the accuracy of such fire should be tested at the same time. The object of the second and third sets of trials would be to find out if dirigibles could drop explosives accurately enough to endanger the safety of 'land magazines, dock gates and storehouses' and 'to discover from what heights such explosives can be discharged with effect'. A fourth series of proposed trials would attempt to 'find out the effect which will be produced on dirigibles, sailing at different heights, by the concussions of heavy charges exploded on the ground below them'. Finally, Kiggell suggested experiments with searchlights to discover how effective they could be in lighting up dirigibles flying at night. Later, in a memorandum submitted to Colonel Seely's Technical Sub-Committee, the General Staff revealed that this programme of trials had been approved.[4]

The Admiralty was first in recognizing that aircraft would be able to bomb important shore installations and vulnerable points. Later, the authorities in the War Office also began to appreciate the vulnerability of certain military establishments to attack from the air. In December 1911 they became concerned about the metropolitan borough of Woolwich, the site of the famous Royal Arsenal and of other factories, laboratories, and workshops involved in the production and storage of warlike stores and equipment.

On 14 December 1911 the Army Council, 'with the concurrence of the Admiralty', sent a letter to the Home Ports Defence Committee requesting it to look into the entire matter and to make recommendations. In July 1912 the Committee prepared a report entitled 'Decentralization of Stores and Manufacturing Plant at

Woolwich and in its Vicinity'. This report recognized the vulnerability of the place to air attack. One of its conclusions was:[5]

> An enemy might succeed in injuring the resources of Woolwich and its vicinity in the following ways:
> (a.) By a raid by a land force.
> (b.) By a raid by a naval force up the Thames.
> (c.) By agents.
> (d.) By aircraft.

The great importance of Woolwich and the surrounding area was emphasized in this report: 'The Navy', it declared,

> is dependent on mobilisation on the resources of Woolwich and its vicinity for stores, the loss of which would deprive the fleet of important fleet auxiliaries and seriously hamper its action. The loss of the victualing stores in particular would cripple the fleet after a certain period. The Army is dependent on Woolwich and its vicinity for essential stores, the loss of which on mobilisation would cripple the Expeditionary Force.

The Committee decided that 'the most probable objectives for an aircraft attack' were the magazines at nearby Plumstead and Purfleet and the laboratories for filling shell and cartridges in the Arsenal itself. They were informed that the magazines at Woolwich 'were reasonably secure against the danger of overhead attack' but at Purfleet 'there was a dangerous accumulation of explosives, chiefly gunpowder, which should be diminished'.

The Committee decided that an attack by airships was a 'certain menace in the near future'. It considered that when the Royal Flying Corps was sufficiently developed and trained 'the Admiralty and the War Office should decide what aerial force is required for the protection of Woolwich'. It did not recommend decentralization:

> Decentralisation cannot be recommended. . . . It would, in the first place, be costly, and, secondly, it would entail the multiplication of the means of defence, which, whether in the form of guns or aircraft, would act less effectively when dispersed than when concentrated for the protection of a single area.

Experiments and trials with guns and aeroplanes for the purposes of air defence now took place. However, as we shall see, in August 1914 when the war broke out the arrangements in each of these spheres or categories were inadequate.

We have already noticed some of Commander Samson's pioneering efforts in the naval air station at Eastchurch. In time, the Military Wing of the Royal Flying Corps also established an experimental branch under the command of a capable officer, Major Herbert Musgrave, RE. This unit carried out experimental work with balloons, kites, wireless telegraph sets, bombs, and also practised co-operation with the artillery. However, it did not enhance the home air defences as much as it might have done. One historian, Neville Jones, has shrewdly pointed out that the 'majority of the specified experiments were intended to improve the tactics of cooperation with the field forces'.[6]

Although the War Office, in this period, continued to claim sole responsibility for the home air defences[7] it did almost nothing to supplement this work of the Military Wing. It failed to provide an adequate number of either aeroplanes or guns for the Air Defence of Great Britain. It concentrated almost all its efforts and resources on the creation of an air force to serve only as a reconnaissance arm for the British Expeditionary Force.

II

Meanwhile, preparation of those gun defences against aircraft recommended by the Home Ports Defence Committee in 1910 proceeded, but not without difficulties. On 4 May 1912 the Admiralty reported to Captain Hankey, by this time Secretary of the Committee of Imperial Defence, that 'careful consideration' had been given to 'the question of the development of an effective gun for use against hostile dirigibles'. The Admiralty communication explained that 'Joint action' in carrying out the necessary experimental work had been agreed upon between the navy and the army 'in order to obviate possible waste of labour through each Service covering the same ground'.[8]

In consequence of this arrangement the Admiralty undertook to develop a 3-inch semi-automatic gun and mounting capable of being carried in 'the smallest cruiser or even in destroyers'. For its part, the War Office agreed to produce a 4-inch gun 'for mounting either in a fixed position or on a travelling carriage'. Oswyn Murray, the assistant-secretary of the Admiralty, made clear in this communication that work on the naval gun was being accelerated 'as much as practicable'. The device had already been ordered from Woolwich. However, Murray wrote, 'My Lords . . . understand that progress with the Army gun is not in such a forward state'.

Murray's conclusion revealed that the heads of the Admiralty were fully alert to the issue of protecting naval dockyards and arsenals from air attack, but they were also aware that the responsibility in this sphere lay not with them, but with the authorities in the War Office. In this period the Admiralty ignored the reports and recommendations of the Committee of Imperial Defence whenever it chose to do so. However, in these circumstances Murray appealed to Hankey for help. He wrote:

> Whilst their Lordships consider that the present menace of Airship attack on warships is not very serious on account of the greater mobility of the latter, they view with some apprehension the defenceless state of Dockyards and Naval Arsenals against this form of attack, and consider the production of a trial gun and mounting is a matter of extreme urgency; they accordingly bring the subject to the notice of the Committee of Imperial Defence for their consideration in connexion with any representations which may be made by them to the Army Council.

Captain Hankey now applied a touch of his own. When he sent a copy of this Admiralty communication to the War Office he pointed out, in his covering letter, that:

> The question of the defence of Magazines, Cordite Factories and other Vulnerable Points against Airship Attack was considered by the Home Ports Defence Committee in 1910 and the views of the Committee at that date are embodied in H.P.D.C. Memorandum 12-M which was approved by the Army Council by their letter General Number 12/45 (M.T.I.) dated 29th August, 1910'.[9]

When the soldiers received these two communications they were annoyed with the procedure adopted by the Admiralty. By this time Sir John French had replaced Sir William Nicholson as Chief of the Imperial General Staff. He at once enquired about the status of the army's trial gun. Sir Charles Hadden, the Master-General of the Ordnance, explained in reply that two mountings for guns to be used against aircraft were being made. One was a field mounting to take an 18-pounder field gun. The gun was already available and the mounting was being made by the firm of Vickers. Hadden expected it to be ready for trial by November 1912. Secondly, a fixed mounting designed to take the naval pattern 4-inch gun was also being made. Hadden did not know when this device would be completed. He declared that the guns presented no problem but the mountings required 'special designs & manufacture'. He also complained about the Admiralty's methods. He wrote: 'I should have thought the Adm. under the joint action would have written us for any information they required, not to C.I.D. in the form adopted'.[10] Eventually, it was decided to send the following message: '. . . the matter referred to by the Lords Commissioners of the Admiralty has not been overlooked by the A(rmy) C(ouncil) and is in hand'.[11]

Unfortunately, this was too optimistic a conclusion. As the months passed the army guns and mountings did not appear but the Admiralty continued to regard the issue as a 'matter of extreme urgency'. Their dockyards and naval arsenals still remained without any protection against attack from the air.

In particular, the officials at the Admiralty were worried about the great Chattenden and Lodge Hill magazines near Chatham which were deemed to be very vulnerable to such an assault. They pressed the War Office to provide guns of some kind to be emplaced at these locations, and as quickly as possible. Colonel Seely and Winston Churchill who were friends always sought to soften or reduce the friction between their two services and on 11 December 1912 Seely wrote to the First Lord with respect to the anti-aircraft gun defences at Chattenden, in the Medway area. His letter reflected the difficulties encountered by the War Office in its attempts to secure artillery for the home air defences. No new mountings or guns were yet available for the purpose. Instead two 6-inch howitzers which would be ineffective against dirigibles were to be sent to Chattenden. The situation with respect to searchlights, Seely revealed, was even worse:[12]

> You will see that as regards anti-aircraft guns we have done all that is now possible: i.e. we shall have the howitzers in

position at Chattenden in a few days. They will terrify but will not hit. By April we shall have guns that may do both.

With regard to search lights we cannot get one with a vertical beam for some time to come.

One important matter arises: the real defence of this & similar places at night may well be found in darkness. At present, as I understand, Chattenden is a blaze of light at night. You will no doubt consider what instructions should be given secretly as to extinguishing all lights when an airship is heard.

Despite every effort and plea by Churchill in his capacity as First Lord no anti-aircraft gun defence system was established by the War Office. Occasionally, various trials and experiments with guns were contemplated and these were sometimes carried out, but not much was accomplished. In 1913 the Air Department of the Admiralty produced a paper entitled 'Aerial Attack on Dockyards'. This document showed how seriously the Admiralty regarded the possibility of aerial attacks at this time. It also revealed the methods the naval authorities thought would be most effective in defence against such attacks:[13]

Aerial Attack on Dockyards

This question has been receiving the close attention which its importance deserves. . . . It is fairly certain that at present attacks are more likely to be expected from Dirigibles than from Aeroplanes. Experiments are being carried out with aircraft guns capable of long range firing at a high elevation and search lights to train vertically. These will be mounted in some prominent position commanding the dockyard. A more suitable method of defence is however now under consideration . . . proposals will shortly be put forward for the early trials of captive kite balloons capable of lifting an armament of several hundred pounds weight; it is considered that guns so mounted would have a wider scope of training etc than those on the ground, also being at an altitude of 2–3000 feet, a correspondingly smaller vertical range would be required. . . .

In addition a certain number of aeroplanes and hydro-aeroplanes will be attached to our Naval ports to assist in scouting and patrol work. Eventually all these will carry an armament which will undoubtedly be used against other aircraft.

Successful experiments have already been conducted with a

maxim gun mounted in an aeroplane and six experimental machines are now about to be constructed, these will be designed to carry specially lightly constructed guns.

The War Office also engaged in experiments with anti-aircraft guns. In August 1913 trials began at Shoeburyness which involved firing at moving aerial targets. Later, similar experiments were made at the Larkhill ranges. There, kites were flown at altitudes up to 500 feet at a speed of about 30 miles per hour.[14] One technical expert later commented: '. . . since enemy aircraft were unlikely to offer themselves as targets at that speed and height the results of the firing were of negligible value'.[15]

When the war began the gun air defences were in a deplorable condition. Winston Churchill, in his paper entitled 'Aerial Defence' which we noticed earlier in this history, revealed that only 33 guns were in position by September 1914. Of this total, moreover, 28 were 1-pounder pom-poms which were of 'very small value'. Churchill explained about these pom-poms: 'Those that have been tried in the field have been found unsatisfactory. No shrapnel is available for them, and the shells provided do not burst on the fabric of a Zeppelin, and almost always fall back to earth as solid projectiles'.[16]

In an appendix to Churchill's paper a chart revealed that in addition to these ineffective pom-poms there was one 4-inch quick-firing gun at Portsmouth; two 3-inch guns at Chattenden and Lodge Hill (the 6-inch howitzers had been withdrawn); a 3-inch gun at Waltham Abbey and another at Purfleet. Aside from the pom-poms these few artillery pieces made up the total of Great Britain's anti-aircraft gun defences in place upon the outbreak of war.[17]

III

The provision of aeroplanes for home air defence was also inadequate. The authorities in the War Office insisted that they should exercise complete control over the home air defences. At the same time they planned to send almost all the aeroplanes of the Military Wing to France instantly, as a part of the Expeditionary Force, as soon as war began.

There remained the aircraft of the Naval Wing. In the pre-war period the War Office and the Admiralty, despite conferences and the exchange of letters, minutes, and memoranda, were unable to decide how these aircraft should be employed in the home defence arrangements.

In August 1912, shortly after the formation of the Royal Flying Corps, the Army Council took up the question of the 'responsibility for coast defence by aircraft'. They decided that such aircraft should be divided into three classes which were: those employed for fleet purposes: those employed with the navy's Coast Defence Flotillas: and a third class of 'Aircraft employed in conjunction with the fixed and mobile land defences of defended ports'.[18]

The Army Council decided that the first two classes of aircraft should be provided and maintained by the Naval Wing. However, they insisted that the aircraft for the third category should be furnished and administered only by the Military Wing. Eventually, the Admiralty agreed with these recommendations and conclusions. Despite this agreement with the Admiralty, the War Office did not establish a distinct defensive air force. They continued to plan to send almost all the aeroplanes of the Military Wing to France for service with the Expeditionary Force. No provision was ever made for aeroplane units responsible for the home air defences.

Winston Churchill was not satisfied with these arrangements and he often complained abut them, but with little effect. For example, he wrote in January 1913:[19]

> The War Office should . . . be asked what arrangements have been made in naval ports generally, particularly at Sheerness and Chatham, for dealing with the arrival of a hostile airship. . . . It would appear desirable that a force of aeroplanes should be stationed in the neighbourhood of Chatham and Sheerness in order that a hostile airship may be attacked by day or night and destroyed. Have any of these arrangements been made?
>
> It is recognised that time is required to make thoroughly satisfactory arrangements, but in the interval we must do the best we can.

By 1913 a number of naval air stations had been established along the east coast so that naval aircraft could be employed to watch the 'German Ocean'. At a conference held at the Admiralty on 19 November it was agreed, for the first time, that aircraft of the Naval Wing based at these air stations would participate in the home air defence arrangements.

By the terms of this agreement between the War Office and the Admiralty the responsibility for aerial defence would be based, henceforward, on several principles. The War Office would be responsible in the British Isles or in any land operations in which the army was involved. The Admiralty would be responsible for all aerial services of the fleet. A third principle declared:

> In order to avoid duplication and overlapping it was agreed that in cases where the naval seaplane stations are close to points of naval importance vulnerable to aerial attack, the Naval Wing should undertake the responsibility for the aerial defence of naval property. Instances of naval seaplane stations so situated are furnished by the Isle of Grain and Eastchurch stations, near Chattenden magazine; the Firth of Forth station, near Crombie magazine; and the Cromarty station, near the Invergordon oil tanks.[20]

In May 1914 Asquith arranged for a sub-Committee of the Committee of Imperial Defence to consider the question of the 'allotment and location of naval and military seaplane and aeroplane stations along or near the coast'.[21] Winston Churchill served as chairman of this body which met for the first time on 25 June.

At this meeting Churchill and Sir David Henderson competed with each other to secure home air defence responsibilities for each of their services. Henderson agreed that the Naval Wing should participate in this activity but he made clear that the Military Wing would require its own aeroplanes 'for home defence'.[22]

This sub-Committee met again on 3 and 21 July. As we have already seen, at the latter meeting General Sir William Robertson, the Director of Military Training, once again claimed for the War Office the sole responsibility for defence against aerial attack, including the air defence of naval ports and other vulnerable points, 'even those of exclusively naval interest'.[23]

A few days later, on 27 July 1914, Sir David Henderson, in preparation for further meetings of this sub-Committee, drew up several 'proposals' for a 'suitable organization' for the aircraft necessary 'for the protection of Defended Ports and Vulnerable Points'. A table he prepared revealed that six squadrons totalling

162 aeroplanes would be required for this purpose, the Air Defence of Great Britain.[24]

According to Henderson's plan, in the Scottish Command one squadron of 27 aeroplanes, with its headquarters at Montrose, would protect Cromarty, Aberdeen, the Tay, and the Clyde. Another squadron of 27 aeroplanes would be assigned to defend the Forth area. No aeroplanes, according to Henderson's calculations, would be needed in the Western or Irish Commands because it was unlikely that 'these points could be attacked from the air for some time to come'. By contrast two squadrons were allotted to the Eastern Command. One of these, with its headquarters at Dover, would defend Harwich, Dover and Newhaven, with a force of 27 aeroplanes. Another squadron, with its headquarters at Orfordness, would be responsible for Shoeburyness and the Thames and Medway areas. Similar organizations were suggested for the Northern and Southern Commands, respectively.

It must be emphasized that these arrangements were only 'proposals'. Henderson and all his colleagues were overtaken by events before any of them were put into effect.

Shortly after the war began the British Expeditionary Force was carried across the Channel; and it then proceeded to its allotted place on the left of the French line. Sir David Henderson now decided he would lead the Royal Flying Corps in the field and he took Sykes with him as his Chief of Staff, and almost all of the aeroplanes of the Military Wing. A junior officer named Sefton Brancker was quickly given a temporary promotion to the rank of Lieutenant-Colonel and the title Assistant Director of Military Aeronautics. In effect, Brancker assumed the position in the War Office Henderson had abandoned for active service in France. Brancker was given 'authority over all branches of the directorate and direct access to the Secretary of State for War'.[25]

The Secretary of State for War was Field-Marshal Lord Kitchener, the brutal but impressive soldier who was selected for the post as soon as the war began. Years later Sir Sefton Brancker, in his unpublished memoirs, described his earliest meeting with Kitchener, in the War Office. His account reveals the deplorable situation Sir David Henderson left behind him in August 1914. There were no aeroplanes of the Military Wing available for home defence. The arrangements were so bad even Lord Kitchener could not mend them:[26]

Lord Kitchener . . . had absolutely no experience of aviation. On almost the first day of his office he sent for me & explained that he had accepted his responsibility to the Cabinet of patrolling with aircraft quite a large portion of the British coast. I think it was from Southampton to the Thames & from Newcastle to the North of Scotland.

My jaw dropped. I pointed out that every man & aeroplane at my disposal at home was required for training & building up new units. . . . K. looked surprised – put on his most terrifying frown – & told me not to talk nonsense but to go away & do what I was told! . . . .

I did nothing. I could think of nothing to do and the next day went to see him again and told him frankly that I had done nothing. . . . K. glared his worst. I expected an explosion but slowly his face relaxed and he said – 'Well – perhaps you are right but you've got to do *something* in order to carry out this bargain although it may be useless'. I sent one pilot & one small & very feeble single-seater to fly in the nighbourhood of the Forth.

IV

On 25 April 1912 the Committee of Imperial Defence met in order to discuss the report on Aerial Navigation of Lord Haldane's Standing Sub-Committee. H. H. Asquith, the Prime Minister, was in the chair on this occasion. He began by stating that he 'had approved this report provisionally in view of the urgency of the matter, and action upon it had been taken'.[27]

During this discussion of 25 April a problem arose. In this period it was very difficult, even for experts, to predict the future development of aeronautics. Winston Churchill was fascinated by the subject and convinced of its importance. As always, his opinions were strongly held. In this instance, however, as technical advances were made, Churchill sometimes changed his mind. For example, at a meeting of Lord Haldane's Standing Sub-Committee held on 18 December 1911 Churchill was sceptical about the military or naval value of airships. He said on that occasion 'he . . . would want a good deal of converting before he would acquiesce in a policy of building dirigibles'.[28] By April 1912 developments in

Germany had convinced Churchill that airships now seemed likely to become 'an indispensable adjunct of the fleet'.[29]

In these altered circumstances Churchill suggested at the meeting of 25 April that Colonel Seely's Technical Sub-Committee should be reconstituted in order to investigate the question of airships. The Prime Minister agreed that the matter should be taken up afresh and on 7 May he requested that 'the Technical Sub-Committee of the Committee of Imperial Defence . . . be reassembled further to consider the provision of Airships for the Royal Flying Corps'.[30]

Seely's Technical Sub-Committee decided to send two of its members, Captain Sueter and Mervyn O'Gorman, to visit Austria, France, and Germany so that they could acquire information about foreign airships. When their trip was over they would prepare a report about their experiences on the Continent. An excursion of this kind had been contemplated for a considerable period before the Technical Sub-Committee was reassembled.

At this time the British Naval Attaché in Berlin, Captain Sir Hugh Watson, RN, was very concerned about the development of aeronautics in Germany. In December 1911 he and his colleague, the Military Attaché, sent a report to London urging that Sueter and Sir Alexander Bannerman should come to Berlin in order to visit the Parseval airship works and possibly the Zeppelin works also. Watson also wrote to Sueter to alert him to this development. In his letter he explained that 'By virtue of dining some of the chief people I am now in a position to be of use to you . . . we believe we can get you to the Parseval works, and through them to Zeppelin perhaps'.[31]

In this letter Watson also explained his views about how a Naval Attaché should carry out his duties:

> My view of a Naval Attaché is that he should investigate as much as possible . . . and then when he has reached the limit of what he can usefully report on . . . get Admiralty to send the Expert, the way for whose visit is paved by many dinners, always useful here! I have done this in the case of Gunnery and Submarines, and now I have worked your branch up to the point where you can usefully step in. Of course, I hope you can say 'We have nothing to learn in England', but even if it is so I suggest a visit . . . the government here don't realise the amount

of leakage that goes on through commercial channels after good Port; this for your own ear only.

Before they left England Colonel Seely's Sub-Committee instructed Sueter and O'Gorman to obtain information about foreign airships; to discover if wireless telegraphy was used in these airships; to enquire about the guns that might be employed for their destruction; and to learn about guns or other weapons that could be used in aircraft. The two agents did not confine themselves to airships. They also investigated the production and deployment of aeroplanes in the countries they visited. They were impressed most by what they saw in Germany.[32]

They arrived in Berlin on 24 June and on the next morning they reported to the British Ambassador, Sir Edward Goschen.[33] Alick Russell, the Military Attaché, arranged for them to visit the Parseval airship works at Bitterfeld.[34] When they returned to London they prepared a long and detailed report for Colonel Seely and included with it photographs and sketches of what they had seen. Their report revealed there were 'no less' than 28 military flying grounds in Germany. In addition, there were six aerodromes for the public but soldiers and sailors were also in training at each of these places. They found 33 large airship sheds and five or six airship factories. Twenty-four firms were engaged in the manufacture of aeroplanes. All these firms, aerodromes and sheds were identified, listed, and located, geographically, in the report.

A high point of their visit came on 4 July when they made a long voyage in a commercial Zeppelin airship, the *Viktoria Luise*. The German authorities had earlier refused to arrange such a journey for them. Eventually, Sueter and O'Gorman disguised themselves as Americans and flew with a party of German civilians and seamen for a distance of 250 miles. They flew over Hamburg, Lubeck, and Kiel. When the journey of nearly six hours duration was over they 'got into conversation' with a German naval lieutenant who was attached to the *Viktoria Luise* for training. He told them he was to be the captain of the first German naval airship which was 'now under construction'.[35]

In their report Sueter and O'Gorman permitted themselves the following reflection:

We were struck with the popular reception given to the airship. On passing over villages, isolated farms etc., everybody turned out and cheered and waved to us. In many hundreds of miles on our small English ships, including trips to London, Farnham; Guildford, Salisbury, and down south to near Portsmouth, no such interest is evoked.

They also pointed out that German airships were able to reconnoitre the entire German North Sea coastline. As a result, in any future war it would be impossible except in foggy or stormy weather for British warships to approach the German coast without their presence being discovered. Unless Britain gained 'command of the air' the idea of such approaches would have to be abandoned. The report also revealed that German airships had covered distances equal to the distance from Germany to the British coast, without refuelling.

Sueter and O'Gorman stated in their report that they had been told that loads up to 1000 pounds had been dropped from airships. They wrote that

> while on this subject it may be remarked that for any nation to have a ton of explosives dropped above their Admiralty, War Office, or Administrative buildings would, to say the least of it, be inconvenient, unless proper alternative underground offices have been foreseen.

V

When Colonel Seely and his colleagues of the Technical Sub-Committee framed their report at the end of July 1912 they began by explaining the principles on which the government's aeronautical policy had been based; and then they set out 'the new factors in the situation'. Their report began by stating:[36]

> Up to the end of the year 1911 the policy of the Government with regard to all branches of aerial navigation was based on a desire to keep in touch with the movement rather than to hasten its development. It was felt that we stood to gain nothing by forcing a means of warfare which tended to reduce the value of our insular position and the protection of our sea power.

> This original policy with respect to aeroplanes had already been

abandoned but such was not yet the case with respect to airships. However, the Technical Sub-Committee had learned several new facts about the recent development of these dirigibles. The new facts were contained in the joint despatch of December 1911 prepared by the Naval and Military Attachés at Berlin; and in other miscellaneous reports including reports from 'A Reliable Source' in Germany. This information had been confirmed and amplified by the Sueter–O'Gorman account: 'Their Report furnishes striking evidence of the progress which has been made on the continent, and particularly in Germany, in airship development'.

The Technical Sub-Committee repeated some of the conclusions offered in the Sueter–O'Gorman report. In its account it then turned to the ability of airships to bomb targets at sea and on the land. Referring to the recent flight of the *Viktoria Luise* the Technical Sub-Committee stated:

> Some idea of the weight of explosives which could be carried may be formed from the fact that in the voyage described in the report thirty persons were carried for six hours. Consequently, torpedo craft, submarines, dock gates, power stations, wireless telegraph stations, magazines, stores of oil fuel, and other vulnerable points, if unprovided with effective means of defence, will be in serious danger from incendiary or explosive charges used by airships. The Sub-Committee wish to draw the attention of the Committee of Imperial Defence to the desirability of further investigating the means of defence of such vulnerable points against aerial attack, and also to the necessity for the immediate provision of a gun capable of being employed against air-craft.

The exploitation of airships in Germany had been so 'remarkable' that the Sub-Committee decided that 'the time has come to abandon the cautious policy which has been adhered to up to the present time, and to adopt a policy of active deployment in this branch of air work'.

It was proposed, therefore, that the Naval Wing of the Royal Flying Corps should be trained to handle airships of the largest size. Initially, since the Military Wing was more experienced in the handling of dirigibles, the Sub-Committee recommended that personnel from the Naval Wing should be attached to the airship company at Farnborough in order to obtain, from the soldiers, as much training as possible. These recommendations were put into effect later in the year when the naval airship section, abandoned after the *Mayfly* disaster, was reconstituted by the Admiralty.

The Seely Sub-Committee report was discussed at two meetings of the Committee of Imperial Defence. As we have seen earlier in this chapter not much was achieved in carrying out the urgent suggestion in the report that it was necessary to secure an efficient gun for the home air defences. However, as a result of these and similar deliberations the Admiralty decided to establish a chain of seaplane and airship stations on the east coast of Great Britain. Eastchurch was considered the first of these seaplane stations. In December 1912 another was established on the Isle of Grain. Similar coastal stations then appeared at Calshot, Felixstowe, Yarmouth and Cromarty, and plans were drawn up for the provision of other stations of the kind.

These places were not designed to serve as defensive fighter-aerodromes. Indeed, when the principal duties of naval aircraft were listed officially in this period reconnaissance was selected to head the list while the prevention of attacks by hostile aircraft on dockyards, magazines, and oil storage tanks was placed last.[37] Nevertheless, even if they were starved of materials and equipment in the pre-war period these coastal air stations were in place when hostilities began in August 1914.

On 6 December 1912 the Committee of Imperial Defence discussed the report of Seely's Technical Sub-Committee, for the second time. The minutes of their meeting reveal the difficulties encountered by the authorities as they worked to prepare their country for this novel form of warfare, war in the air. Asquith was in the chair on this occasion but Colonel Seely opened the discussion about airships. He pointed out that it might become necessary for the sake of British security to 'maintain command of the air' but for the moment, he said, 'what was most wanted was to evolve a type of airship in no wise inferior to the best available in foreign countries . . .'.[38]

Winston Churchill said the matter was 'important' and 'urgent'. He feared that the development of aircraft was certain to involve the nation in heavy and continuous expenditure. At the moment he did not believe Britain was in actual peril 'but he did not feel very confident about it, and it was a matter of great anxiety'. 'Our dockyards', he said, 'were absolutely defenceless against this form of attack if an enemy airship were to succeed in reaching a position whence it could attack'. Magazines were also vulnerable to such an assault.

At this, the Prime Minister enquired if the threatened magazines could be put underground. Seely replied that the expert view was that the consequences of an underground explosion, if one took place, were so much more serious that it was better to have the magazines in the open. Churchill explained that various suggestions had been made and were being tried, such as 'cover from view by planting trees'; or employing torpedo netting for the purpose; and the establishment of dummy magazines. Lloyd George asked if guns could not be used against attacking airships. It remained for Seely to say: 'there were some very great difficulties in the use of guns yet to be got over'. No British gun had yet been made that could be relied upon to hit an airship. Furthermore, it had been found very difficult to construct a fuse that could burst a shell upon striking the gas-bag of an airship. Difficulties had also been encountered in making an incendiary projectile to ignite the gas in a dirigible.

It remained for Churchill to explain what was being done. He said:

> they were trying guns and mountings – a 3-inch gun for the Army and a 4-inch for the Navy. They would have an all-round arc of fire . . . and though still only in an experimental stage he thought it imperative to provide some at once. . . . We were very seriously behindhand. The failure of the first naval airship had been a bad set-back. Without ships we could not train the personnel to handle them. The Admiralty had purchased an airship of the Astra-Torres type in France, and were endeavouring to buy one of the Parseval type in Germany, but they were doubtful of getting it. He had no doubt that we should eventually recover our lost ground, but in the meantime they could not but feel very anxious, and if events took an untoward turn the matter might become one of very grave reproach.

Another problem of moment arose in the course of this discussion of 6 December. It was one that had plagued the British planners for months past. Admiral Sir A. K. Wilson had, by this time, resigned as First Sea Lord but he continued to serve as a member of the Committee of Imperial Defence. He now intervened to say he had 'gone very thoroughly into this question'.

Wilson thought the danger to Britain from German airships was 'much exaggerated'. He argued that at sea naval guns would be able to hit dirigibles without too much difficulty since the 'target offered by an airship was larger than that offered by any battleship'. When Churchill said he had not been considering the dangers to a fleet at sea, the Admiral replied that 'defence by aeroplanes' could supplement the gun defences of dockyards, magazines, and other vulnerable points at home. Earlier, as we know, Wilson had not thought much of the effectiveness of aeroplanes but now he revealed that, like Churchill, he had changed his mind. He argued that aeroplanes could destroy a dirigible by means of a 'squib fired out of a pistol at close range' or 'by ripping her up with a fishing line and hooks'.

Winston Churchill would not give way in this argument. He pointed out that the latest airships were equipped with a platform on the top from which the crew were able to fire, in their own defence.[39] Asquith now intervened to say he thought Wilson's criticisms were pertinent. He suggested 'the matter needed further consideration' since the Committee of Imperial Defence did not possess enough experience or information to enable sound judgements to be made.

As a result, the meeting adjourned consideration of the Technical Sub-Committee's report on airships. The Sub-Committee had concluded, in its report, that the best way to defeat enemy airships would be to employ a superior force of such vessels against them. In December 1912, however, the British authorities found themselves unable or unqualified to come to any decision in the matter. Like their contemporaries in other countries they found it difficult to predict which engine of war, airship or aeroplane, would finally dominate. Earlier, at a Committee of Imperial Defence meeting held on 1 August 1912, when the report of Colonel Seely's Technical Sub-Committee was first discussed, Winston Churchill had lamented: 'We could not afford to wait until the relative value of airships and aeroplanes was settled, and we must therefore seriously study airships'.[40] Four months later the issue was still a subject of bickering among Britain's highest leaders.

VI

In October 1912 an incident occurred at Sheerness that fixed the attention of the entire country upon the condition of Britain's air defences. Sheerness was a garrison town and naval seaport in the Isle of Sheppey on the right bank of the Medway estuary at its junction with the Thames. It was defended by strong modern forts at Garrison point and Barton's point. These forts commanded the entrances of both the Thames and the Medway. The dockyard at Sheerness was used for naval repairs and there were also extensive naval stores and barracks in the place. On the evening of 14 October 1912 the sounds of an aircraft engine were heard over Sheerness. The result was an alarm that spread to every part of Britain. It was widely believed that a Zeppelin, under cover of darkness, had paid a visit to this strategically important area.[41]

On 14 November Captain Sueter submitted a minute to the Third Sea Lord that set out the details of this incident:[42]

> Commander Samson reported . . . that an airship had been heard over Sheerness and he thought it might have been a German Zeppelin.
> On the evening in question a telephone message was received at Eastchurch[43] from Sheerness that an aircraft was in the air over Sheerness. Flares were lit but no landing was made and the aircraft disappeared.
> Commander Samson reported that no naval machines or any of the aero club machines were out.

On the same day that Sueter prepared this minute C. G. Grey produced an article about Sheerness in his magazine, *The Aeroplane*. Count Zeppelin and a party of his associates were flying in the new Zeppelin *L1* on the night of 14 October and Grey published this fact. Grey pointed out, in his article, that Britain possessed no weapon that could attack airships effectively. Modern aeroplanes could serve this purpose, he explained, but the War Office and the Admiralty, he argued, still refused to 'guarantee' that British firms producing high speed aeroplanes would ever obtain orders for their machines. Never, he concluded, was the ' "Wake up England" spirit more immediately of importance than it is at the moment'.

This was only the beginning of an agitation that lasted for

months. Count Zeppelin at once denied that his airship had approached the English coast, and his disclaimer was published in the *Daily Mail*. Later, a reader of *The Aeroplane*, a Dr C. Abel-Musgrave of Brighton, sent a cutting of *The Aeroplane*'s article of 14 November to Zeppelin. The Count promptly replied with a cablegram 'denying that any of his ships had crossed over England. . . .'[44]

These disclaimers, of course, produced little or no effect on the developing situation. On 18 November the *Daily Mail* published a long article about the episode at Sheerness. All the considerable resources of the paper were employed to discover what had occurred there. A special correspondent was sent to Sheerness, Eastchurch, and nearby Queensborough to make enquiries, at the scene. The *Daily Mail* correspondents in Hamburg and in Berlin were ordered to investigate the German end of the puzzle.

As a result of these efforts, the *Daily Mail* reported on 18 November that many 'responsible witnesses' in Sheerness and Queensborough heard an aerial engine over both towns on the evening of 14 October. The *Daily Mail* also discovered that the authorities at the Eastchurch naval air station had received a report about the engine noises by telephone. In the belief that an aeroplane was in distress they lit flares on the flying ground, but no aircraft appeared. The Hamburg correspondent of the *Daily Mail* reported that a new German naval airship, with Count Zeppelin and 21 men on board, passed over Borkum and Norderney and then disappeared in the mists over the North Sea on the afternoon of 14 October. The ship returned to Kiel at 2 a.m. on the following morning. The paper also announced, in this article, that William Joynson-Hicks, the Unionist Member for Brentford, proposed to ask a question about the Zeppelin airship in the House of Commons.

Joynson-Hicks and his friends in the House of Commons now began to harry the Liberal Ministers about their aeronautical policies. As a matter of course, Colonel Seely and Winston Churchill were the primary targets in these attacks. When Churchill, on 21 November, made an evasive reply in the House the issue was taken up by the *Daily Mail*, again. In a leading article of 23 November the paper explained that Sheerness was 'one of the most important of British dockyards'. It declared:

The matter cannot be left where Mr Churchill's answer . . .

leaves it. Of what nationality was the mysterious craft? Whence did it come? If the Admiralty tells the country that it does not know, the country's reply will be that it is the business of the Admiralty to find out.'

As the agitation continued many people in Britain came to the conclusion that it did not matter if a Zeppelin airship had visited Sheerness on the night of 14 October. The great point was that the Germans were capable of such a foray and there was nothing the British could do to prevent it. As the technical magazine *Flight* put it in an article on 23 November:

Without committing ourselves to any definite opinion as to whether or not 'L1' has in fact flown over Sheerness, we say that there is not the slightest reason why she should not have done so. . . . It is not only possible for such craft as the Zeppelin to cross the North Sea and get back in safety, but such flights as this may be said to fall quite within the ordinary compass of these vessels. To Great Britain as a nation the moral is so obvious as to need no emphasis.

On 27 November 1912 the battle in Parliament was resumed when the Tories decided to attack Winston Churchill. Joynson-Hicks began by asking the First Lord if Britain possessed any airship equivalent in size and power to a Zeppelin. Churchill could only reply in the negative. Next, Bolton Eyres-Monsell, the Conservative member for South Worcestershire, and a former naval officer, asked Churchill:[45]

(1) whether he proposes to use aeroplanes with the view of preventing an airship from proceeding to London or elsewhere; (2) what will be the relative position of Great Britain and Germany with regard to fighting airships (rigid) at the end of 1912, 1914 and 1915; and (3) whether it is the policy of the Admiralty to employ non-rigid airships for engaging rigid airships?

Churchill could only say that with respect to rigid airships 'Germany has won a great pre-eminence'. As to the fighting value of such vessels, Churchill said: 'opinions differ and experience is lacking ... the whole subject is receiving the attention its undoubted importance demands'.

Colonel Charles Warde, the Conservative member for the Mid. Division of Kent, then asked: 'Whether our dockyards and repairing docks are provided with searchlights and howitzers capable of repelling aerial night attacks'. Churchill refused to discuss this subject in public but he explained that its importance was 'recognised' by the Admiralty.

When Joynson-Hicks enquired if the Committee of Imperial Defence had considered what effect the development of aeroplanes and airships had had upon the defences of the country, the Prime Minister himself intervened. Asquith blandly informed the House that the Air Committee, a permanent Sub-Committee of the Committee of Imperial Defence, 'keeps the question of aerial navigation constantly under review'.

These replies could not satisfy a public that was becoming more and more insecure as Britain's weakness in the air began to be compared with German aerial strength and power. Critics demanded that the Liberals provide more funds for naval and military aeronautics in order to redress this deadly imbalance. Although he could not say so in public, this was also Winston Churchill's opinion. When the Committee of Imperial Defence discussed the report of Colonel Seely's Technical Sub-Committee on 6 December, Churchill told his colleagues:[46]

> There was very little doubt that the airship reported recently to have passed over Sheerness was a German vessel, and this incident had renewed anxiety. Some form of overhead protection would have to be provided for magazines, and guns for use against airships designed and supplied.

On 19 December *The Aeroplane* published a special article, 'Dirigibles and the Nations', by a technical authority of the day, W. E. de B. Whittaker. He gave clear expression to the fears aroused by the incident at Sheerness and then proceeded to attack the Liberal Government because it put its social policy ahead of the needs of national defence:

> Recent events ... ought to convey a warning to those in whose hands lie the destinies of the British Empire. The Sheerness affair ... was ... an actual fact. We cannot identify the

airship beyond knowing that it came from the Continent, but its visit points a moral. We, the officially supreme nation of the earth, the arbiter of world policy, are unable to return a call made on us by a powerful rival. What one can do is possible to others. . . .

We, as a country allot £6,000,000 in the bestowal of old age pensions. Nothing could be more admirable in itself; but if, in the desire to find the money, the country's defences are neglected, then in the future it is possible that in a state of suzerainty we shall be unable to pay even a fraction of these subventions for the maintenance of the antique poor.

VII

Liberal Ministers were attempting to shore up the aeronautical defences of the country but little could be done in this direction in a short space of time. The provision of adequate funds for the purpose was not the only problem. The technical difficulties were immense. Efficient guns, gun-mountings, and searchlights could not be procured in an instant. Moreover, British builders lacked experience so that the mysteries and technicalities of airship development continued to confound them. It was widely believed in Britain that the French and the Germans possessed better aeroplanes while it was known that they controlled far greater numbers of these machines than did the Royal Flying Corps. Moreover, officers like Henderson and Sykes, and some counterparts in the Royal Navy, continued to put the greatest emphasis upon the reconnaissance functions of aircraft when they made their plans. There now grew up in some sections of the British public a feeling of outrage. Men began to say that if war came and if it then turned out that the home air defences had been neglected, those responsible would be hanged from lamp-posts in Whitehall. Indeed, it is fair to say that this connection between the home air defences and lamp-posts became something of a theme of recent British life. An early example of the idea appeared in *Flight*, the official publication of the Royal Aero Club. On 1 February 1913 *Flight* discussed the subject of 'Our Aerial Fleet':

> In dealing with the matter of our aerial defences we have at all times endeavoured to steer clear of alarmist tendencies and

to give the naval and military authorities due credit for what they have accomplished in the teeth of Government apathy. . . . But there comes a time when it is necessary to talk plainly. . . . That time has come now. We have waited to see the awakening, and we have seen nothing but a continued policy of discouraging apathy, which has left this country hopelessly behind its rivals, without an air fleet worthy of the name, and almost entirely at the mercy of the first aerial power which cares to launch its air squadrons on a mission of destruction across the North Sea. . . .

This is plain enough speaking, and the pity is that it is true to the letter. We repeat that we are not alarmists . . . The political outlook is darker than it has been for many years. . . . And if war should come suddenly, we most certainly shall not have command of the air – but the lamp-posts of Whitehall may have unfamiliar ornaments. And well it might be under the circumstances.

In this connection it may be mentioned here that in July 1917 German aeroplanes successfully bombed London. In the panic that followed it was decided at once to combine all the elements of the defence, in the air and on the ground, under one commander. His responsibility would be to see to it that such air raids were met and turned away before the metropolis could be bombed again. The man chosen for this post was a brilliant officer, Major-General E. B. Ashmore, RA. At the time he was commanding the artillery of the 29th Division in the line north of Ypres, and his dug-out there was regularly drenched with poison gas. His new combined command was to be called the London Air Defence Area (LADA). Ashmore was a brave man but he also realized the significance of this change in his responsibilities, especially if he failed in the new post. He wrote: 'The fact that I was exchanging the comparative safety of the Front for the probability of being hanged in the streets of London did not worry me'.[47]

Lord Beaverbrook experienced a similar reaction when he became Minister of Aircraft Production in 1940 during the Battle of Britain. His task, at that time, was to see to it that the aircraft needed for the defence of the country were turned out in adequate numbers. He was terrified by the responsibilities of the post. He once said: 'I was TERRIFIED. If I failed I knew it meant a lamp-post for me. I took a sleeping draught every night. . . . After three days I understood the job'.⁴⁸

11

The Air Panic of 1913

Churchill and Seely Disagree with Admiral Wilson – German Pride in Their Fleet of Airships – Sir John Jellicoe's Opinions – Asquith's Decision – Anti-Aircraft Gun Defences Recommended – An Order of Priorities – The Airship Panic of 1913 – Historians and Lord Northcliffe – Flight – The Aeroplane – Alfred Stead – The Supply Debates Provide an Opportunity to Criticize – Colonel Seely's Preparations – His Audience with the King – His Terrible Blunder – Severe Criticism of Seely – Seely is a Goose – The German Capital Levy – Funds for a German Air Fleet – British Alarm – A Meeting at the Mansion House – Northcliffe's Attitude – The Mansion House Meeting – A New Stage.

I

The incident at Sheerness caused alarm in every part of the country. Early in 1913 the Press sought to concentrate the attention of its readers upon the issue of the nation's air defences. Lord Mayors, city councillors, and other local dignitaries wrote letters to the Press in order to give expression to their concern about the possibility of an attack from the air upon British cities. As this agitation developed, the responsible authorities in the Committee of Imperial Defence and in the War Office and at the Admiralty continued to work at the technical aspects of the problem.

By the end of 1912 Winston Churchill and Colonel Seely were convinced that Britain needed a fleet of large dirigible airships in order to prepare, efficiently, for war. Their attitude had been challenged, in the Committee of Imperial Defence, by Sir Arthur Knyvett Wilson. He argued that the guns of warships at sea could drive dirigibles away easily enough, and that aeroplanes could be employed to defend dockyards, magazines, and other vulnerable points ashore against them.

Wilson's views won the support of Asquith, the Prime Minister. The result, for the British technical authorities, was an impasse or impediment in their planning. Moreover, it was realized that an anti-aircraft gun system, the second aspect of the air defence problem, would also have to be planned for, and put into place.

On 7 January 1913 the Committee of Imperial Defence met in order to take up again their discussion of the report of Colonel Seely's Technical Sub-Committee, adjourned earlier on 1 August and 6 December 1912. Churchill at once explained that Sir Arthur Wilson's opinions were being studied closely at the Admiralty. The Second Sea Lord, Sir John Jellicoe, would deal with the subject but he had been appointed only recently and had not yet been able to consider the matter, adequately.

Colonel Seely now pointed out that the question had been under consideration for a long time and that it was most desirable to resolve it at last. He urged that Britain should at once develop an efficient airship, in no way inferior to the best foreign airships. The Prime Minister refused to agree. Asquith declared he had been impressed by Wilson's criticisms of the report and he now wanted to be supplied with the considered views of the Admiralty, before any decision was made. Because of this attitude, the discussion was adjourned yet again and no further progress was achieved as a result of this meeting.[1]

Winston Churchill now acted vigorously in order to influence Sir A. K. Wilson's outlook. On 24 January he suggested that Wilson might like to see a confidential report about 'Dirigible Airships in time of War', prepared by the Naval and Military Attachés in Berlin, dated 7 December 1912. This report was sent to Admiral Wilson a few days later. It explained that the German Government had probably subsidized the Zeppelin Airship Company and other firms engaged in the construction of dirigibles. This meant that on the outbreak of war the German military and naval authorities would have at their disposal not only all the government airships, but also a number of dirigibles belonging to private firms, 'fully manned and equipped and ready for instant service'. The attachés believed the German authorities would, in such circumstances, be in a position to deploy a force of between 21 and 23 airships. Their report continued:[2]

> A number of vessels in this formidable array of airships would be capable of sailing from Germany to Sheerness, Woolwich, or any other desired point in England and return without the necessity of an intermediate descent to the earth. Though none of these dirigibles are [sic] as yet capable of maintaining average

speeds of 60 miles an hour, only very unfavourable winds could . . . prevent the passage of the North Sea being carried out.

The attachés emphasized, in their report, 'the pride and patriotic satisfaction' felt by the German nation because of their airships. The idea of an aerial fleet that could 'reduce or neutralize' the superiority of the Royal Navy was responsible, they argued, 'for the abounding enthusiasm of the German people over their dirigible airships'.

They also revealed that in a recent lecture at Kiel a German naval officer, Kapitän zur See von Pustau, had explained to an enthusiastic audience that airships could carry the violence of war to England in a way no other weapon could. This officer suggested to his listeners that bombs dropped from airships on targets in England would force the English 'to speak with us on quite different terms . . .'. The lecturer further explained that several aeroplanes were stationed along England's south coast, but 'these craft cannot manoeuvre at night and can, therefore, afford no protection against airships'. The attachés concluded that:

> These passages from a lecture . . . are merely given as indications of the hopes entertained in German hearts of the valuable assistance to be expected from their aerial fleet in the event of a conflict with Great Britain.

When Sir A. K. Wilson read this account he was not impressed. He showed no interest in the Teutonic enthusiasms and ambitions the Germans entertained for their fleet of airships. Instead, he concentrated on technicalities. The attachés in Berlin had reported that the Zeppelin airship *Viktoria Luise* had carried out a test flight on 28 November 1912. During the flight the airship flew over Coblenz, Mainz, and Frankfurt. In a docket dated 2 February 1913 which gave his opinion of their report Admiral Wilson wrote:

> The distance travelled by the Viktoria Luise in her test flight appears to have been a little under 100 sea miles or less than half the distance from the nearest point of the German frontier to Chatham. The information is of little value without knowing the weight carried and the time taken to perform the journey.[3]

This response of Wilson's to the report of 7 December meant

that his disagreement with Churchill and with Seely would continue. In these circumstances the Committee of Imperial Defence could still not decide upon its airship policy.

As a next step, Wilson and Churchill exchanged letters on the question of the validity of airships in war. On 3 February 1913 Churchill sent a long missive to the Admiral setting out the 'chief points' of the subject 'which have been made by my advisers'. In response, Wilson wrote his own 'marginal comments' on the Admiralty's arguments. The two sides could still not agree upon the value of airships in war. The impasse continued.[4]

The Committee of Imperial Defence met again on 6 February 1913. Admiral Sir John Jellicoe was now fully prepared to take part in the deliberations and he took up the attack upon A. K. Wilson's point of view as forcefully as he could. Jellicoe, at this time, was extremely interested in the capacities of aeroplanes and airships but he was convinced that the radius of action of aeroplanes in early 1913 was so small that they possessed little value for scouting purposes. For this reason he favoured the development of airships for the Royal Navy.[5] Furthermore, he insisted at this meeting of 6 February that the newest Zeppelins were, without question, a 'menace to our magazines, oil-fuel tanks, and dockyards . . . if war should break out'.[6]

He also stated, baldly, that a German airship of the Parseval type had 'flown over Sheerness and back to Germany'. He further explained that at night an aeroplane was still so inefficient that it could not cope with an attacking airship. The aeroplane would find it very difficult to rise above an airship and night landings by aeroplanes were still 'very dangerous'.

For his part, Sir Arthur Wilson continued to refuse to accept the validity of these arguments. He insisted that airships would prove to be of little value in war. Nevertheless, as a result of these exchanges, Asquith now decided, after months of controversy, that the proper course with respect to airships would be to continue to confine 'our present efforts to experimental work'. Therefore, it was at last agreed that the recommendations contained in the Report of Colonel Seely's Technical Sub-Committee should be approved. Now, the army would concentrate all its attention on aeroplanes while experiments with airships would be left to the Royal Navy. As we shall see, this course of action exposed the Liberal Government to bitter criticisms later on in the year.

II

While they argued about the relative value of aeroplanes and airships in war, the responsible authorities did not neglect the importance of anti-aircraft gun defences when they analysed the arrangements that would be needed for an efficient home air defence system. In this period several officers paid serious attention to the idea that guns stationed on the ground could be employed to destroy aircraft flying in the air. These soldiers wrote articles in the technical and popular Press, they spoke at meetings of professional societies, and they discussed the subject among themselves. A leader in this new field was Brigadier-General F. G. Stone. He published articles about anti-aircraft gun defences in *The Aeroplane* and in other technical journals and also delivered lectures at the Royal Naval War College and before the Royal United Service Institution.

Stone was not the only soldier to concern himself with this novel problem of modern warfare. Some officers concentrated their attention upon the gun defences that would be required by troops in the field. Some concerned themselves with anti-aircraft guns needed to protect fixed defences, fortresses, dockyards, and magazines. Several of their conclusions were based only on speculation. Others published the details of experiments already carried out in the field. By 1913 a corpus of articles dealing with anti-aircraft gun defences had been built up in the periodical literature. These were regularly supplemented by the publication of translated accounts, originally printed in French or German technical journals.

For example, Major H. T. Hawkins, RA, wrote an article at this time entitled 'The Attack of Airships', published in the *Journal of the Royal Artillery*. In it he argued that the aeroplane had improved so much in recent months that dirigibles would be 'at its mercy' in daylight. He concluded, therefore, that the 'only possible use' for dirigibles would be in 'night raids' designed to destroy a large target like a dock or a railway station. He went on, in the article, to discuss the kind of guns that would be most valuable when deployed against dirigibles and aeroplanes. He argued that shrapnel would be more effective than high explosive shell when firing was practised from the ground to the air. He urged that experiments should be undertaken to find out if his ideas were correct. He pointed out that aircraft in the future were certain to be armed with bombs and that 'these weapons will be used against terrestrial targets . . .'.[7]

III

In January 1913 the Army Council began to discuss the anti-aircraft gun defences that would be required for the protection of magazines, cordite factories, and other vulnerable points 'against airship attack'. On 3 January 1913 the Army Council applied, by letter, to the Home Ports Defence Committee in order to ask them to consider *'at an early date'*:[8]

(a) The classes of places and vulnerable points which should be provided with anti-aircraft defences,
(b) The places, if any, at which they recommend that these defences should be provided *in the immediate future*, and the order of importance of the places to which guns should be supplied.

The letter went on to point out that the War Office had already made temporary arrangements to guard the Chattenden and Lodge Hill magazines by stationing 6-inch howitzers there, together with the necessary personnel. It further explained that special semi-automatic guns, 'with mountings adapted for use against aircraft', were on order, and that in time they would replace the howitzers. The War Office took care to send a copy of this document to the Admiralty to soothe and reassure the officials there, and to make certain they understood that the department responsible for the home air defences was carrying out its duties in this sphere.

In February 1913 the War Office communicated again with the Home Ports Defence Committee. They sent it a schedule of government factories, stores, and magazines 'the destruction of which would seriously affect our military position'. For its part, the Home Ports Defence Committee decided to apply to the Admiralty for a list of 'naval vulnerable points'. When the Admiralty supplied such a list, the Committee would then be in a position to draw up a 'comprehensive programme which would satisfy the more pressing requirements of either department, and at the same time obviate the duplication of anti-aircraft defences'.[9]

When the Home Ports Defence Committee worked out the order in which the anti-aircraft guns were to be provided, they considered two separate factors. First, they thought about the 'intrinsic value' of each of the establishments that had to be defended. Secondly, they bore in mind the distances of these establishments from the British coast. As they saw it, 'it may be assumed that an enemy aircraft will hesitate to choose for its objective a locality necessitating a double journey over hostile territory, where it is liable to be attacked both by aeroplanes and artillery'.

After deliberation, the Home Ports Defence Committee produced an order of preference for the allotment of anti-aircraft guns to particular localities. This schedule, a remarkable document, reflected the priorities of contemporary planners as they came to grips with a military problem that was without precedent.

The Chattenden, Lodge Hill, and Upnor magazines in the Chatham district were placed first on the list. It was felt they were so close to the sea frontier they were especially vulnerable to an aerial attack, flying in from the 'German Ocean'.

Second on the list was Waltham Abbey in Essex. The royal gunpowder factory was in the immediate vicinity of the place while west of the town there were government works for the manufacture of cordite. Waltham Abbey was also the site of guncotton and percussion-cap factories. Woolwich and Purfleet, already dealt with by the Committee, followed in this roster.

The Admiralty itself appeared next on the list and was singled out for particular comment:

> The Committee are aware that it is impossible to fire projectiles into the air from the vicinity of the Admiralty buildings without risk of injury to life or property, but they are informed that the protection of these buildings against hostile aircraft is imperative.

Dockyards in the Chatham area followed in the list of priorities and then came the magazines and dockyards in the Portsmouth District. Guns would thereafter be allotted to defend the Elswick Ordnance Company Works and the firm of Vickers, Son and Maxim at Barrow-in-Furness. Barrow was on the west coast of England and its inclusion at this point seemed to violate the rule that enemy aircraft would hesitate to risk a long double journey over hostile territory. The Committee took care to explain this aberration:

> The Committee have made an apparent departure from one of their principles in the comparatively early allotment of anti-aircraft guns to Barrow; but exceptional action is justifiable in this instance, owing to the very extensive ship-building programme which is being carried out at present for the Admiralty by Messrs. Vickers, Son and Maxim.

Next in order of importance were the oil-fuel tanks at Sheerness and at Thameshaven. Last in this catalogue of places that were to be supplied with anti-aircraft guns, on a priority basis, was the Wireless Telegraph Station at Cleethorpes, on the coast of Lincolnshire. As a final recommendation, the Home Ports Defence Committee suggested that the Admiralty and the War Office should confer, where necessary, when selecting the sites for these guns.

Here was a far-ranging and comprehensive plan for the establishment of an anti-aircraft gun defence system. A few days after the plan was prepared the War Office, on 4 April, took care to write to the Admiralty to point out that the howitzers temporarily allotted to the Chattenden and Lodge Hill magazines were now mounted in their emplacements, and that ammunition had been made available for them: 'They are therefore in a position to open fire without delay'.[10]

This was an immediate arrangement. The War Office, in its letter, also revealed that the Army Council had received an advance copy of the report of the Home Ports Defence Committee, dealing with the distribution of anti-aircraft guns on a more permanent basis. The War Office assured the Admiralty that these guns would be distributed in conformity with the recommendations of the Home Ports Defence Committee 'as they become available'. Unfortunately, as we already know, an adequate supply of these guns was not secured in the period before the outbreak of war even though these plans for their deployment had been worked out many months before August 1914.

In this letter of 4 April the War Office further explained to the Admiralty that the recommendation 'that a force of aeroplanes be stationed in the neighbourhood of Chatham or Sheerness will be dealt with as part of the general question of anti-aircraft defence'.

However, no such aeroplanes were stationed near either of these places in the period between April 1913 and August 1914.

IV

The Sheerness incident was responsible for the famous 'airship panic' of 1913. While the naval and military planners laboured at the technicalities involved in the establishment of an adequate home air defence system, sections of the public and the Press gave themselves up to a kind of hysteria.

Early in the year and for weeks after airships were seen at several places in the night sky over England. On 27 February 1913 the *Daily Mail* reported lights in the sky had appeared in many localities and that the observers of the lights were convinced they belonged to airships. According to the *Daily Mail* account such lights were seen in Yorkshire, Bristol, Hull, Grimsby, Dover, Leeds, Ipswich, Portsmouth, and in other places.

Historians have sometimes blamed Lord Northcliffe for the 1913 panic. According to their analysis he was capable of any act, however shameless, that might increase the sales of his newspapers; and a panic of this order would achieve such a result for him.[11]

Even the great historian, George Dangerfield, adopted such an argument in his dazzling work, *The Strange Death of Liberal England*. According to his account stories of airships 'hovering at midnight' over the east coast of England suddenly appeared 'in the *Daily Mail* and other sensational papers'.[12] Unfortunately, the airship panic of 1913 was more complicated than Dangerfield allowed and, for once, his brilliant quips and asides fell wide of their target.

The airship stories were printed in all the British journals of the day, not merely in the sensational ones. Furthermore, although the *Daily Mail* offered new and handsome aeronautical prizes in this period, it did not take the lead in orchestrating the panic of 1913. The *Review of Reviews*, C. G. Grey's *The Aeroplane*, and *Flight*, the official publication of the Royal Aero Club, played more prominent roles in the agitation.

Indeed, in its account of 27 February the *Daily Mail* took care to point out that the stories of airship sightings over England should not be believed. According to the *Daily Mail*: 'The very multiplicity of these reports discredit them'. Furthermore, the *Daily Mail* of 27

February also gave prominence to a Reuters' account of an article published in the *Cologne Gazette* on the previous day. According to this article, as published in the *Daily Mail*, no German airship had sailed over England and if such a secret visit were undertaken 'care would be taken to show no lights'. The idea that German airships were flying over the English countryside was dismissed, by the *Cologne Gazette*, as a 'monstrous suggestion'.

The attitude of Lord Northcliffe, and of many other journalists concerned by Britain's obvious weakness in the air, was well expressed in a leading article published in the *Daily Mail* of 25 February. According to the argument of this leader it did not matter if German airships were flying over England in secret night-time forays. The point of consequence was that they were capable of such flights and the British Government, with all the forces of the Crown at its disposal, could do nothing about it. This article emphasized that a foreign airship had flown over Sheerness but, as we have already seen, this was also the opinion of the Second Sea Lord, Admiral Sir John Jellicoe, and no one has ever argued that this officer was given over to sensation-mongering. The first leading article in the *Daily Mail* of 25 February declared:

> Whether or not we accept the circumstantial reports that a strange airship was seen hovering over British territory . . . it must be taken as certain that this country has recently been visited by foreign aircraft. As to the Sheerness visit there is now no doubt in official quarters. The serious fact is that, if these vessels in peace can fly over our territory, they can also come in time of war. . . . Our present position is one of humiliation and danger.

Lord Northcliffe also employed *The Times*, which he owned, to give prominence to German denials that their aircraft were flying over England. A special article was published in the paper by *The Times'* Berlin correspondent on 27 February. It made clear that the German authorities looked upon the accounts of airships over England as a serious and disturbing issue:

> In German official quarters it is stated most emphatically that the mysterious aerial apparitions in England cannot possibly have proceeded from Germany. The German Government seems to be really anxious to dispose of all the legends on this

subject, and is inclined to treat the matter as a potential peril to Anglo-German relations. The *Cologne Gazette* declares . . . that no German military or naval airship has ever visited England. . . . This is undoubtedly an official denial which ought to be accepted as such.

The writer of the inspired article goes on to prove that a secret cruise to England and back is an impossibility. . . . Germany could have nothing whatever to gain by such an adventure, which could, at most, provide only material for the agitation against Germany.

Even the *Morning Post*, one of the savage enemies of the Liberal Government in this era of British political history, did not take partisan advantage of the airship panic. In a remarkably thoughtful leader, published on 1 March, the *Morning Post* declared there was a 'widespread feeling of uneasiness in the country' with respect to aeronautics. This fear was responsible for the 'extraordinary credence attached to the doubtful stories of nocturnal voyages in England by airships'. The *Morning Post* concluded that since foreign nations were building aeroplanes and airships it was 'of prime importance that we should do likewise'.

V

The technical publication *Flight* adopted a somewhat different argument. In its editorial for 1 February 1913 *Flight* explained that it did not take the reports of German airships flying over England 'too seriously'. Nevertheless, the paper argued, 'there is nothing in the wide world, least of all British aircraft, to prevent them so doing whenever those directing them are inclined'.

Flight's purpose was to use these reports of the mysterious airships in order to make it clear that Britain possessed no defence against them. The situation could only be remedied, *Flight* argued, if the Liberal Ministers now, at last, provided an adequate supply of funds for the development of a national air force.

Flight urged people to rise up and 'insist that the safety of the country shall take precedence of schemes of so-called social reform, which no one wants and which are frankly designed to catch the votes of the unthinking populace . . .'. This was the beginning of a campaign to force the government to spend money on the air service even if its programmes of social reform suffered in consequence. This was the object of those who kept the airship panic alive in these weeks.

By 1 March *Flight* claimed that its efforts to arouse the British public to a 'sense of the terrible danger the country is incurring, by the neglect of our aerial defences by the Government', were beginning to have an effect. *Flight* explained that as a result of its articles a 'simultaneous campaign' in almost the whole Press of the country had begun against the supineness of the Liberals in the sphere of aeronautics. *Flight* noted that the Army and Navy estimates would shortly be presented to Parliament and it urged a continuation of the campaign until substantial results were achieved: 'The thing cannot be allowed to rest until we see in the Estimates the provision of a sum adequate to gain for this country supremacy in the air as absolute as that which we enjoy at sea'.

Similar arguments were adduced by C. G. Grey in *The Aeroplane*. On 6 March he wrote that the daily Press in Britain had at last woken up and was demanding 'with quite surprising unanimity' that the sum of one million pounds should be spent on military aeronautics in the coming year. Grey attributed this agitation in the Press to the reports about 'certain strange aerial visitors' which had been hovering over England in recent months, beginning with the dirigible heard at Sheerness in the previous October. Grey believed the incident at Sheerness was real enough but he dismissed the other sightings with contempt: 'Undoubtedly some of the roving lights seen at night have been those of fire balloons sent up either by practical jokers or by people who seriously intended to wake up the obtuse English'. Indeed, Grey revealed he knew about two or three such devices, shaped like dirigibles, and released at night in order to fool anyone who saw them. The main point for Grey, however, was that those who wanted serious expenditures for military aeronautics would have to challenge Lloyd George, the Radical Chancellor. In Grey's view, Lloyd George was determined to carry out his own schemes of social reform in the new Budget for 1913–1914 even if this meant he would have to neglect the nation's 'defence forces'.

At the same time that such arguments were being offered in the technical and daily Press, a more forceful champion entered the lists. This was Alfred Stead, Editor of the *Review of Reviews*. His famous father, W. T. Stead, had perished when the *Titanic* sank in the previous year and Alfred had succeeded to the control of the magazine. W. T. Stead, a man of extraordinary vigour and spirit, had been one of the inaugurators of the 'new journalism' in England. His lively editorial campaigns had been responsible for significant changes in British life. In 1884, for example, he published a powerful series of articles on 'The Truth about the Navy' which awoke the British people to the fact that the Royal Navy was not so capable as they thought it was. Lord Morley once called Stead 'the most powerful journalist' in England. Now, Alfred Stead hoped to repeat his father's triumphs by producing a series of exciting articles about the country's weakness in the field of military aeronautics.

The first of these articles appeared in the *Review of Reviews* for February 1913. It was written in strident tones, designed to cause alarm. Alfred Stead was capable of writing the sort of 'rouser' that had made his father famous:

> The inhabitants of these islands before King Alfred created his fleet were no more defenceless against alien attack than we are today.... To command the sea is no longer to put an impenetrable barrier in the way of invasion.... Today the danger to this country is not on the sea, nor on the land; it is very decidedly in the air.
>
> The great Continental nations have realised this and acted upon it.... Today those responsible for the government of this country have not only made no effort to join the dominion of the air to British sea supremacy, but have absolutely handed the air over to foreign countries, leaving every portion of the British Isles completely defenceless from aerial attack.

Alfred Stead argued in this article that Britain was 'absolutely ... at the mercy of ... German airships'. These vessels, he explained, had been constructed for one purpose only, an air attack upon the British Isles. The airships were strategically based at Heligoland, Kiel, Cologne, and Wilhelmshaven. Each was capable of arriving above a British target 'without any warning'. Stead cautioned the Liberal Ministers, in the fiercest tones, about what would happen to them if they continued upon their present uncaring course:

> If they any longer neglect their duty towards the nation, they will be betraying the future of this country in the most cold-

blooded and unpardonable fashion. In the past the mistakes of Ministers have been retrieved and this country has muddled through; but with regard to a possible attack from the air there will be no possibility of muddling through, and the disorganised and panic-stricken survivors of the population of London will have the sole, although sorry, satisfaction, before passing under German domination, of hanging the guilty Ministers.

Alfred Stead argued the country needed a large fleet of aeroplanes as a first line of defence in the air. He urged that hundreds of thousands of British citizens should subscribe, voluntarily, so that the nation could obtain the funds to acquire these machines. When this was achieved it would be impossible for the Liberal Cabinet to 'hold its hand'. The Cabinet would be forced to take up and continue the work:

A great voluntary giving for aeroplanes is imperative. . . . Let each county, each great city or town, each collection of villages in the homeland and the Empire, give one or more aeroplanes to the State. . . . Let every mayor, every lord-lieutenant of a county open a list for subscriptions.

Alfred Stead continued this campaign in the months that followed. In the March edition of the *Review of Reviews* he published the responses of several Lord Mayors and mayors. The Lord Mayor of Liverpool declared he was prepared to talk the matter over with his friends but the Lord Mayor of York believed it would be best if taxes produced the funds required for the 'necessary aeroplanes'. S. W. Royse, the Lord Mayor of Manchester, agreed that the subject was one of 'great importance'. However, he was convinced it was receiving the full and careful consideration of the government. Many statements in the Press, he wrote, 'have created some disquietude in the public mind' but he was certain the 'military departments' were preparing to deal with the matter in short order. This was the opinion of several of the local officials who responded to Alfred Stead's patriotic appeal.

It is thus clear that by the beginning of March 1913 observers awaited, with novel interest, the appearance of the Liberal Government's army and navy estimates. These military budgets would make clear the government's response to the existing situation in the air.

VI

According to the established practice of the time, the estimates for the army, navy, and civil service became the subject of separate votes when the House of Commons resolved itself into the Committee of Supply. These estimates were worked out by the appropriate departments in the late autumn, revised by the Treasury, and by custom submitted to Parliament about the middle of February. In the case of the service estimates, it was the practice to vote in March the grants for the pay and wages of the men for the entire year.

The general debate upon the motion to go into Committee of Supply upon the estimates afforded an excellent chance to attack the government. Essentially, such debates were not discussions of financial issues. Instead, they furnished an opportunity to criticize the policy of ministers, and an opportunity to condemn the conduct of the departments under their control. In March 1913 everyone realized the Liberal Ministry lay open to easy attack in the field of military aeronautics. The supply debates were recognized as a political opening for the enemies of the government.

The chief target upon this occasion was Colonel Seely and his aeronautical policy at the War Office. Although Winston Churchill, the First Lord of the Admiralty, was despised by the Conservative hosts in the House of Commons, his devotion to the Naval Wing of the Royal Flying Corps was recognized, and appreciated.[13] For the moment, the Tories looked elsewhere than at the Admiralty when they sought to give expression to their feelings of outrage at the Liberal Government's failures in the sphere of aeronautics.

Colonel Seely was not the most capable politician in Asquith's ministry. Nevertheless, he tried, however feebly, to prepare himself for the ordeal of presenting the army estimates of March 1913. In mid-December 1912 he applied to the Prime Minister for help in the matter. In a letter of 9 December Seely explained to Asquith that his experts in the Financial Branch at the War Office had told him that decisions about the estimates 'had always been decided before this'. Seely wrote to Asquith:

I should be grateful if you could arrange for a decision on Army expenditure to be reached without further delay. There are many matters of detail of more than ordinary importance in themselves about which it is urgently necessary to deal with [*sic*].

Seely's letter continued:[14]

For example
(i) Vertical fire guns for protection against hostile aircraft
(ii) Vertical observation searchlights for the same purpose

As the weeks passed curiosity in the Press and in Parliament with respect to the army estimates became more and more intense. The airship 'scare' of the time fixed particular attention upon the provision for aeronautics decided upon by the War Office in its estimates for 1913–1914. The King himself developed a keen interest in the subject. Early in March he arranged an audience for Seely and they discussed the entire question of military aeronautics. On 14 March Seely wrote to Lord Stamfordham, private secretary to King George V, to place upon the record all that had taken place at the interview. His account reveals that he was too optimistic in his analysis of the situation:[15]

I think the following substantially represents what passed at my Audience with the King. . . .
I told His Majesty that the total strength of the Military Wing, Royal Flying Corps, and Central Flying School, was now 51 Officers and 598 Men, of whom all the Officers except three are qualified aeroplane flyers: eight men have also been trained as flyers.
With regard to aeroplanes, I said that we hoped to be in possession of 100 by the 31st March and 150 by the 31st May.
With regard to Dirigibles, I said that I had come to a working arrangement with Mr Churchill and that the demarcation between the Admiralty and the War Office should be that the former should be responsible for all aircraft lighter than air (an exception of course being made in the case of hydroplanes) while the latter should be responsible for machines heavier than air.
His Majesty impressed upon me the importance of paying

due attention to the potentialities of dirigibles which presented certain advantages not possessed by aeroplanes and particularly offered a more stable platform for guns. I undertook to discuss this matter with Mr Churchill. . . .

I then gave the King some account of my visit to Shoeburyness where I had seen experiments with anti-aircraft guns. The difficulties of manufacturing suitable weapons had, I explained, been somewhat exaggerated, and the problem was more susceptible of easy solution than we had reason to think. I said that we expected to have within four weeks a considerable number of guns of the pom-pom type capable of vertical fire and a supply of larger guns towards the end of the year. We intended eventually to have two guns in each brigade of Field Artillery which could be used for ordinary and vertical fire.

VII

Colonel Seely presented the army estimates for the year 1913–1914 to the House of Commons, sitting as the Committee of Supply on 19 March. His speech which lasted for more than two hours included a lengthy and 'optimistic' discussion of Britain's position in the air. He 'enthralled' his audience with his aeronautical account while Lord Haldane, his predecessor as Secretary of State sat above the clock, 'his face beaming with satisfaction', as he observed the scene below him with obvious pleasure.[16]

Despite Haldane's reaction Seely committed a terrible, and unnecessary, blunder in the course of these exchanges of 19 March. In his speech he said with respect to aeroplanes: 'It will be asked, "What have you got?" and I propose to tell the House. Last year when I introduced the Estimates, we had seventeen, but today we have in possession of the War Office 101 capable of flying . . .'. He was at once challenged by several Conservatives who were certain the War Office did not possess so great a number of aeroplanes.

Seely, in his enthusiasm, had fixed upon a number that was incorrect but in the days that followed he refused to alter his tally of aeroplanes. On 24 March W. Joynson-Hicks, a severe Tory critic of the government's aeronautical policy, challenged Seely on the subject of military aviation. He began his speech in the House by giving expression to one of the chief charges levelled against the

War Office at this time, that official policy 'starved' British manufacturers of aeroplanes. Joynson-Hicks said:[18]

> We must buy a certain number of foreign aeroplanes, for improvements are being made in the construction of machines by France, Germany, and America . . . and it is important that we should have the best types. But in the event of war breaking out we shall have to rely on English manufacturers, and I am anxious that the policy of the War Office should be that our manufacturers should have a continuous supply of orders, so that their work can be kept going. I am afraid that during the past year they have been very largely starved. I think the whole industry has been starved by lack of orders from the Government.

Joynson-Hicks went on to say that Seely was boasting when he claimed the War Office possessed 101 aeroplanes. Seely responded at once. He declared: 'I said so. I did not boast. Let us clear this way. I say we have got 101 aeroplanes'. He added: 'I say on my full responsibility as a Minister that we have got 101 aeroplanes which are flying. . . . We certainly have got 101 aeroplanes'.[19]

Years later, Sir Sefton Brancker offered his own amused comments upon this incident. Brancker recalled, in his memoirs, that Seely at once ordered him to buy 'eight heterogeneous machines "off the peg" ' to 'make up our strength to 101 aeroplanes'. Brancker pointed out that a mission headed by Joynson-Hicks actually toured the country counting aeroplanes at different aerodromes controlled by the War Office. 'I'll swear', he wrote of Seely's machines, 'that they were all there. Some were not very vigorous, others never flew again, but there *were* 101 of sorts!'[20]

The effect upon contemporaries was more serious. Lord Haldane, when he was responsible for the aeronautical programme of the Liberal Government, had contrived to keep the issue largely outside the realm of partisan party politics. Owing to Seely's parliamentary blunder, however, there now followed an outburst of Conservative rage that altered the situation Haldane had contrived.

By 1913 the Conservatives had been out of office for so long that their frustration was, itself, a significant factor in the fierce politics of the day. In particular, they despised the Liberals as incompetents in the sphere of national defence in a period when the menace of Imperial Germany seemed a more dangerous threat than any faced by the nation since the time of Napoleon. In 1909 the Conservatives were responsible for the greatest naval panic in recent British history when they moved a vote of censure against the government because of its inadequate naval policy. Now, they hoped to make military aeronautics another issue they could employ to assault the Liberals, in Parliament and in the country.

The new situation was explained, clearly, by C. G. Grey, in *The Aeroplane*. On 20 March he wrote:

> The Army Estimates are now public property, and it is at once evident that the Secretary of State for War has betrayed his trust. So far from putting our aerial defences on a sound footing during the past year, he has actually left them in a worse relative condition than they were a year ago.

A week later Grey wrote in *The Aeroplane* of 27 March that:

> Not the least unpleasant feature of this tragic occurrence is the fact that aerial defence has become a phase of party politics. I am not a Conservative, nor am I a Liberal, but I am compelled to admit against my will that national defence in general and aerial defence in particular seems to have become entirely the concern of His Majesty's Opposition.

Grey's contempt for Colonel Seely was terrible. In his article of 27 March he wrote:

> Those of us who are . . . in touch with Service matters hear . . . that the pressure of work in the office of the Secretary of State for War has been so great . . . that Colonel Seely . . . is breaking down under a task which is too big for him.
> By hardly any other possible explanation can one account for the astounding statements made by him when introducing the Army Estimates. . . . From start to finish his speech, so far as aviation was concerned, was such an astonishing fabric of distorted facts, that one could not imagine a responsible Minister with any knowledge of the truth giving utterance to it.

VIII

Attacks like this one continued for weeks and months after Seely's introduction of the army estimates in March. Tory stalwarts like Arthur Lee, William Joynson-Hicks, Lord Montagu of Beaulieu and others interested in the issue of aerial defence regularly condemned Seely for his ridiculous claim that the War Office possessed 101 aeroplanes.

Joynson-Hicks and a colleague, G. J. Sandys, the Conservative member for the Wells Division of Somerset, undertook a thorough investigation of the matter. They visited Salisbury Plain, Farnborough, and other places in order to count for themselves the aeroplanes available to the War Office. At the end of July 1913 they produced a report which destroyed the position Colonel Seely had taken up. In their investigation they exploited documents furnished them by the War Office itself. They came to a remarkable conclusion. They calculated that the Military Wing of the Royal Flying Corps controlled about 50 efficient aeroplanes. The editorial comment on their revelations was almost without precedent, even for this bitter era of British political history. On 9 August the technical publication *Flight* declared:

> What must be thought abroad of the cynical effrontery of a Minister of the Crown who so lightly regards the truth as to allow himself to be convicted in the face of Parliament and the country of what can only be construed into a deliberate attempt to cloak neglect and to mislead the country. . . . At one time it was the proud boast of this country that its Ministers were above the petty attempt to deceive. . . . After the shameful disclosures which are contained in the report made by Messrs. Joynson-Hicks and Sandys, it is quite impossible that we can look the world in the face and say with the Pharisee that at least we are better than others. To our way of thinking, anything more shameful, more deplorable than the cynically light-hearted manner in which Col. Seely has dealt with the whole question of the Military Wing would be hard to find in the records of British public life.

C. G. Grey in *The Aeroplane* was even more blunt. On 7 August he wrote:

After such an exposure, any politician with any sense of decency would retire into private life. . . . Discredited as a politician, dishonoured as a gentleman, Colonel Seely is now stripped of every shred of respect anyone may ever have had for him. . . To say that only a knave could have made his original statements, and that only a fool would have stuck to them, would be fair comment if one believed him to be in full possession of his mental powers. . . . For practical purposes Colonel Seely is 'down' and 'out'; and he may depart at any moment, unwept, unhonoured, and unsung.

It is no wonder that Lord Esher later wrote of Seely that 'he is a goose, and always was'.[21] Seely's management of the aeronautical programme of the Liberal Government in the House of Commons may be reckoned a minor disaster for the Asquith Ministry. After March 1913, owing to Seely's ineptitude, it became impossible to attempt to continue Haldane's strategic course of keeping the subject of military aeronautics outside the realm of partisan politics. It now became yet another bone of contention between the parties.

IX

The situation between Britain and Germany suddenly became more tense at this time. The German Government had been preparing a new Army Bill since the previous January and this was formally introduced by the Chancellor, Bethmann Hollweg, at the end of March 1913. The new bill was unprecedented since the armed forces, by its terms, were to be financed by a capital levy of 1000 million marks or £50 000 000.

Historians have paid serious attention to this remarkable departure of 1913. A. J. P. Taylor wrote of it:

The Lloyd George Budget of 1909, which had caused a constitutional crisis in Great Britain, had only raised taxes by fifteen million pounds. Germany, a poorer country, had to provide an extra fifty million pounds within eighteen months. This was an effort that could not be repeated. In the summer of 1914 German preparations for war would be at their height.[22]

Sir Robert Ensor looked upon this capital levy or *Wehrbeitrag*, as it was called, in a similar way. He felt it was a triumph of the power of the soldiers in Germany over the civilian ministers. Ensor wrote in his classic history: 'All the different arrangements for collecting and spending this utterly unprecedented sum converged towards a common date – the late summer of 1914 . . . the mastery of the soldiers in Berlin was complete'.[23]

The historians neglected to notice an aspect of this bill which greatly alarmed British contemporaries. Gigantic sums were set aside in it for the provision of a huge new German air fleet of dirigibles and aeroplanes. Those in Britain who were concerned with the subject of air defence now began to argue that the supineness of the Liberal Government had produced a situation in which the British would find it impossible to catch up with their German rivals in the air.

The alarm was sounded first by J. E. Mackenzie, the Berlin Correspondent of *The Times*. In an article published in *The Times* on 29 March he explained that

> The 'levy on property' . . . has been given the name of 'Wehrbeitrag' . . . The basis of the levy is to be a subscription of one-half per cent of the value of property, supplemented by a special levy on large incomes. . . . There is to be no graduation of the levy and it is to fall upon all property of the capital value of more than £500.

The issue of the *Wehrbeitrag* and military aviation was then taken up by Colonel Repington, the famous Military Correspondent of *The Times*. He prepared an entire series of articles that began in *The Times* on 31 March. He wrote on that day:

> We now know the best and the worst of the Government policy concerning aerial warfare, and are in a position to realize the full effects of the neglect of the King's Ministers in this branch of defence.
>
> We are aided in this unpleasant task by the news which we publish from Berlin . . . showing that the Germans propose to allocate nearly four millions more to their aerial fleet, bringing up the total sum available for this purpose to between six and seven millions sterling. . . .
>
> . . . It is the Government as a whole on whom the responsi-

bility rests . . . for their inability to understand the importance of this new branch of warfare. . . .

It was obvious to every looker-on that when M. Blériot crossed the Channel a new chapter was opened in the military history of the British Isles. . . . Nevertheless we took no action worth mentioning.

Another article in *The Times* of 31 March, written by Mackenzie, drew attention to a supplementary estimate for the German Navy, 'tucked away' at the end of the Army Bill. This supplementary estimate was concerned with naval aeronautics. It provided further large sums, amounting almost to an additional million pounds, for new naval airships, aeroplanes, and air stations. A memorandum attached to the supplementary estimate made several significant points: 'the new arm (aeronautics) provides for naval purposes a valuable extension of the sphere of tactical and strategic reconnaissance, and in certain circumstances, can usefully be employed as a weapon of attack'.

These various articles were the subject of a leader published in *The Times* on 31 March. The leader was an exposition but it also drew some sober conclusions:

About £50,000,000 is to be raised by the 'Wehrbeitrag', or the levy on capital and large incomes. . . . It is certainly a startling departure from what have been generally regarded as among the first principles of national finance. . . .

To ourselves the portions of the German scheme which relate to aviation . . . are of special moment. . . . The pregnant remarks in the German official communications on the employment of the new arm for purposes of naval attack will be particularly noted. . . . We have no wish to use the language of alarm on this or on any other point of our defence. But it is plain that, while we have hitherto done but little in preparation for warfare in the air, the 'next greatest naval Power' has done much, and intends forthwith to do a very great deal more. That is a grave fact which it would be the height of folly to ignore.

It must not be supposed that only the Conservative Press took alarm at the *Wehrbeitrag* of 1913. Liberal journals also commented on the huge German military expenditure. For example, on 31 March a long article in the *Manchester Guardian* drew attention to the 'deeply interesting' details of the new German Army Bill. A headline above the article stated 'HEAVY OUTLAY ON AIRSHIPS' and in the text attention was called to the 'very heavy expenditure on airships' which was 'something of a surprise'.

Publication of the details of the new German Army Bill produced serious tensions in England in the days that followed. The Liberal Government began to be blamed more and more severely for the country's weakness in the air. Major-General H. T. Arbuthnot, Chairman of the Aerial League of the British Empire, wrote to the Press to declare that Germany was building an 'invincible air fleet' which was a 'serious menace to the stability of our Empire'. He announced that the Aerial League now planned to launch a programme of lectures aimed at the 'leaders of public opinion in every locality'. The object of the campaign would be to force the Liberal Ministers to realize that 'no sacrifice shall be deemed too heavy to combat the peril while it is not yet too late'.[24]

Colonel Repington adopted a similar tone of alarm in his series of articles in *The Times*. The details of the new German aerial programme, he explained in an article published on 9 April, made it clear that Britain was far behind in the sphere of aeronautics. In Repington's view, someone bore the responsibility for this terrible situation. He suggested that those responsible should be hanged: 'If naval or military authorities have given bad advice to the Cabinet . . . these authorities should be gibbeted; but if they have given good advice and it has not been taken, than a vote of censure would be a legitimate course for the Opposition to adopt'.

These anxieties produced a result. It was announced in the Press on 10 April that the Lord Mayor of London, Sir David Burnett, proposed to hold a meeting at the Mansion House, early in the following month. The purpose of the meeting, designed to enlist the help of public figures in all walks of life, regardless of party affiliation, was to consider the weak condition of Britain's air defences, and to recommend ways in which it could be improved.

X

Lord Northcliffe was very concerned by Britain's position of inferiority in the air but the attitude he adopted at this time was a cautious and not a sensational one. On 5 March he had a long talk with his old friend, Orville Wright, who had been in Germany to confer about his aeroplane business there. Orville, in Northcliffe's opinion, was a 'very large minded man' who possessed great knowledge of aeroplanes and airships. He conveyed Orville's technical opinions to Geoffrey Dawson, Editor of *The Times*, and also cautioned him that the British should move slowly in the new sphere: 'There is no frightful hurry because neither the French nor the German machines are very practical yet'.[25]

Northcliffe then commissioned H. G. Wells to write a series of three articles dealing with 'the war of the future', and these were published in the *Daily Mail* on 7, 8 and 9 April 1913. The *Daily Mail* described Wells as an 'independent and daring thinker' while it emphasized that the ideas he adduced in these articles were his own and did not 'necessarily' reflect the editorial views of the paper. Northcliffe's object was to perform a public service by providing Wells with a forum.

Wells argued, in his articles, that Dreadnought battleships, the great contemporary symbols of maritime power, would not determine the outcome of naval battles in the future. Britain's enemies, he wrote, would employ 'unsportsmanlike devices against our capital ships'. These would include submarines, torpedo-boats, seaplanes, and aeroplanes. The Royal Navy, if it was to avoid defeat, would have to beat an enemy 'in the air above and in the waters beneath'. The burden of his message was that new technical devices would determine the outcome of all conflicts in the future.

Wells also argued, in his articles, that in recent years British imagination in naval and military matters had undergone a 'complete arrest'. Britain, he said, suffered from a terrible gap in her defences. As he put it, 'She is short of minds'. He argued that 'we need a new arm to our service; we need it urgently, and we shall need it more and more, and that arm is Research'.

Northcliffe was so impressed by these ideas that he sent the articles to several important politicians. On 8 April he wrote to the Conservative leader, Andrew Bonar Law, and urged him to read what Wells had written.[26] On 11 April he wrote to Colonel Seely in the most civil terms and enclosed the articles with his letter. Northcliffe's letter to Seely was sympathetic in tone while it expressed regret that the subject of aeronautics was now becoming an issue between the political parties. In it, he made suggestions designed to improve Britain's aeronautical position. It was not the missive of one who sought to exploit a panic:[27]

The Air Panic of 1913

> I do hope that you will be able to induce the Treasury to do more than they have done. There is no inducement to the inventive brain to leave profitable forms of work for the nebulous rewards of aeroplane planning. The thing now looks like becoming a Party matter. I think it is a great pity that we do not use our friendship with France for finding out what they have done and know, for in this matter they know more than the rest of the world put together. . . .
>
> . . . I venture to send three rather remarkable articles by a man whose books show that he has a very unusual gift of foresight.

Meanwhile, further details of the Mansion House meeting were made public. It was revealed in *The Times* of 10 April that when the Lord Mayor decided to hold such a meeting he did so in response to a request made by the Aerial Defence Committee of the Navy League. The object of the meeting, which was to be 'entirely non-party in character', was to impress upon the British people and their government the 'vital necessity' of an adequate aerial defence. In the *Daily Mail* of 12 April it was announced that Arthur Balfour, Field-Marshal Lord Roberts, Lord Rosebery, and other notables would address the Mansion House gathering. At the meeting resolutions would be put urging the preparation and introduction of supplementary estimates that would provide for 'at least a two to one standard of superiority over the airships and aeroplanes of the next strongest naval power'.

Lord Northcliffe did not approve of the enthusiasm for the policy of the Navy League shown in the *Daily Mail* of 12 April. He held strong ideas of his own on this particular subject. On that day he addressed a stiff note to Thomas Marlowe, Editor of the *Daily Mail*. It was the kind of missive that always gained the instant attention of his employees:[28]

> I do not advocate 'Two to One in the Air'.[29] It is a preposterous scheme. The headings this morning make it appear as though it is the policy of the 'Daily Mail'. It has never been mine, and I wrote a strong letter to the Navy League this week, objecting to their policy. What we want is five to one in aeroplanes and waterplanes.

XI

The Mansion House meeting, scheduled for 5 May 1913, and the entire subject of aerial defence, now began to receive serious attention in the Press and in Parliament. A remarkable agitation developed in the weeks before the fifth of May.

In its first leading article for 13 April the *Observer* called the meeting a 'national conference on aerial defence'. The paper looked upon the planned gathering as the beginning of an 'overwhelming movement' that would force the Liberal Government into a more active course with respect to aeronautics. The *Observer* urged that the help of 'a few more patriotic Liberals' should be secured so that the movement would be a national, and not a partisan, issue. The *Observer* warned that '. . . the command of the air around these islands is as absolutely vital to us as ever was the command of the sea'.

Several days later Lord Montagu of Beaulieu took up the same themes in a speech in the House of Lords. He began by insisting that he was not speaking as a political partisan. His object, he said, was to call public attention to the importance of aeronautics. In particular, as a technical authority, he desired to make clear that before long airships and aeroplanes would be used not merely for intelligence work and reconnoitring, but also for purposes of offence. Large dirigibles were now capable of lifting weights of up to ten tons and if these were dropped at inconvenient places 'it would produce a state of dismay and a scene of ruin which could hardly be paralleled . . .'. Modern civilization, said Lord Montagu, was now confronted by a new and extremely grave problem. Air bombing could produce so serious a state of panic it might result in a revolution, 'a social uprising . . . which would be difficult to quell . . .'. He stressed that there was no town in England, even those far distant from the coast, which was not now vulnerable to serious damage 'from explosives dropped from above'. He insisted he did not want to exaggerate the nature of the problems involved but he urged they should be studied by the War Office and by the Admiralty. In his opinion, there was only one way to deal with the situation. Britain must be provided with an ample number of airships and aeroplanes for the purpose of aerial defence. 'We must realise', he concluded, 'that we are only at the beginning of huge developments in this direction, and His Majesty's Government will be well advised if they give the whole subject . . . their most careful and serious consideration'.[30]

Those in the vanguard of this agitation insisted they were not acting as political partisans, concerned only to score debating points at the expense of the Liberal Government. They explained that their purpose, a patriotic one, was to make clear the implications and the consequences of a new and terrible national problem. However, all who took an active part were critical of the Liberal Ministry's failure to proceed more vigorously in the matter of aeronautics and air defence.

For example, Colonel Repington wrote a special article, full of praise for Lord Montagu's speech in *The Times* of 1 May. Repington argued that Britain's aerial forces were so far behind those of her continental neighbours that parity could be obtained only if the government made 'great efforts' to catch up. However, the Liberal Press reacted completely differently to Lord Montagu's remarks. The political parties and their adherents in Britain were so bitterly divided in this period that aeronautics now became a political issue, a subject of dispute between the politicians on each side and between their supporters in the Press. In a leading article published on 30 April the *Manchester Guardian* defended the government's record in aeronautics while it condemned those critics who demanded that more money should be spent even though, like Montagu, they were unable to 'indicate in what direction further provision should be made'. Referring to airships and aeroplanes, the *Manchester Guardian* declared: '. . . there is no apparent reason to believe that the War Office and the Admiralty are less interested in them, or less alive to their importance, than any of those who are now clamouring for more to be done in some undefined way'.

Some sections of the Liberal Press now began to condemn the enthusiasts who urged a more active aeronautical policy upon the government as 'scaremongers'. This outraged the Editor of the *Observer*, J. L. Garvin. On 4 May the *Observer* made the deteriorating situation between the two factions worse when it published a leading article about the forthcoming Mansion House meeting that was as violent and as pugnacious as it could be:

> It was once an axiom that any business which was endorsed by a Mansion House meeting was as good as done. . . . Men of both the great parties will be on the platform . . . Those who

are afraid of being ridiculed as 'scaremongers' will stay away. The parrot-cry in this case is only worthy of the brains of birds. . . . Zeppelins . . . have a radius of action of over fourteen hundred miles. Does it need to be a 'scaremonger' to see that our mobilization arrangements can be read like an open book, our signal stations, depots, railways, and bridges destroyed unless we have the power to meet and beat these intruders in their own element? If Englishmen are striving to awaken their fellow countrymen to the danger, is it necessary or fitting to attribute to them the ignoble motive of personal gain or the poor, base lust for party triumph?

This slime of vituperative faction smirches our national life on every side.

On 5 May the *Daily Telegraph* published a more moderate appeal in its first leading article, which was devoted to the Mansion House meeting, scheduled to take place later on that day. The *Daily Telegraph* explained that the meeting would mark the opening of a new era in 'warlike preparations'. It emphasized the occasion would be 'non-political in its character. There will be no appeal to party passions'. Men of all parties would come together in the Mansion House to deal with the emergency that had arisen in the air. 'The Government and the Treasury', the *Daily Telegraph* argued, 'must be convinced that the country will tolerate no delay . . .'.

XII

Unfortunately, the Mansion House meeting was not the triumph its supporters had hoped for when they began their campaign. The great Egyptian Hall of the Mansion House was not filled to capacity. The announcement that Arthur Balfour, Lord Roberts, and Lord Rosebery would put in an appearance proved to be incorrect since they were not present at the meeting. The speakers at the Mansion House were eminent and distinguished but they were not men of the very first rank in British political life. Moreover, the reaction to the meeting in the Press followed partisan lines. The Conservative journals praised the speakers and organizers while the Liberal and Radical newspapers condemned the meeting, in harsh terms. The attempt to organize a non-partisan or non-party national movement in support of a more positive aeronautical policy was not a success.

The Lord Mayor began the proceedings by stating that this was 'one of the most important gatherings that had ever been held in the Mansion House'.[31] Other speakers including the Duke of Argyll, Admiral Sir Edward Seymour, and Lord Desborough moved several resolutions. Eventually, it was decided to establish a National Aeronautical Defence Association to arouse and educate public opinion on questions affecting the aerial defence of the country. One speaker, Admiral Sir John Hopkins, said that 'owing to the extraordinary progress made in aviation, England was no longer an island'.[32]

The reaction in the Press on the following day was intense. It reflected the fierce political animosities of the day. The Radical *Daily News and Leader* stated that the 'really important persons who were advertised to attend yesterday's Mansion House meeting in support of the latest newspaper panic had the good sense to stay away, and the list of those who attended is . . . unimpressive'. The paper dismissed the entire campaign as a 'temporary, if . . . expensive, sensation'.[33] The *Manchester Guardian* of 6 May offered a clear comment: 'we are disposed to deprecate the mere clamouring for expenditure unrelated to any definite and well-thought-out plan of action'. According to the *Manchester Guardian* the country required 'clear thinking' and not 'free spending' when it worked out its aeronautical programme. By contrast, the Conservative *Morning Post* praised the organizers of the Mansion House meeting and welcomed the formation of the National Aeronautical Defence Association. The *Morning Post* savagely attacked Colonel Seely, the Liberal Government, and the democratic system itself in a leading article of 6 May:

> The other day the Secretary of State for War gave some misleading figures as to our aeroplane fleet. There is no such fleet of working and up-to-date aeroplanes in existence as would appear from that statement, and it is a disastrous thing to lull the country into a false security by exaggerating and falsifying its real resources. . . . We hear a great deal about Committees which are considering this matter, but not enough of things actually done and resources actually prepared. It is the great danger of Democracy that it is apt to content itself with shams instead of realities, and with things said rather than with things

done . . . if we lose our Empire and are defeated by our rivals it is because we are growing accustomed to substitute words for preparations.

It will be recalled that when the House of Commons held its first major debate on aeronautics, the historic debate of 2 August 1909, *The Times* had declared that the subject would have an 'assured place' in all future Parliamentary proceedings.[34] At that time and for a period thereafter contemporaries tried to deal with the new technology in a non-partisan way. By the spring of 1913, however, a new stage had been reached. Aeronautical policy had become a subject of serious disagreement between the two great political parties.

12

Preparing for War with Germany

The Problem of Invasion – Two Schools of Thought – The Inquiry of 1913 – An Admiralty Memorandum – Lord Roberts and his friends – The greatest change in the art of war since the discovery of gunpowder – Aircraft and Invasion – The High Level Bridge – Winston Churchill's Interest in it – The Time Table of a Nightmare – Decisions of the High Level Bridge with respect to Air Defence – Winston Churchill's Ideas about Aeroplanes – Activity of the Naval Wing – Systematic Trials – Churchill mentions Hornets – His Boast in the House of Commons – Commander Samson and Captain Scarlett work out an Air Defence Arrangement – Formation of The Royal Naval Air Service – A Knock-out Blow Against London – Air Attack on London Considered – Sir Percy Scott's Opinions – Captain Hankey wants an Inquiry into the Requirements of the Air Defence of Great Britain – Churchill on the Alert.

I

The beginning of the year 1913 witnessed a development in Britain that reflected the extreme tensions of the pre-war period. This development fused together a fairly ancient and a more recent theme of British life. In January 1913 Asquith, the Prime Minister, ordered the appointment of a Standing Sub-Committee of the Committee of Imperial Defence to consider the problem of the invasion of the British Isles.

This was the third formal examination of the invasion question within a decade. The membership of this Standing Sub-Committee of 1913 was so formidable that one expert has called it, with reason, 'one of the greatest gatherings of British military and political talent in the twentieth century'.[1]

The British people, and their leaders also, looked upon the insular condition of their country as a blessing of divine providence. The English Channel or 'silver streak', as it was variously called by contemporaries, was a frontier that had spared Britain the problems which plagued the various states of the European continent for centuries past. It was widely believed that this fact of geography was responsible, in good measure, for those differences in history, culture, tradition, and political and constitutional development that set the British apart from their less fortunate European neighbours.

From ancient times the British always feared an invasion of their island home from the European continent. This fear of invasion was a major theme of British life throughout the ages. In modern times the problem of an adequate defence against invasion threw up two schools of strategic thought.

On the one hand were those who placed their faith in the Royal Navy as the supreme barrier against invasion. On the other was a military school whose adherents argued that the Royal Navy was not invincible; that the defending fleets might be rendered ineffective by storms, by defeat in battle, or by some ruse or trick that would lure them away from the decisive point at a time chosen by some calculating and ruthless enemy. This second school urged that fortifications should be built at key places along the coasts and that a home army should be organized behind them. Its task would be the destruction of an invading force.

Since the funds available for national defence were strictly limited a never-ending quarrel developed. The competition for these funds pitted the military leaders of the country against its great sea captains in a recurring battle.

In the nineteenth century a 'Bolt from the Blue' school of strategic thought arose. Its members suggested that the Royal Navy might not be able to deal with a 'lightning invasion' and that a second line of fortifications, together with a home defence army, was needed in order to render the country safe. Later, a 'Blue Water' school challenged these views. The 'Blue Water' theorists argued that so long as the Royal Navy was maintained in adequate numbers it would dominate the seas and make a successful invasion impossible. The dispute between these factions flared intermittently, thereafter.

It was Arthur Balfour who sought to solve this problem. In December 1902, shortly after he became Prime Minister, he tentatively reconstituted the old Defence Committee of the Cabinet and began to turn it into the now more familiar Committee of Imperial Defence. One of his objects was to employ the new institution as a kind of supreme command that might eventually eliminate the inter-service rivalry of earlier years. Lord Esher, who helped Balfour in this work, hoped the Committee of Imperial Defence could be used for exactly this purpose.

The first subject submitted to the Committee of Imperial Defence for its consideration was the invasion question. In February 1903 a special sub-committee began to investigate the matter. The result was a triumph for the 'Blue Water' school. Balfour took care to reveal its technical conclusions to the British public. He explained, on more than one occasion, that so long as the Royal Navy was maintained adequately, the British people had no reason to fear the possibility of an invasion. Moreover, when his government was replaced by Sir Henry Campbell-Bannerman's Liberal Ministry in December 1905, the 'Blue Water' doctrines of the previous administration were accepted at once by the new Secretary of State for War, R. B. Haldane.

In the years that followed a new factor began to bear upon this situation. The fear of Germany became so strong in Britain that Field-Marshal Lord Roberts and his friends, advocates of military conscription, decided it was their patriotic duty to force the Liberal Government to reopen the invasion question, and to examine it anew.

In November 1907 Campbell-Bannerman agreed to their demands and in that month the Committee of Imperial Defence, with H. H. Asquith in the chair, began its second invasion enquiry 'on the assumption that Germany might be the enemy'. The result of their deliberations, which lasted for months, was another victory for the 'Blue Water' school. Asquith and his colleagues came to the conclusion that so long as Britain's naval supremacy was assured an invasion of the United Kingdom was an impracticable operation of war.[2]

We saw earlier that at the time of the Agadir crisis in 1911 it was discovered that the War Office and the Admiralty had worked out their plans for war with Germany in isolation. The war strategies adopted by each of the services at that time were mutually exclusive. Historians, we noticed, tended to blame H. H. Asquith for this almost incredible situation.[3]

The 1913 invasion enquiry revealed a similarly deplorable confusion of aims. In 1913 the War Office hoped to send the six divisions of the regular army to France, immediately upon the outbreak of war. This was the key to the continental strategy of the soldiers. However, by this time the Admiralty had become so nervous about the prospects of an invasion of the British Isles that they wanted to retain several of the regular divisions at home to cope with an invading force from Germany. The admirals had modified and changed the 'Blue Water' theories of an earlier time.

In 1911, Asquith had been unable to force his generals and admirals to agree upon a strategic policy acceptable to each of their great departments. In 1913 the military and naval leaders continued to disagree about the proper course to be followed at the beginning of a war with Imperial Germany.

Indeed, one authority has, by implication, criticized Asquith for even setting up the invasion enquiry of 1913. This authority referred to the problem as the 'moss-grown question of the invasion of England' and he wrote also that the 'problem of invasion remained a hardy perennial' in the years between 1905 and 1914.[4]

Nevertheless, we must be clear that Asquith acted in strict conformity with established policy in 1913. In 1903 Arthur Balfour had explained to his Committee of Imperial Defence that the subject of invasion was of such overmastering importance that no decision about it could ever be considered final. He argued at that time that 'In a matter which concerns national existence the arguments on either side must be made the subject of constant revision and reconsideration . . .'.[5]

When Asquith's Standing Sub-Committee of 1913 prepared its final Report they referred to these explanations of 1903 and added by way of further clarification:

> The present inquiry was undertaken with precisely the same objects in view as those mentioned by Mr Balfour, and in pursuance of a definite policy of submitting the question of invasion to thorough and exhaustive re-examination every few years.[6]

By 1913 several new developments had taken place which, in Asquith's view, made it imperative to reopen the matter. When he announced, on 7 January 1913, that an invasion enquiry would be taken up yet again, he called attention to recent progress in naval architecture; to the growing naval strength of other powers in the North Sea and the Mediterranean; to significant developments in the field of wireless telegraphy; and to the new technical abilities of aircraft that might have an effect upon the invasion issue.[7]

It will be realized at once that aircraft played little or no part in the earlier deliberations of the Committee of Imperial Defence on the question of the 'Attack on the British Isles from Oversea'. By 1913, however, it was recognized that airships and aeroplanes had now become factors of consequence in any analysis of the subject.

II

In March 1913 the Admiralty prepared a memorandum for the information of the Standing Sub-Committee. This paper dealt with the subject of aircraft in connection with the 'problem of Oversea Attack from Germany'. Most of the conclusions offered by the Admiralty on this occasion were cautious and circumspect. They reflected the aeronautical opinions of Sir John Jellicoe who was the Second Sea Lord, at this time.[8]

In its memorandum the Admiralty pointed out that Germany possessed several airships capable of long flights. On the other hand, the United Kingdom did not own any airships that could make an 'extended voyage'. However, Britain had made a 'better start than Germany with respect to hydro-aeroplanes'. Airships, it was explained were suitable for 'independent distant work' while aeroplanes were only capable of operations conducted at no great distance from their bases.

Modern German airships had a radius of 1500 miles which meant, in the opinion of the Admiralty, that they could 'visit Scapa Flow, scout down the entire East Coast of Britain . . . and return to Germany'. The Admiralty estimated that the Germans possessed eight airships with such capabilities.

The Admiralty memorandum revealed that aeroplanes might be worked from cruisers or from specially constructed ships in order to carry out scouting patrols. However, the range of aeroplanes was so limited in 1913 that the 'parent' ships would have to steam to an extremely advanced position where they would at once lie open to attack and destruction. Such aeroplanes would also find it difficult to return to a 'parent' ship, even in clear weather. For these reasons, the Admiralty paper argued that airships were more suitable for scouting than were aeroplanes.

Nevertheless, the Admiralty paper also explained that an aeroplane patrol of Britain's East coast was being established and that this patrol, flown from the new naval air stations, 'should do much to exclude the possibility of surprise landings in that locality'. The conclusion was offered that: 'The command of the air would add greatly to the safety of the United Kingdom in regard to oversea attack'.

However, the firmness of this statement was at once mitigated by other comments. In calm weather, it was explained, the hydro-aeroplane patrol of the East coast could observe the sea for a distance of about 120 miles from the coast: but emphasis was also placed upon the facts that adverse winds or fog could nullify the efficiency of such patrols. The memorandum offered as a conclusion: '. . . aeroplanes may fail to give warning of approaching danger when the wind force is strong and off shore. In fog both scouting and disembarkation are impossible'.

A far more spirited document dealing with aircraft and invasion was submitted to the Standing Sub-Committee by Field-Marshal Lord Roberts and his friends, at exactly this time.

Lord Roberts and his allies were convinced that only a programme of national military service or conscription could prepare Britain for the great continental war with Germany that was bound to come in the next few years.

They were also certain that Germany possessed the resources needed to carry out a successful invasion of the United Kingdom.

Since 1909 Roberts had feared that the Germans might strike at Britain from the air, with crippling effect, as a prelude to a full-scale invasion.[9]

Now, in 1913, Roberts, Colonel Repington, Lord Lovat, and Sir Samuel Scott carried their agitation to the Committee of Imperial Defence itself. They produced a memorandum which argued that a German invasion of Britain was a feasible operation of war and they argued that aircraft could play a significant part in contributing to the success of such an invasion.

In a section of their memorandum entitled 'Aircraft' they pointed out that Germany possessed about 28 dirigibles and that Britain had 'nothing to set against them'. These dirigibles were a powerful new factor in the situation. In bold terms, the memorandum declared:[10]

> We consider that our situation in regard to aeronautics is extremely serious, believing as we do that aerial navigation is the greatest change in the art of war since the discovery of gunpowder.

Without doubting for a moment the very great progress which may take place during the next few years in the adaptation of heavier-than-air machines to the purposes of war . . . we consider that for the moment, and for the special purposes of the Committee's reference, the dirigibles represent Germany's offensive arm. . . .

We see no reason to doubt that some German dirigibles . . . will be able to reach England continually in war, and to return to Germany after an attempt . . . to injure our ships, docks, arsenals, and magazines. The Rhine stations are only 250 miles from Chatham . . . There is no longer any reason to doubt that aircraft will prove to be extremely formidable engines of war. . . .

They will prove invaluable in facilitating an act of invasion. . . . We can come to no other conclusion . . . but that the German aircraft may do us great hurt. . . .

. . . we can come to no other conclusion . . . but that Germany's superiority in the air will prove a grave disadvantage to us in war, facilitate surprise and invasion, and endanger our naval supremacy.

The Standing Sub-Committee continued its investigations for months after it received this paper. As a direct result of its work a Conference was held at the Admiralty on 19 November 1913. Winston Churchill, Colonel Seely, and their highest technical advisers from the Admiralty and the War Office took part in these deliberations. They considered proposals dealing with several matters connected with the subject of invasion. These included the manning of Coast Defence batteries at ports on the East coast of Britain; the arrangements for watching the coast; and 'the distribution of responsibility for aerial defence'.[11]

III

The Standing Sub-Committee finally produced its Report in April 1914. Generally, its conclusions were similar to the earlier Committee of Imperial Defence reports dealing with the subject. In 1914 the Sub-Committee decided that at least two divisions of regular troops should be kept in the country in the earliest stages of a war, in case there was an invasion. This was insisted upon even though the War Office continued to demand the instant despatch to France of all six divisions of the Expeditionary Force. With respect to aeronautics, three conclusions were offered. None was especially positive.[12]

Indeed, the first conclusion in the Report that dealt with aircraft was extremely tentative. For some time the Admiralty had planned to use big ships, on station in the North Sea, to look out for the German fleet in case it ever embarked upon an invasion of the British Isles. However, the recent development of the mine and the torpedo made it impossible to maintain a close watch on the exits from the Heligoland Bight with heavy ships. According to the Report if this were attempted it would result in a 'serious wastage' of heavy units which, if prolonged, might 'alter the balance of naval power'. If smaller torpedo craft were employed for the purpose they could be overwhelmed at any time by the sudden attack of a larger enemy force. The Report suggested that the development of submarines of 'ocean-going capacity' might modify this situation in Britain's favour. It also argued that aircraft might be used for this work but it did so in the most hesitant terms: 'the use of aircraft may also help in this direction, but without extended exercises to ascertain their capabilities this cannot be counted on with any certainty'.[13]

The Report further argued that an enemy fleet might employ aircraft as scouts to give warning of the presence of British ships at sea. Since the British, owing to the recent improvements in mines and torpedoes, would have to withdraw their blockading line 'the evasion of our fleets by the enemy' would become 'less difficult than formerly'. 'It is possible', the Report stated, 'that the evasion of our ships might be facilitated by the employment of aircraft . . .'.[14]

The third conclusion dealing with aircraft was somewhat less negative. The Report explained that since 1908 a naval system of coastal patrols had been developed along Britain's East coast. These patrols would be among the first units to attack an enemy invasion force. In order to obtain the best results from the patrols they would have to be 'worked in conjunction with a complete system of coastal intelligence and communications'. The Sub-Committee recommended that the Admiralty and War Office should at once concert arrangements for the provision of such a system along the East coast. In this connection, the Sub-Committee also pointed out: 'The completion of the Admiralty's sea-plane stations should add materially to the efficiency of the coastal intelligence system'.[15] This was the final reference to aircraft in the Standing Sub-Committee's Report of 1914 dealing with the issue of an 'Attack On The British Isles From Oversea'.

IV

While the Standing Sub-Committee was carrying out this examination of the invasion problem, Winston Churchill and Colonel Seely embarked upon an initiative of their own. Everyone concerned with the issue of national defence at the highest level was upset by the continued failure of the Admiralty and the War Office to prepare their plans for war in concert. Years later Lord Hankey complained about a 'system of Supreme Command which allowed the two Services to work out their plans independently without providing for any central control'.[16]

Churchill and Seely decided to address this issue, in their own way. The two men were warm personal friends. They resolved to try to reduce or do away with the friction that had characterized the relationship between their two great departments for years past. The device they hit upon to achieve this object was a joint committee of the Admiralty and the War Office, known as the 'High Level Bridge'.

A series of these 'High Level Bridge' conferences was held either at the War Office or at the Admiralty between August 1913 and May 1914. The 'High Level Bridge' could not take up every issue of mutual or joint concern to the War Office and the Admiralty. However, they did address questions connected with home defence, colonial defence, and also the problem of the air defence of Great Britain. Indeed, as we shall see, Churchill used these 'High Level Bridge' meetings to achieve a particular goal of his own, an increase in the Admiralty's responsibility in the area of home air defence.

The highest service leaders took part in these 'High Level Bridge' conferences. The Secretary of State for War and the Chief of the Imperial General Staff attended regularly. The Admiralty was represented by the First Lord and the First Sea Lord. At their discretion, certain technical authorities or experts were invited to join in the deliberations. Asquith insisted that Captain Hankey should act as the secretary of the 'High Level Bridge'. In this way he would be kept informed of its activities and there would be no duplication of work already allotted to the Committee of Imperial Defence.[17]

In later years Seely claimed that he had given the name 'High Level Bridge' to this joint committee.[18] Nevertheless, the initiative in its employment was often taken by Winston Churchill. His general attitude with respect to the 'High Level Bridge' was revealed in a letter he wrote to Seely in December 1913. At that time Churchill explained to Seely that there were so many different questions

> connected with the watching of the East Coast, so much intermingling of naval and military functions . . . that nothing less than the highest authority in the two Departments can frame the outlines of a good and thorough scheme.[19]

Churchill went on to declare that it would be useless to delegate work of such importance: 'I propose to you', he wrote to Seely,

> that it shall form the subject of regular and special enquiry on the high level bridge at which you and I, Sir John French, and Prince Louis, with Hankey as recorder shall personally deal with it. Any officers that we require can be in attendance. . . . The meetings could be held alternately in each Department. . . . What do you say?[20]

In Churchill's opinion the role to be assigned to the new naval seaplane stations 'and the Air Service generally' was one of the important questions to be decided upon at the 'High Level Bridge'.[21]

In this period Churchill was convinced that aircraft would play a significant part in any invasion of the British Isles. One of his great objects at this time was to do what he could to increase the role of the Naval Wing of the Royal Flying Corps in the air defence of the country.

His lively concern with these matters was demonstrated in April 1913 when he prepared a memorandum he called 'The Time Table of a Nightmare'. This document, designed by the First Lord to jolt and shock his admirals, was an account of a hypothetical landing by German troops on the East coast of England. According to Churchill's scenario, the Germans invaded at exactly the time the

British Expeditionary Force was being rushed to France, upon the outbreak of war. They forced themselves ashore at the port of Harwich and proceeded to march upon London by way of Colchester and Chelmsford. After they took Chelmsford the Germans occupied 'the line Harlow–Ongar–Billericay' where they paused to prepare for a final thrust against the metropolis.[22]

Rioting now broke out in London and a terrified populace there sought to prevent the British regular soldiers from leaving England for their assigned places in France. The leader of the opposition in the House of Commons demanded to know from the government what steps were being taken to protect London. He wanted a portion of the British Expeditionary Force diverted from France so that it could be employed for the purpose.

Meanwhile, by the terms of Churchill's account, the British mobilization at Chatham was interrupted when five German airships appeared over the dockyards and basins there and dropped from fifteen to twenty tons of explosives on their targets below. Several ships were sunk and some of the largest workshops in the place were wrecked. Panic, involving townspeople and dockyard labourers alike, followed; and there was a general exodus from the town. One of the airships was brought down by a 'lucky shot from an improvized high-angle howitzer'; but the 'whole of Chatham and its resources . . . were seriously crippled for a week'.

One of Churchill's admirals, Sir Henry Jackson, dismissed this paper as 'sensational', 'alarmist', and 'satirical'. The First Lord, however, would not budge. He countered by writing of his memorandum: 'It was written with the intention of raising certain very serious issues . . . which will be apparent the more the facts . . . are studied'.

Some of these 'serious issues' were among the subjects dealt with at the 'High Level Bridge' when it began its deliberations a few months later, in August 1913.

V

The 'High Level Bridge' turned to the subject of aeronautics at its second meeting, held at the War Office on 15 October 1913. David Henderson, the Director-General of Military Aeronautics, was invited to attend the conference on this occasion.[23]

At the meeting the committee took up the question of the transfer of airships from the control of the War Office to that of the Admiralty. It will be recalled that this transfer had been recommended by the Committee of Imperial Defence. The 'High Level Bridge' now decided that the details of the exchange should be worked out by a Joint Committee presided over by the Second Sea Lord, and composed of the Director-General of Military Aeronautics representing the War Office, and the Director of the Air Department, representing the Admiralty.

Later, the question of 'The Responsibility for Aerial Coastal Defence' was also examined. We have discussed this subject earlier in our history but it should be looked at again in this particular context. Winston Churchill was present at this meeting of 15 October. He was determined to enlarge the part played by the Naval Wing in the air defence of the country. In the pre-war period he achieved at least a portion of this goal as a result of decisions taken by the 'High Level Bridge'.

Churchill was not alone in his attitude about the employment of naval aircraft. The enthusiasts in the Naval Wing of the Royal Flying Corps always sought to increase the scope of their duties and responsibilities. In August 1912, for example, Captain Murray F. Sueter submitted a paper to his superiors at the Admiralty. In it, he listed the tasks he felt naval aircraft should be asked to carry out. The eighth item in his list read as follows: 'Preventing attacks on Dockyards, Magazines, Oil Storage Tanks, etc. by hostile Aircraft'.[24]

This submission was rejected at once by Sueter's superiors. One of them wrote:

> The estimate of the . . . naval duties of air-craft is concurred in, except item 8. The prevention of attacks on establishments on shore has always been a responsibility of the Army hitherto, and should remain so whether the attack is by air-craft or sea-craft.

This opinion, as we already know, was entirely in accord with established policy and tradition.

However, at the meeting of the 'High Level Bridge' on 15 October 1913 a subtle change in this position was revealed. While the members of the conference agreed that the War Office was responsible for all 'aerial services ancillary to fortresses and fixed defences' they accepted another conclusion also:[25]

> In view of the fact that the Admiralty has already established seaplane stations at or near several points of vital importance which are vulnerable to aerial attack, *e.g.*, the Isle of Grain, and Eastchurch, near the Chattenden Magazine; the Firth of Forth, near Crombie Magazine; Cromarty, near Invergordon, etc. it was agreed that the Admiralty should now undertake the responsibility for the provision of such aircraft as may be necessary for reconnaissance in connection with the local defence of naval property, except in the vicinity of the permanent stations of units of the Military Wing of the Royal Flying Corps.

The 'High Level Bridge' met next at the Admiralty, on 19 November 1913. Several matters were taken up and then the discussion of 15 October, dealing with the responsibility for aerial defence, was continued. This time the 'High Level Bridge' came to a more substantial conclusion:[26]

> The Admiralty are establishing seaplane stations along the coast. . . . In order to avoid duplication and overlapping, it was agreed that in cases where the naval seaplane stations are close to points of naval importance vulnerable to aerial attack, the naval wing should undertake the responsibility for the aerial defence of naval property.

As we have seen in an earlier chapter, the military authorities later challenged this conclusion and again demanded that sole responsibility for the aerial defence of the country should be assigned to them.[27] Nevertheless, this decision of the 'High Level Bridge', arrived at on 19 November 1913, proved to be of substantial help to Winston Churchill in organizing those home air defence arrangements the country was forced to rely upon, at the outbreak of war in August 1914.

VI

Churchill and Colonel Seely were convinced that war with Germany was bound to come, sooner or later. Each, in his own department, worked as hard as possible to make certain the country would be ready when the terrible day arrived. They were constantly aware of their awful burdens and responsibilities. They concurred in the view that aircraft would play an important role in the warfare of the future.

Much ingenuity was demonstrated in each of the services in carrying out experiments with aircraft; and in devising ways to attack them, from the ground or in the air. It is not unfair to say, however, that more imagination was exercised in the Naval Wing than in the Military Wing of the Royal Flying Corps in the pre-war period.

When David Henderson became Director-General of Military Aeronautics his new Directorate was placed under the personal supervision of the Secretary of State for War. As a result, Henderson enjoyed the right of 'direct access' to Colonel Seely; and this enhanced his ability to get things done in the War Office. He encouraged all kinds of experiments in the Military Wing but he continued to emphasize the importance of reconnaissance as the principal function or task of the Wing's squadrons.

Frederick Sykes, the Military Wing's commander, agreed with his chief's opinions about the future employment of aircraft in war. Sykes further argued that some aeroplanes would have to be equipped to fight and win battles in the air so that reconnaissance missions could be carried out, unhampered by opposing enemy aircraft.

Asquith had appointed Winston Churchill First Lord of the Admiralty in order to make certain that the Royal Navy would be ready for war with Germany, whenever it came. This meant that Churchill's chief concern, after 1911, was with the ships, men, and naval bases of the fleets. Nevertheless, in this pre-war period he also devoted an extraordinary amount of his time and energy to the development of aeronautics in the navy.

At the end of October 1913 Churchill prepared a minute which revealed something about the policy he proposed to follow with respect to the employment of naval aircraft. His ideas, as revealed by this minute, were more imaginative than anything entertained by his contemporaries in the War Office.[28]

Preparing for War with Germany 275

In his minute the First Lord recommended three new types of aeroplanes for future employment with the Royal Navy. These were an 'oversea fighting seaplane' to operate from a ship as a base; a 'scouting seaplane' to be employed with the fleet at sea; and a 'home-service fighting aeroplane' whose task would be to repel enemy aircraft whenever they attacked vulnerable points or other targets in Britain. This minute dealt also with the 'menace' of Zeppelins. These should be attacked in the air, Churchill argued, by dropping a string of bombs or fireballs on them from above.

This minute reflected the results of a great number of trials and experiments carried out by the Naval Wing, in this general period. These trials were described in a document prepared for the Admiralty War Staff, at the end of December 1913. Entitled 'Naval and Military Aeronautics – 1913', the Report was compiled in the Intelligence Division and the Air Department of the Admiralty.[29]

The report revealed that the Royal Navy, at the end of 1913, possessed 30 seaplanes, 32 biplanes, and four monoplanes. A further 54 machines were on order, at this date. There were 108 trained pilots in the 'Naval Air Service'.[30]

The report dealt, also, with the arming of aeroplanes. It stated that Maxim guns had been mounted in and fired from aeroplanes at Eastchurch, and elsewhere. The firing of these guns, it was explained, had no effect upon the stability of the machines in flight. A new light Maxim gun that weighed only 28 pounds was being considered for use with naval aircraft. A conclusion offered in this connection was that the 'mounting of Maxim guns in aeroplanes does not . . . offer any serious difficulties on the score of weight'.

An entire section of the report was devoted to 'Offensive Operations Against Airships'. Significantly, this section began by stating: 'Some method of destroying these craft is one of the most vitally needed things in the Naval Air Service'. The report revealed there was a 'very prevalent idea' that the hydrogen in airships was easy to ignite. This was incorrect. Great difficulties had been experienced in attempting to set this gas on fire and 'numberless' experiments had been carried out, without much success. The report suggested that dropping a succession of bombs or grenades on airships in flight might prove to be a feasible way of destroying them.

The report then turned to the subject of dropping bombs – on targets at sea, and on the land. An aircraft, the report said, would probably not choose to attack a ship at sea if it was armed with anti-aircraft guns. In an attack on a land target bombs dropped at random might have a 'demoralizing effect' but they would probably not do much 'vital damage'.

The report also dealt with the employment of anti-aircraft guns. It referred to the army's experiments with a 4-inch gun and mounting which had been carried out at Shoeburyness. A similar gun had been procured for naval use. Twelve were now available for arming ships of the 'Iron Duke' class. 'Very good shooting', the report declared, 'is possible with this gun'. The navy had also taken over from the army twelve 1-pdr. pom-pom guns and conversion of their mountings to high-angle fire was being carried out. It was proposed to mount guns of this type in some of the most modern destroyers, two in each ship. The firm of Vickers had designed a high-angle mounting for 3-pdr. guns and these, too, might be supplied to ships for use against aircraft. The report came to the tentative conclusion that the most suitable projectile for use against aeroplanes would probably be a 'high explosive one' but difficulties in designing a fuse for such a device had already been encountered. In the case of airships, it was proposed to employ a high explosive shell, fitted with a special fuse. A fuse of the type required had been designed and was undergoing trials.

Finally, the report revealed that since 1912 the Admiralty had been interested in experiments involving wireless telegraphy sets for use in aircraft. It had been decided to fit all 'war seaplanes' with transmitting sets, and all naval air stations with complete transmitting and receiving installations.

Another step in these already extensive arrangements was taken at the end of December 1913. The post of Inspecting Captain of Aircraft was created, and in time Captain F. R. Scarlett, RN, later an Air Vice-Marshal, was appointed to it. Churchill wanted an officer of 'extensive flying experience and knowledge of aeronautics' for this position.[31] The Inspecting Captain of Aircraft was in charge of all naval air stations, under instructions from the Admiralty Air Department. He was also required to make certain that all naval aircraft, ashore and at sea, were in a state of readiness for use. In order to deal with the increased volume of aeronautical work in the navy Churchill further arranged that an office known as the Central Air Office should be established in Sheerness dockyard. This now became the headquarters of the Naval Wing of the Royal Flying Corps.

VII

In the new year Churchill saw to it that the Naval Wing paid careful attention to the problems of home air defence. In January 1914 systematic trials, later called one of the 'most important events connected with the Naval Wing at this time', were carried out with a Vickers machine-gun firing from an aeroplane. It was the duty of Lieutenant R. H. Clark-Hall, RN, the Wing's Armament Officer, to examine the results of these trials and to offer his superiors conclusions about them. After study, Clark-Hall decided that 'machine-gun aeroplanes' would be needed to drive off enemy aircraft 'approaching our ports with the intention of obtaining information or attacking with bombs our magazines, oil tanks, or dockyards'. As a result of the trials Clark-Hall recommended that a 'gun-carrying machine' should be developed which could fly at a very great speed and possess 'good climbing power to enable it to get up to the attack quickly'.[32]

Churchill, at this time, was preparing the naval estimates for the year 1914–1915. On 17 March he made his formal statement about them in the House of Commons. He devoted a substantial portion of his remarks to naval and military aeronautics, and also to the issue of the air defence of Great Britain. He began upon the subject by explaining his special and particular interest in 'the Air Service'.[33]

> I now come to air. I cannot help feeling a special interest in the Air Service, because it has come into full activity during my tenure. When I came to the Admiralty two and a half years ago there were four aeroplanes and five pilots. Now we have 103 aeroplanes.

The First Lord proceeded to explain to his audience the several functions which the seaplanes and aeroplanes of the Naval Wing were expected to carry out. Firstly, they would perform as scouts at sea, and in two ways. Their scouting missions might begin from an air station on land or from a 'seaplane ship', a special vessel equipped to carry seaplanes out with the fleet, thus enabling them to reconnoitre from a 'moving base'.

In the second place the aeroplanes of the Naval Wing would serve the 'extremely useful function' of watching the coasts. Patrol flotillas of British ships were already grouped together at strategic points along the East coast so that they could attack an enemy invasion fleet at the shortest possible notice. The naval aeroplanes could render very valuable assistance to these flotillas by bringing them timely information about the enemy's ships. 'Of course', Churchill continued, 'the heavy seaplanes which we are developing now will carry formidable explosives, which could be dropped on to transports and disturb the landing even before the patrol flotillas could arrive'. He offered further information about these heavy seaplanes. In ordinary weather they could fly by night as well as by day. They carried wireless telegraphy sets that enabled them to send effective signals up to a distance of 120 miles, and 'quite recently' some of them had 'even been able to receive a message while in the air'.

A third function involved the home air defences. Churchill's exposition continued, when he dealt with this subject, in some detail. It was his own considered analysis, the result of study and reflection:

> In the third place, the seaplanes and aeroplanes will be of great value for the defence of vulnerable points. Oil tanks, magazines, docks, workshops, power houses, all these nerve centres of naval power have in the last few years been exposed to the indefinite menace of aerial attacks. Passive defence against such an attack is perfectly hopeless and endless. You would have to roof in the world to be quite sure. Something may be done, and something has been done, by the provision of guns which fire upwards, and by searchlights which train throughout the entire arc, but the only real security upon which sound military principles will rely is that you should be master of your own air. Nor is this unduly difficult. The war aeroplane flying in its own country, unhampered by floats, and close to its base, must be a far more efficient fighting instrument than any similar craft that could come across the seas . . . the home aeroplane will have decided advantages over the intruder.

Now Churchill proceeded to involve himself in a boast about

Britain's home air defences that later caused him a great deal of trouble. When the country was being bombed from the air in the First World War this boast of 17 March 1914 was remembered by his political enemies and was used by them in savage attacks upon him and upon his pre-war administration of the Admiralty. Referring to the aeroplanes of the Naval Wing of the Royal Flying Corps, Churchill declared:

> With the Military Wing, with which great progress has been made, we are already in a position of effective strength, and any hostile aircraft, airships or aeroplanes, which reached our coast during the coming year would be promptly attacked in superior force by a swarm of very formidable hornets. This is the true military protection of vulnerable points.

In April 1914, in conformity with these ideas, Commander C. R. Samson and his pilots of the Naval Wing carried out a series of air operations designed to test the defences of the Nore. Several lessons were learnt from these exercises and they were taken note of at each of the naval air stations and also at the headquarters of the Naval Wing in Sheerness. Commander Samson decided, as a result of these tests, that 'defending aeroplanes should not be lured from their posts by the attacker'; that an observer should be carried in each aeroplane engaged in a defensive patrol; and that 'efficient observers' were needed on the ground to communicate with the defending machines in the air. Samson concluded that at least five aeroplanes were needed at each 'defended point'. He decided his pilots should 'endeavour to attack with the sun or against a dark background' and that the defending aircraft should not be painted white.[34]

In the same month Captain Scarlett, writing from his headquarters in Sheerness, sent a report to the Commander-in-Chief at the Nore. His report dealt with the question 'of the means to be taken for the protection of Dockyards, Sheds at Grain, Kingsnorth Airship Station, Magazines at Chattenden & Lodge Hill, Oil Tanks & Floating Docks, Ships in harbour . . . against an attack from the air'.[35]

Scarlett calculated that such an attack might be made by airships or by aeroplanes. He revealed he had discussed the matter with Commander Samson 'on many occasions'. He proposed that 'War Flights' composed of 'fast armed machines that can climb rapidly' should be established at Eastchurch naval air station. According to his plan, patrolling seaplanes and observers stationed at different points along the coast would be employed to flash word to these War Flights upon the approach of enemy aircraft. The duty of the 'land machines' at Eastchurch would be to attack and destroy the enemy's aircraft 'before they have attained their objective'.

Scarlett stressed he had received valuable advice and assistance from Samson in working out his scheme. He urged that seven machines 'ready for instant action' should be stationed at Eastchurch while others would be employed to fly the patrols. These patrols would commence at daybreak and would be maintained in relays until dark. All the seaplanes attached to these patrols, Scarlett reported, 'are fitted with W.T.'

Winston Churchill soon interested himself in these arrangements, and added several ideas of his own. In May 1914 Churchill prepared a memorandum for his colleagues in the Admiralty. As a result of recent 'experiences', he now wanted a squadron of his 'hornets' put into place, as soon as it could be done. He wrote in his minute that it was important that a 'war squadron' of ten fighting aeroplanes should be created at Eastchurch, 'as quickly as possible'. This squadron would provide for the aerial defence of Chatham dockyard, the Chattenden magazines, and the oil-fuel tanks of the area. The First Lord explained that the proposed squadron should consist of two flights of four machines each, with one in reserve. 'The design of these aeroplanes', he wrote, 'should be reconsidered in the light of the latest experience'. He wanted all the aeroplanes of the squadron to be identical in pattern, each produced by the same maker of aircraft. Each flight should be provided with two spare engines. The aeroplanes of the squadron were to be kept 'quite separate' from the school machines used for training and practice at Eastchurch and 'eight of the ten should always be ready to fly'.[36]

VIII

Colonel Seely's colleagues did not look upon him as a particularly effective minister. In March 1914 he so blundered in his duties as Secretary of State for War that he seriously embarrassed the government; and the Prime Minister was forced to repudiate his actions in the House of Commons. His resignation from office followed at once.[37] Winston Churchill took advantage of this development and soon embarked upon a new departure for the Naval Wing, a change he had contemplated for some time past.[38]

In the original scheme approved by the Cabinet provision had been made for a single Royal Flying Corps, divided into two wings, one naval and one military. A Central Flying School, on Salisbury Plain, was also set up where the art of flying was to be taught in the first instance, to be followed by specialist training in separate military and naval facilities. Under these arrangements the Aircraft Factory at Farnborough was to be used as the principal establishment for the design, manufacture and repair, of all service machines.

Almost from the first, however, several naval officers argued it would be best to divorce the Naval Wing from its military counterpart, and to form a separate air service. Originally, Winston Churchill was opposed to this idea. In a minute dated 12 March 1913, addressed to the First and Second Sea Lords, he declared:[39]

> I am not prepared at present to accept . . . the idea of a separation from the War Office so far as the Royal Flying Corps is concerned. The decision to work in common was not taken without much thought, and to upset it would not only imply a change of policy but would break down the closer co-operation between the Army and Navy which has been so important a feature in recent Admiralty administration. The attitude which the War Office is showing in many matters connected with the air service is one of most friendly co-operation, and my own relations with the Secretary of State are of the closest and pleasantest character.

As time passed Churchill's experiences forced him to change his mind. There was friction between the two departments because of a dispute about the airship sheds at Farnborough. Although he was a close friend of Colonel Seely they sometimes competed with each other for the funds available for the air services. Many naval officers believed their pilots could be trained more cheaply and more efficiently at their own school at Eastchurch, instead of being sent to the Central Flying School at Upavon, on Salisbury Plain. Eventually, Churchill found he agreed with this opinion. He was angered by the treatment shown to the Admiralty by the Royal Aircraft Factory. Machines ordered for the Naval Wing were delivered late or sometimes were not delivered at all. Above all else the rivalry between the two great departments caused him annoyance, and even distress.

Churchill brought all these points out in a letter he sent to the Prime Minister in July 1914. The letter complained in the strongest terms about the lack of cooperation between the War Office and the Admiralty. This failure to work in harmony extended far beyond the sphere of aeronautics.[40]

> There ought to be a full and free interchange of technical information between the Naval and Military Wings. I have given directions that every facility in our power is to be given to the War Office, and that no secret is to be withheld, it being absurd that Departments of the Government should treat each other like foreign Powers. It is obviously wasteful and wrong that the Navy and the Army should develop types and designs of aeroplanes in strict isolation and without pooling their knowledge and communicating regularly with each other.
>
> There are a number of other small points apart from the Air Service which indicate a revival of that departmental friction and jealousy which until recent times led to such a complete isolation of the War Office and Admiralty interests . . . in my time we tried resolutely to overcome this. Matters were agreed upon between Ministers and others at the head, and the Departments were made to feel that narrow views and departmental rivalries would be severely repressed. I earnestly hope that if such conduct or such an attitude can be detected in any part of the Admiralty administration, you will have me informed so that the persons concerned may be brought to their senses without delay.

In these circumstances Churchill and his admirals decided to break up the original organization of the Royal Flying Corps so that they could chart an aeronautical course of their own. In June 1914 the Admiralty issued a series of regulations which were to take effect in the following month. By the terms of these regulations the Naval Wing became a distinct branch of the Royal Navy and was designated 'The Royal Naval Air Service'.

The regulations explained that the Royal Naval Air Service would be administered by the Admiralty. It would consist of the Admiralty Air Department; the Central Air Office; the Royal Naval

Flying School; the Royal Naval Air stations; and all the seaplanes, aeroplanes, airships, seaplane ships, balloons, kites, and any other type of aircraft that might from time to time be employed for naval purposes. All ranks and ratings of the Royal Naval Air Service would be 'borne on the books of one of His Majesty's ships and will serve under the provisions of the Naval Discipline Act . . .'. It has been pointed out that the Royal Flying Corps was formed, originally, as the result of a Cabinet decision. However, this partition of the Corps in the summer of 1914 was the result of actions taken by the Board of Admiralty, alone; and it was never questioned or criticized.[41]

Winston Churchill, as we have seen, deplored the inability of the War Office and the Admiralty to work together in a reasonable and co-operative way. The new arrangement was no solution to the difficulty. After it was put into effect the friction between the two departments, in the sphere of aeronautics, increased; and during the First World War it was responsible for very serious problems, especially in the area of home air defence.[42]

IX

During the course of the year 1914 people in Britain became more and more concerned about the possibility of an air attack upon their country in case of war with Germany. The subject was discussed, regularly.

One expert in this sphere was Colonel Louis Jackson, a retired officer of Royal Engineers. In April 1914 he delivered a lecture at the Royal United Service Institution and this was published in the Institution's Journal in the following June, under the title 'The Defence of Localities Against Aerial Attack'. This lecture at once attracted a nation-wide interest. An exchange of letters dealing with the lecture was published in *The Times*, and there was comment about it in other journals, also.[43]

Colonel Jackson was prompted to deliver his lecture, he explained, because the public, for years past, had been disturbed by a vague menace, the menace of an attack from the air. His purpose was to define this danger more precisely, and to suggest the steps necessary to guard against it.

By the year 1917, he surmised, France and Germany would possess airships capable of dropping a ton of explosives. These vessels, he predicted, would be armed with light, quick-firing guns and would operate at considerable heights in order to avoid defensive fire from the ground. Since they possessed the ability to stop in the air over their targets they would be able to drop their projectiles accurately. At night, they would descend to much lower elevations and thus bomb their targets even more precisely. The aeroplane of this period, he calculated, would be armed either with a machine-gun or with bomb-dropping apparatus.

If these assumptions were correct, he argued, a great many vulnerable points in Britain would soon be exposed to attack from the air. These vulnerable points were the 'localities' to be defended and they included coast batteries, dockyards, magazines, ammunition factories, oil reservoirs, wireless stations, and great centres of population.

Guns would probably be employed in the defence but their capabilities for this purpose were still not known. Some authorities felt they would be ineffective against aircraft but Colonel Jackson believed that a properly designed 'vertical fire gun' would be of use. 'As for aeroplanes', he said, when a suitable projectile was designed 'shooting at them will be just like game shooting'. The guns, he argued, would have to be situated close to the 'localities' they were protecting: 'They should be placed centrally, along the highest ground available'.

Colonel Jackson was particularly concerned by the vulnerability of London to air attack. In this context he recognized that London was unique and different from other cities:

> London is for us the prime object of consideration. Destruction and panic in the largest of provincial towns would cause trouble, but need not affect our national policy. London in this respect stands alone, that it is not only the habitat of a large fraction of our whole population, but the seat of Government, the centre of our financial and business systems, and the nerve centre of our military and naval forces. A serious blow aimed at London would be more effective against the national life than in any other capital of the world.

Jackson declared that the British were now face to face with a 'new era' in warfare. He then referred to certain statements made by a French officer, published a few months before in the *Daily Mail*. This officer had written: 'Even admitting that a Zeppelin were to pass over the English countryside, it is not easy to see what result would be effected, for even in time of war it would not be permissible to drop explosives into unfortified towns'. Jackson's comment was: 'I have no wish to be an alarmist or to make anyone's flesh creep, but I am not prepared to accept this dictum . . .'.

Colonel Jackson was convinced that in wars of the future unfortified towns might be bombed from the air. London, in particular, would become an important target for such attacks. His comments, in this connection, were remarkable. He believed London might be exposed to a 'knock-out blow' from the air:[44]

> If a Zeppelin dropped half a ton of guncotton on to the Admiralty or the War Office . . . what would be the result in disorganization and discouragement? What would be the effect of cutting off the water supply of the East End, or sinking food-ships in the Thames? These things seem incredible to us, who have only known of wars on the frontiers. I confess I am reluctant to go the length of my own arguments, but . . . such action will soon be possible; and this is the age of the 'knock-out blow' in everything. Would any ruler harden his heart to such action? Who can say? . . . If it seemed probable that such panic and riot would be caused as to force the Home Government to accept an unfavourable peace, then perhaps it might be done.

Colonel Jackson next took up the question of the defence of London against such an attack. He believed those responsible for the defence were confronted with serious problems. He offered the opinion that the best way to drive off aerial attackers would be to employ large numbers of 'armed aeroplanes' against them. The task of these machines would be to 'hunt their quarry out of existence'.

During his lecture Jackson asked: 'Can any student of international law tell us definitely that such a thing as aerial attack on London is outside the rules: and further, that there exists an authority by which the rules can be enforced?'

Two days later he received an answer from one of the most eminent jurists in the country. Thomas Erskine Holland, a Fellow of All Souls College, Oxford, and formerly Chichele Professor of International Law and Diplomacy in the University, wrote to *The Times* in order to reply.[45]

Holland began his reply by calling Jackson's lecture an 'interesting and important address'. He pointed out that Article 25 of the Hague Convention of 1907 stated: 'It is forbidden to attack or bombard *by any means whatever* towns, villages, habitations, or buildings which are not defended'. The words 'which I have italicised' were added deliberately, Holland explained, 'in order to render illegal any attack from the air upon undefended localities; among which I conceive that London would unquestionably be included'.

Colonel Jackson, however, was unable to accept the validity of this argument. He replied with a letter of his own and sent it to *The Times* on 28 April. In it he stated that aircraft, a 'new factor in warfare', were now able to attack London. The Hague Convention, however it was worded, did not provide 'an adequate safeguard' against such an assault.

Professor Holland found that he was not prepared to challenge this conclusion. Therefore, he wrote to *The Times* yet again in order to stress that London should not rely for its protection upon Article 25 of the Hague Convention. He praised the principle involved in the article and hoped it would win full acceptance from the international community, but he also declared: '. . . in the meantime, let us not for a moment relax our preparation of vertical firing guns and defensive aeroplanes'.

In this way Holland expressed his agreement with Colonel Jackson's opinion that London now faced the possibility of a novel threat, the possibility of an attack from the air designed to cause so much disturbance in the capital that it would have a serious effect upon the ability of any British Government to wage war.

X

Several people in Britain, impressed by recent advances in technology, began to argue that drastic and radical changes would have to be made if the safety of the country was to be assured. Since 1911, for example, Admiral Lord Fisher had stressed the high importance of Zeppelin airships and submarines in any valid defensive arrangements.[46] As we have seen, H. G. Wells argued at some length in the *Daily Mail* that battleships were now out of date since submarines, seaplanes, and aeroplanes would dominate the naval battles of the future. In the early summer of 1914 a minor sensation was caused by the publication in *The Times* of a letter written by Admiral Sir Percy Scott. In his letter Scott declared, in the boldest terms:[47]

> Submarines and aeroplanes have entirely revolutionized naval warfare, no fleet can hide itself from the aeroplane eye, and the submarine can deliver a deadly attack even in broad daylight.
> Under these circumstances I can see no use for battleships and very little chance of much employment for fast cruisers. The Navy will be entirely changed. . . . In war time the scouting aeroplane will always be high above on the look-out, and the submarines in constant readiness, as are the engines at a fire station.

Sir Percy Scott's technical abilities had won him the respect of the British public. He was the Royal Navy's foremost gunnery expert. He had been responsible for nothing less than a 'revolution' in naval gunnery in the years since 1899–1900. By employing new methods and new technical devices, some of which he himself invented, he greatly increased the accuracy and rapidity of fire of the ships in which he served. His work attracted public notice and in 1903 he was appointed to the command of HMS *Excellent*, the gunnery school at Whale Island. He turned this institution into a model barracks and training establishment. He was one of Admiral Fisher's warm supporters and in 1905 he became Inspector of Target Practice. Scott was convinced of the validity of his own ideas. He never hesitated to urge his opinions upon anyone who would listen to him.

Although he retired from the navy in 1909, the admiral maintained a lively interest in naval developments. He also believed the British public would benefit if it was kept informed of his views. Therefore, on 31 May 1914 he wrote a brief note to the Editor of *The Times* in which he claimed that 'many people' still applied to him for his technical opinions. These had been set out at some length, he explained, in a letter he had written on 15 December 1913, and he enclosed this with his note. Each of these documents was published in *The Times* of 5 June 1914. They caused a controversy as soon as they appeared.

The Times introduced the two letters by stating that they 'must attract general attention both at home and abroad'. Scott, the paper explained, was a naval officer who had shown the ability to 'think ahead of his contemporaries'. In its first leading article *The Times* argued that Scott's opinions deserved to be read with respect but it came to the conclusion that the 'British Empire is not to be risked for a new theory'. *The Times* offered its own opinion that battleships were not yet 'doomed'.

On 6 June the *Daily Mail*, in its first leading article, declared that Scott's letter raised a 'question of great national importance'. While the admiral's ideas were entitled to 'great respect' they were very largely 'speculative' and it would be irresponsible to base future naval policy on them.

In the *Observer* of 7 June a sterner analysis was offered. Sir Percy Scott's distinguished services to the navy were well known, the paper declared, 'but he does not enhance them by the tone and method of his letter in "The Times" '. The *Observer* explained there were other naval experts concerned with these very problems and they believed the 'offensive possibilities' of aeroplanes and submarines had been over-estimated.[48]

For its part, the Liberal Press looked with more favour on Scott's opinions. Organs like the *Daily News* and the *Daily Chronicle* argued that if his analysis was correct the great expenditures required to build and maintain a fleet of battleships would no longer be needed; and the 'bloated armaments' of their day could be reduced, as a result.

This Scott affair of June 1914 was a reflection of the nervousness and uncertainty of the pre-war period. The British fear of Imperial Germany which so marked out this general era was sometimes enhanced and made worse when contemporaries contemplated the technological changes and advances that were being made in the armed forces of every country. During the Scott controversy, although they did not mention Germany by name, the various protagonists had in mind German submarines and German airplanes and German battleships but there was no need to remind British readers of the source of their apprehensions.

XI

In June 1914 some of the responsible authorities in Whitehall were very concerned with technological developments. Captain Hankey, in particular, was worried by recent foreign advances in aeronautics and he hoped he could persuade H. H. Asquith to take some action that might improve the situation for their country. On 16 June he prepared a minute, addressed to the Prime Minister. In it, he explained that Colonel Seely, before his resignation, had 'pointed out that our aeroplanes were not particularly well equipped for dealing with enemy airships'.[49]

There had been some discussion between Seely and the Prime Minister about an enquiry into the matter. However, nothing was done. Hankey now proposed that an enquiry into the entire question of the Air Defence of Great Britain should be undertaken. He prepared the draft of a paper that would set his plan in motion, and submitted it to Asquith for his consideration. His draft read as follows:[50]

SECRET   TERMS OF REFERENCE

The Prime Minister desires that the Air Committee should consider the liability of the United Kingdom to damage by foreign aircraft, having regard to the present position of aviation abroad, and make recommendations as to any further measures of aerial defence or aerial defensive organization against attack from the air which may be required in this country.

Asquith rejected the idea at once. On the next day he wrote in his own hand on Hankey's original minute: 'The attention both of the naval & military experts is now being given to this matter, and I do not think there is any necessity at the moment for a special inquiry. H.H.A. 17/June. 14'.[51]

Meanwhile, the War Office ordered that all the squadrons of the Military Wing should be concentrated at the airfield at Netheravon for a period of 'combined training' during the months of June and July. Lieutenant-Colonel Sykes, the Wing's Commander, was in charge. Later, he wrote: 'I always look on those two previous months as a godsend. . . . Rapid mobilization would have been almost impossible but for this concentration'.[52]

There was also a 'concentration' of the aircraft of the Royal Naval Air Service when the King held a review of the fleet at Spithead, from 18–22 July, 1914. The pilots of the Royal Naval Air Service took the opportunity of the review to practice cooperation with one another. When the review ended all the aeroplanes, seaplanes, and airships returned to their bases except for Wing Commander Samson's flight of aeroplanes.

Samson and his pilots embarked upon a tour of aerodromes

in different parts of the country. However, their journey was interrupted by an urgent order directing them to return at once to their base at Eastchurch. They arrived there on 27 July. On the same day the seaplanes were assembled at the naval air stations at the Isle of Grain, Felixstowe, and Great Yarmouth so that they could patrol the coast in the event of war.

Winston Churchill was responsible for these arrangements. As the diplomatic situation in Europe worsened, he became more and more concerned by the threat of an air attack upon the British Isles. He was now determined to employ the aeroplanes and seaplanes of the Royal Naval Air Service as a home defence air force. These were the first steps he took to put his resolve into effect.

13

Winston Churchill Takes Charge

The European Crisis – Policy of the Military Aeronautics Directorate – Winston Churchill's Attitude – He orders a New Policy for Naval Aircraft – The Royal Flying Corps is Mobilized – Anti-aircraft Gun Defences Alerted – No. 4 Squadron Diverted to Eastchurch – London Will be Bombed from the Air – The Admiralty Wireless Installation – An Anti-aircraft Gun for the Admiralty Building – Fear of Civil Disturbance in Case of War – Churchill and Lord Kitchener – Coastal Air Patrols – The Passage of the Army to France – Churchill's Paper on 'Aerial Defence' Looked at Again – Lack of a British Aircraft Industry – Sir Charles Ottley's Secret Report – Situation in the Royal Flying Corps in August 1914 – Its Commanders Hate Each Other – Lady Henderson's Account – Royal Flying Corps in Action – Dover and the Thames Defences Become Vulnerable – Winston Churchill sends the Eastchurch Squadron to Dunkirk – Its Task is to Attack Zeppelins – Asquith and Venetia Stanley – A Major Change – Churchill's Second Thoughts.

I

The gravity of the European crisis in the summer of 1914 was not recognized quickly by Britain's political leaders. Their attention was fixed upon the terrible problem of Ireland which had poisoned their lives for two years past, and still threatened the very foundations of the British State. The dispute over the issue of Home Rule for Ireland was so fierce that it dominated all political activity in the country.

By 24 July, however, even as they struggled to work out a compromise solution to the Irish question, some British leaders at last realized that the European powers might soon involve themselves in a great continental war. When this happened the responsible authorities in the War Office and at the Admiralty sprang into action. They could not know if their country would be involved in the war that might come but it was their duty to be ready to fight in case the British Government decided to intervene.

In these circumstances the Directorate of Military Aeronautics in the War Office at once embarked upon a war policy of its own. The course it chose to follow at the end of July was completely different from that pursued by the Admiralty and the Admiralty Air Department at this time.

The staff of the Military Aeronautics Directorate worked night and day in order to bring the four aeroplane squadrons of the Military Wing destined for service with the Expeditionary Force up to 'war strength'. It was planned to send each of these squadrons to the continent as quickly as possible so that they could begin to carry out reconnaissance missions for the Expeditionary Force in case it was ordered abroad.

During this pre-mobilization period the question of appointments, in case of war, was settled. Brigadier-General Sir David Henderson was selected to command the Royal Flying Corps in the field, with Lieutenant-Colonel F. H. Sykes as his Chief of Staff. The command of the Military Wing at Farnborough was now assigned to Major Hugh Trenchard, previously the Assistant-Commandant at the Central Flying School. Major Sefton Brancker was left in charge of the Directorate of Military Aeronautics but he was so junior an officer he was made an Assistant Director only, with the temporary rank of Lieutenant-Colonel. These two latter officers were made responsible for the home command of the Flying Corps while their superiors prepared to leave for the scene of action in Belgium and France.

Winston Churchill at the Admiralty was dominated by other considerations when he thought about the wartime employment of the aeroplanes at his disposal. He was worried about the home air defences.

In a series of conferences that took place in June and July 1914 he expressed these anxieties to David Henderson, General William Robertson, and other representatives of the War Office. At these meetings Churchill made it clear that the Admiralty was making its own preparations for the air defence of certain vulnerable points along the coast.[1]

It was on 24 July that Churchill realized a European war might break out. On that day he attended a Cabinet meeting where the Irish issue was discussed, at some length but with no result. At the end of the Cabinet Sir Edward Grey read out to his colleagues the terms of the Austrian ultimatum to Serbia. Churchill was instantly alarmed. He returned to the Admiralty from Downing Street and told his colleagues there that a serious danger of war existed. He now embarked upon a programme of preparations designed to make certain that the Royal Navy would be ready in case Britain became involved in the war.

Of course, his greatest concern was with the condition of his various fleets, flotillas, and squadrons but he did not neglect the subject of aircraft. On 28 July he asked the First Lord to obtain a report from the Director of the Air Division about the 'exact positions of the aircraft which were concentrated yesterday in the neighbourhood of the Thames estuary'. He also wanted to be certain that there was a 'complete understanding' between these aeroplanes and the military authorities in charge of the aerial gun defences. 'This', he sagely explained, 'is of the utmost importance if accidents are to be avoided'.[2]

On the next day Churchill came to a bold decision of his own about these aeroplanes of the Royal Naval Air Service. He decided that their primary task was to fight enemy aircraft and thus provide the country with a defence against aerial attack.

This was a major change in Admiralty policy. Only a few years earlier a First Sea Lord, Sir Arthur Knyvett Wilson, had offered the opinion that the needs of the Royal Navy could be met with two aeroplanes. Even the enthusiasts of the Naval Air Service had found it necessary to exercise caution and prudence in connection with the subject. They often stressed the importance of reconnaissance functions when they contemplated the 'uses of naval aircraft'. It will be recalled that in 1912 even that great champion, Captain Murray F. Sueter, had placed the duty of 'preventing attacks . . . by hostile aircraft' last in his list of tasks for which naval aircraft should have responsibility. Now Churchill altered these priorities, completely.

On 29 July 1914 he prepared a minute in his own hand. It was addressed to the First Sea Lord, the Fourth Sea Lord, and the Director of the Air Department. It ordered the new policy to be put into effect, at once:[3]

1 S.L.
4 S.L.
D.A.D.

In the present stage of aeronautics the primary duty of British aircraft is to fight enemy aircraft & thus afford protection against aerial attack. This should be made clear to Air officers, C. in C. Nore and A. O. P.[4] in order that the machines may not be needlessly used up in ordinary scouting duties. After the primary requirement is well provided for, whatever aid is possible for coastal watch & extended defence scouting shd. be organised. But the naval aircraft are to regard the defense against attack from the air as their first and main responsibility. They must be carefully husbanded.

W.S.C. 29.7.

We must be clear, in this connection, that the offensive equipment of the aeroplanes of the Royal Naval Air Service still reflected the experimental nature of the force. The armament of these aircraft was not impressive. Only two aeroplanes were armed with machine-guns. The main weapon carried in the great majority of the machines was the 0.45-inch rifle which was supplied with incendiary bullets, and also with ball ammunition. In addition, it was planned to issue the pilots with the Marten-Hale rifle grenades. These devices contained a few ounces of explosives. Their extreme range was about 350 yards. Often, after the war began the pilots flew their patrols armed only with a rifle or a shot-gun.[5]

Churchill's order of 29 July was passed on to the responsible naval authorities at once. On that day a cypher telegram was sent from the Admiralty to the Commander-in-Chief Nore, to the Admiral of Patrols, and to the Inspecting Captain of Aircraft. In effect, the telegram created a rudimentary fighter defence air force:[6]

From – Admiralty   29.7.14
To   C-in-C Nore   A.O.P.   I.C.A.

Inform I.C.A. for the present the duties of Aircraft are to be confined to affording protection against hostile aircraft. Scouting and patrol duties in connection with hostile water craft are to be considered secondary to this duty. All machines are to be kept tuned up and ready for immediate action.

When the war began, as we shall see, the Admiralty found it necessary to organize a system of aerial patrols around the coasts. Furthermore, it was the task of the Royal Navy to protect the Expeditionary Force when it was carried across the Channel in mid-August. The Royal Naval Air Service was employed to watch over this passage, from above, and patrols of aeroplanes and dirigibles were then organized for the purpose. In these circumstances it was found necessary to alter and modify Churchill's arrangement of 29 July; but his minute and this Admiralty cypher telegram reflect the way in which he thought at the time the threat of war seemed about to develop into reality.

II

While Winston Churchill began to prepare the Royal Navy for action, Sefton Brancker in the War Office also became alarmed by the international situation. He realized a crisis was at hand as early as 25 July. At once, Brancker alerted David Henderson, his small staff in the Directorate of Military Aeronautics, and other concerned officers. On 26 July Lieutenant-Colonel Sykes called a conference of his senior officers at Farnborough and decided to begin mobilizing the Royal Flying Corps. In Sykes's opinion the 'war was upon us'.[7]

The great object of these officers was to make certain the four squadrons of the Royal Flying Corps would be ready to depart for the front at the same time that the Expeditionary Force left for the continent.

In addition to these concerns, Sefton Brancker was one of the officers responsible for the condition and readiness of the home anti-aircraft gun defences. As early as 27 July cypher telegrams were sent from the War Office to the various military Commands in Britain ordering them to 'man anti-aircraft guns with regulars'. On the next day 'recognition diagrams' of British and German airships were sent to each of the Commands. On 29 July orders were sent to mount anti-aircraft guns at Purfleet and Waltham Abbey. At this stage it was the policy of the War Office that airships were to be fired upon, but fire was not to be directed against aeroplanes or seaplanes because the two latter classes of aircraft might be British machines: 'To Commands. . . . Fire to be opened at all airships which do not answer signals . . . which are specified. . . . Aeroplanes and seaplanes not to be fired at'.[8]

People in Britain knew the Germans possessed several Zeppelin airships capable of attacking London. As the diplomatic situation became more tense in the summer of 1914 it was feared that the metropolis might be bombed from the air, at any moment. At the end of July a rumour spread that Zeppelins were about to raid Woolwich and London.

The rumour was taken seriously. On 30 July the Army Council agreed to send No.4 Squadron of the Royal Flying Corps to Eastchurch, as a reinforcement for the naval aeroplanes already stationed there.

No one in the War Office welcomed this diversion from their plan to send all the aircraft they could muster to France. It was looked upon as a mere temporary aberration in policy.

Lieutenant-Colonel Sykes gave expression to the military point of view when he wrote, later: '. . . No.4 Squadron . . . was delayed by a rumour of a Zeppelin raid . . . and dispatched to Eastchurch, where it was obliged to stay until the alarm proved to be false'.[9]

III

Despite this false alarm people remained convinced that the Germans would bomb London from the air. The feeling was widespread in the period immediately before the war began and it continued to grow after Britain declared war on Germany on 4 August.

An example of this attitude appeared in *The Aeroplane* of 12 August when the journal explained to its readers that:

> So far as Great Britain is concerned there seems to be no reason to fear a German invasion. . . . But there is . . . always the chance of one or two German ships raiding the coast. . . . Also, there is still more probability of a raid by air. . . . London would, of course, be the objective. . . . Even a ton or so of explosives dropped in small instalments on London at night would have a . . . disconcerting effect on the strongest nerves, and it is hard to judge whether it would result in a panic-stricken mob rushing to Buckingham Palace clamouring for peace.

John Ledeboer, the respected editor of the technical publication *Aeronautics* was convinced that London would be bombed from the air, and for a good reason. In his opinion an air raid on London would not cause much 'material damage' but it could still result in 'grave mischief'.[10]

He feared the Germans would try to bomb the wireless apparatus on the roof of the Admiralty and thus throw the country's 'naval arrangements out of gear'. In Ledeboer's opinion if the Germans could disrupt the Admiralty's communications with the various fleets at sea, even for a short time, they would have accomplished a great deal.

This wireless installation on the Admiralty roof was a subject of much discussion between military and naval officers in the period before the war and also after the beginning of hostilities. Indeed, Winston Churchill mentioned it prominently in one of the dramatic passages in his *The World Crisis* when he described how the Fleet took up its war station, on 29 July 1914. Churchill wrote:

> If war should come no one would know where to look for the British Fleet. Somewhere in that enormous waste of waters to the north of our islands, cruising now this way, now that, shrouded in storms and mists dwelt this mighty organization. Yet from the Admiralty building we could speak to them at any moment if need arose. The king's ships were at sea.[11]

However, we should notice that the danger John Ledeboer wrote about had been foreseen. Winston Churchill's reference to the wireless installation in the Admiralty was not incorrect but it was misleading. The responsible naval authorities had taken care to make certain they would not have to rely solely upon the wireless apparatus in the Admiralty whenever they desired to 'speak' to their ships at sea. There were alternatives.

The Admiralty building possessed 'landline' communications with a number of wireless telegraphy stations situated at strategic points along the East coast. These stations could be employed to communicate with the fleet by wireless whenever it became necessary to do so. The most important of them was the Wireless Telegraph Station at Cleethorpes in Lincolnshire.[12] These technical arrangements were explained by Rear-Admiral E. F. B. Charlton, the Assistant Director of Torpedoes, in a minute he wrote in April 1914. He was not too worried by the threat of an air attack upon the Admiralty's wireless station:[13]

> In any event there are many alternatives possible should Whitehall W/T station (Admiralty) be irreparably damaged. It is hardly conceivable that the landlines to Cleethorpes, Horsea,

Clifden and Poldeu would all be destroyed. Even then a route by wireless to Cleethorpes or Horsea is available via Marconi House and Chelmsford. . . .

As regards their comparative value to the Fleet as W/T Stations damage to Cleethorpes Station would be more serious than any damage to the Admiralty W/T Station.

IV

Charlton prepared these remarks at a time when the responsible authorities were planning to install an anti-aircraft gun in the vicinity of the Admiralty. On 3 April 1914 a party of high-ranking officers from the War Office, in company with representatives of the Office of Works, met Rear-Admiral Frederick Tudor, the Director of Naval Operations, in order to inspect the Admiralty building and to pick a site for the gun. Eventually, these experts decided the gun should be put in place on top of the Admiralty Arch. However, a problem soon arose which threatened the entire project.

Tudor reported to the Board of Admiralty that it would be impossible to fire projectiles into the air from the vicinity of the Admiralty building without the risk of injury to life and property. Bits of shell casings and shards from the projectiles were certain to fall over a wide area whenever the gun was fired. In these circumstances, the experts who met on 3 April wanted the whole question reconsidered. Tudor explained that some of the most famous places in London could be put at hazard:[14]

> In the opinion of the officers who have inspected the site, this risk amounts to a certainty that some of the most important buildings in the Metropolis, such as Buckingham Palace, Houses of Parliament etc. would be seriously damaged if shell were fired from a gun mounted at the Admiralty.
> In addition, the knowledge that such a gun was mounted or to be mounted would be almost certain to raise very serious public and possibly political comment and protest.

In these circumstances the Board of Admiralty decided to re-examine the matter. They now had to decide if the gun was really needed at the Admiralty when its presence at that site could cause so much damage in the very centre of London. The conclusion was reached that the Admiralty building was not important enough to be provided with an anti-aircraft gun because of its wireless telegraphy equipment.

However, the Admiralty as the 'central governing authority of the Navy' did require the installation of such a gun. This had been recommended, initially, by the Home Ports Defence Committee.[15] When Admiral Tudor looked into the question more closely he prepared another minute, dated 21 April 1914: 'It was understood during the discussions at the H.P.D. Committee meetings that it was the Admiralty as the centre of Naval Administration rather than the wireless installation which required protection'.[16]

The Board of Admiralty decided to apply to the War Office for an anti-aircraft gun; and on 15 July 1914 they were informed that the Army Council would hold in readiness an 'anti-aircraft gun, with a view to its use in the protection of the Admiralty buildings, on the inauguration of the Precautionary Period'.[17] Shortly after the war began the Army Council fulfilled this promise of July 1914. Years later, Captain L. S. Stansfeld, RN, was asked to prepare a history of the Royal Naval Volunteer Reserve Anti-Aircraft Corps. He produced a document of great historical value. Referring to the date 8 August 1914 he wrote in his account as follows:

> On this date Major Molesworth, R.G.A., 2 N.C.O's and 30 gunners, reported themselves at the Admiralty to man three 1 Pdr. Pom Pom guns, that had been mounted on Admiralty Arch, The Foreign Office, and the Office of the Crown Agents for the Colonies, 4 Whitehall Gardens.[18]

V

British preparations for war were not limited to plans aimed at the German enemy. The responsible authorities also made arrangements to deal with their own people in case of riots, strikes, or other civil disturbances. Such outbreaks were expected and it was held they might result from several causes, including air-raids against population centres. It was foreseen that there might be looting or other types of criminal activity after the war began.

In 1913 a series of conferences between representatives of the General Staff and of the Metropolitan Police was held. The purpose of the conferences was to work out the principles upon which it was proposed to base a scheme 'for preventing Civil disturbances in London in time of war'.[19]

The War Office did not wait for hostilities to begin before acting in this sphere. On 2 August 1914 Reginald Brade, the Assistant Secretary in that department, sent a letter to the General Officer Commanding, London District. In it, Brade declared:

> I am commanded by the Army Council to request that, in conjunction with the Commissioner of Police of the Metropolis and the Commissioner of Police for the City of London, you will draw up a scheme for preventing civil disturbance in London in time of war . . . it should be borne in mind that it is essential to employ the fewest possible number of troops on these duties.

Brade then listed the units that might be employed for the purpose. They included: the Household Mounted Brigade; the London Mounted Brigade; 'F' Battery, Royal Horse Artillery; XXXV Brigade, Royal Field Artillery; and two Battalions of Foot Guards.[20]

A scheme of the kind desired by the Army Council was prepared at once. By its terms, military guards were to be stationed at certain selected buildings 'the destruction of which would constitute a serious danger to public safety'. The authors of this scheme were impressed by the seriousness of the situation. They wrote: 'As the emergency will be great, armed guards will be given definite orders to fire to protect the premises they are watching or to prevent crime or riot'.[21]

There was reason for the British authorities to show such vigilance at the beginning of the war. Shortly after it began riots broke out in London, and also in the provinces. Shop owners with German-sounding names were attacked and their premises looted. Often, the rioters took advantage of the situation and robbed any store they chose, in the confusion of these riots.[22] Later, when the Germans began to bomb English cities and towns from the air, similar rioting broke out in the wake of the air-raids. Sometimes, troops had to be summoned to the aid of the civil authorities in order to control the rioters.

Meanwhile, in the very first days of the war it was the urgent and imperative business of the War Office and the Admiralty to cooperate with each other in discharging certain joint tasks and duties. The relationship between Winston Churchill and Field-

Marshal Lord Kitchener now became a factor of consequence. The two had met first in 1898 on the field of Omdurman when Churchill was a lieutenant in the 21st Lancers and Kitchener was Commander-in-Chief. It was, Churchill wrote later, 'a case of dislike before first sight'.[23] Fortunately, this attitude changed with the passage of time. When Kitchener was appointed Secretary of State for War early in August 1914 the two were able to work together easily.

They quickly applied themselves to their mutual aeronautical concerns but a lack of harmony between the two departments became apparent, almost at once. Thus began what two technical experts have called, with good reason, 'the remarkable prolonged wrangling between the Admiralty and the War Office over air defence . . .'.[24]

On 4 August the Admiralty asked the War Office to furnish a defensive flight of four Royal Flying Corps aeroplanes, for the defence of London. It was suggested that the unit designated for this purpose might be stationed in Hyde Park.

The Army Council replied that the aeroplane 'would not be a suitable form of defence for use over London' because there would be too many casualties if aeroplanes were ordered to 'take action in such an area'. The Army Council suggested also that Hendon, rather than Hyde Park, might be a better location for such a flight. The soldiers further argued that a more efficient employment of aeroplanes would be to station them on the East coast where they would be in a position to intercept dirigibles before these latter aircraft even reached London.[25]

Next, Winston Churchill and his colleagues in the Admiralty began to prepare a scheme of coastal air patrols to guard the entire East coast from air or sea attack. These patrols were to operate from the Moray Firth in the north of Scotland to Dungeness, on the south coast of Kent.

This was a major defensive undertaking. The authorities in the Admiralty hoped the War Office would cooperate with them in supplying pilots and machines for the purpose. Nevertheless, the lion's share of this work was assigned to the Royal Naval Air Service. In the original defensive scheme its pilots were awarded the patrols protecting the most vulnerable areas of the coast, from the Firth of Forth to the Thames. The scheme assigned the northern and southern extremities of the coast to the Royal Flying Corps. Later, Churchill had reason to be annoyed with the War

Office because the part played by the Royal Flying Corps in these arrangements did not amount to very much. On 8 August the Admiralty ordered these patrols into action. They were based on eight seaplane stations and two airship stations.

We have to notice, however, that even before the patrolling began, Churchill was dissatisfied. On 7 August he took care to explain the entire plan to Lord Kitchener. In his letter he made it clear he was eager to cooperate with the War Office but he also let it be known that the burden placed on the Royal Naval Air Service was, in his opinion, too heavy:[26]

> I send you herewith a copy of the scheme worked out between the Naval and Military Wings for the patrol of the East Coast. I think that the part assigned to the Naval Wing is more than we have machines enough to do. A better arrangement would be for the military to take everything North of the Humber as well as the neighbourhood of Dover, and leave us the section from the Thames to the Humber. We would, however, put three or four seaplanes in the neighbourhood of the Firth of Forth and the Tyne. Experience shows that unless numbers are available no regular patrol can be maintained by aeroplanes day after day.

After this patrolling began the War Office had to formulate a policy for the anti-aircraft gun defences. The orders were sent to the various Commands by B. B. Cubitt of the War Office on 18 August:[27]

> I am commanded by the Army Council to inform you that British aeroplanes and seaplanes will in future be freely used for coastal patrol and reconnaissance duties in Great Britain; the probability of German aeroplanes arriving in this country at present is, however, remote.
> It has therefore been decided that until further orders no aeroplanes or seaplanes shall be fired upon by troops employed in the United Kingdom. . . .
> As regards airships – The only class of German airship which is likely to reach this country is the Zeppelin. Copies of a recognition diagram are enclosed and it will be seen that this type is perfectly distinctive and should be easily recognized.
> Arrangements should therefore be made to employ infantry as well as artillery fire against any airship of the Zeppelin class

which may come within range of the troops detailed for the purpose.

VI

In these early days of the war Churchill and Kitchener cooperated closely in working out the final plans for the passage of the Expeditionary Force to France. This was carried out successfully in the period 9–22 August. In this enterprise not a ship nor a man was lost. Indeed, the concentration of the army on the continent was completed three days in advance of Sir John French's original promise to General Lanrezac, the French commander most nearly involved in the arrangement.

We must be clear, however, that this was a time of great stress at the Admiralty and in the War Office. It was feared the German enemy might begin the war by delivering a smashing but unexpected blow at the British that could destroy the army before it reached the continent.

Churchill, in particular, worried the Germans might throw a military force against the British Isles themselves or else launch a naval raid into the Channel, designed to permit German warships to break in among the transports as they passed from England to France. The bulk of the army was in transit from 12–17 August and, as Churchill put it later, it was at this time that 'the strategic tension reached its climax'.[28]

Formidable naval forces were deployed to make certain the transports reached their destinations safely. The Eastern approaches to the straits of Dover were patrolled by cruiser squadrons and flotillas from Harwich and the Thames. In the straits themselves were British and French destroyer flotillas. These were supported by submarines under the command of Commodore Keyes. In addition, early in August a Channel Fleet of 19 battleships was organized. Its task was to cruise the Western reaches of the English Channel, and to support these other forces when occasion required it.

The aeroplanes, seaplanes, and dirigibles of the Royal Naval Air Service were also pressed into service in these August days. Their task was to guard the crossings from above. In connection with this last arrangement, Winston Churchill became very annoyed with his colleagues in the War Office. They failed to help him, as he felt they should have done, in providing air cover for the passage of the army.

On 9 August a seaplane patrol was begun, and was continued on a regular basis, to protect the Eastern flank of the transports. The seaplanes flew from Westgate-on-Sea close to the North Foreland in Kent, to Ostend in Belgium. These seaplanes of the Royal Naval Air Service followed each other at intervals of two hours so that an almost continuous watch was maintained over the waters where the German Navy was most likely to appear. On 13 August a temporary base for these seaplanes, under the command of Flight Lieutenant E. T. Chambers, was set up at Ostend. The base was maintained in operation until 22 August when the passage of the army was completed and Ostend itself threatened by the advance of German forces.

Two airships of the Royal Naval Air Service began patrolling the Channel on 10 August. This patrol was continued throughout the month. The airships flew directly over the transports and also scouted far out over the North Sea searching for any approach of hostile ships.

After the Expeditionary Force was safely ashore these routine patrols were no longer carried out but the burden on the Royal Naval Air Service continued to be a heavy one. Its aircraft carried out reconnaissance flights along the coast at dawn and at sunset, whenever the weather permitted.

In this connection we must look, again, at the reaction of Winston Churchill to these events of August 1914. The significance of his memorandum of October 1914, 'Aerial Defence', has not been recognized, and for a good reason.[29]

'Aerial Defence' clearly reflected the 'strategic tension' Churchill felt at a time of great national hazard, the time when the British Expeditionary Force was carried to France. In it, Churchill wrote of the War Office demand that it should be entrusted with the 'sole responsibility' for the home air defences, even the air defence of naval ports and vulnerable points ashore 'exclusively of naval interest'. The memorandum continued:

> Notwithstanding these views, the War Office had not up to the time of the declaration of war provided any aeroplanes for home defence . . . practically all the army aeroplanes were sent abroad. Not only were none available for guarding the vulnerable points, but none could be found even for the temporary

Winston Churchill Takes Charge 305

purpose of coast-watching during the passage of the army to the Continent.

Later, in October 1915, Churchill produced a very different analysis of this failure of the War Office to discharge its air defence responsibilities, adequately. In this memorandum Churchill argued that the War Office was correct in its decision to send all its aeroplanes to France upon the outbreak of war. This second account was more bland in its exposition than was 'Aerial Defence':[30]

> Up to the outbreak of war, the War Office was responsible for all forms of home defence, including even the aerial defence of naval ports. But owing to the fact that aviation was a new arm and that General Seely had to fight for every penny in his Estimates, the Army Air Service . . . had only up to the outbreak of war been able to prepare the flying squadrons required for service with the Expeditionary Army and they had not yet reached the point when any steps had been taken for aerial home defence. I consider that the War Office was quite right to do the most important work, namely, the equipping of our armies in the field, first.

Later still, Churchill published an exposition similar to this one in *The World Crisis*.[31] The book, like his memorandum of 25 October 1915, did not in any way reflect his anger at the War Office because of its lapses in the sphere of aerial defence. How can we explain these differences?

In the one account Churchill was critical of the soldiers because they claimed and demanded air defence responsibilities which they would not give up even though they had made no preparations, whatsoever, to meet or discharge them. In the second he claimed that the War Office was correct in its policy and that he, as First Lord, concurred in what was done by the military leaders.

Until October 1914 when he wrote 'Aerial Defence' Churchill was one of the most powerful politicians in the country. He was at the peak of his political career. The war was run, at that time, by a Triumvirate. Its membership included the most important war leaders in the country. These were: Asquith, the Prime

Minister; Lord Kitchener, the Secretary of State for War; and Winston Churchill, First Lord of the Admiralty.

After all, it had been revealed to the world in August 1914 that the Royal Navy was prepared and ready to discharge all its many functions and duties. It was shown that Churchill's three years at the Admiralty had been a successful stewardship. Even his Conservative enemies joined in applauding his achievement. He had ensured that the fleet was ready for war.

The first battering blows to this reputation as a war minister came in October 1914. When the key port of Antwerp was threatened by a German advance, Churchill rashly suggested he should go there in order to report on the situation. Once in the town he became excited by the military activity he observed and desired to remain there as the commander of its defences. Antwerp fell to the Germans on 10 October. When Churchill returned to London he was greeted by a storm of hostile criticism. Men began to ask why Churchill, a civilian minister, had rushed from his post in Whitehall to the scene of battle. It was widely stated that he had shown himself to be unfit for office in a war-time administration. Churchill was momentarily cast down by these criticisms. He contemplated resignation.

Nevertheless, despite these attacks, he could still speak or write to his colleagues without fear or the need for undue restraint, whenever he desired to express himself about any aspect of war policy. His opinions still counted in the deliberations of the government. He could still influence the conduct of the war.

His situation in October 1915 when he prepared his second air defence memorandum was completely different. By that time his political career was in ruins. In May 1915 he had been forced to retire from the Admiralty because his Tory enemies, and others also, blamed him for the bungled campaign at the Dardanelles. Antwerp and the Dardanelles, in combination,, seemed to prove his total lack of ability to serve as a Cabinet Minister in war. He became Chancellor of the Duchy of Lancaster in the Coalition Government formed at that time. In his new post he hoped Asquith might offer him some worthwhile employment but when this failed to happen he decided to resign. He was allowed no influence in the conduct of war policy.

It was in these entirely altered circumstances that he wrote his second air defence memorandum. It was a defence of his air policy as First Lord, prepared because his enemies had circulated inaccurate or 'partial' accounts of what he had done. On this occasion, he could afford to offend no one. As his biographer put it when he wrote of this memorandum of October 1915: 'No appeal to history could help him to regain the influence he had lost in May'.[32]

VII

At the beginning of the war Britain's first-line strength in aeroplanes was modest. The Royal Flying Corps disposed of 63 aeroplanes while the Royal Naval Air Service possessed 50 effective machines. There were several others, at the Central Flying School, and at other places, also, but they were not deemed battle-worthy for a variety of technical reasons. France at this time possessed 120 aeroplanes that were ready for action and Germany 232.[33]

In August 1914 the country was almost without an aeroplane industry. There were some firms involved in 'the trade', as it was called, but they were relatively inexperienced and were not capable of rapid expansion. The Admiralty had encouraged them with a few orders but the over-all effect of these purchases was not especially significant, by this stage.

For its part, the War Office relied, to a greater extent, upon the output of the Royal Aircraft Factory at Farnborough. Sir Frederick Sykes explained this situation with some care when he wrote of the B.E., Farman, Blériot and Avro aeroplanes with which the Royal Flying Corps was equipped in August 1914:

> The Royal Aircraft Factory could only turn out two machines a month, and private firms could not standardize their plant. The consequence of this was that when the war broke out we had only a hundred machines of various types, and less than a hundred pilots. But we made up for lack of numbers by keenness and sound organization and training.[34]

The situation with respect to the supply of aircraft engines was even worse. One expert has called it 'disastrous'.[35] There were no suitable British engines. The Admiralty and the War Office were each forced to rely, largely, on foreign sources when they sought to acquire aircraft engines.

In the days immediately before the war began, Major Brancker ordered his subordinates in the Military Aeronautics Directorate to travel to every part of the country, and also to France to purchase all the available aeronautical material they could find. The French authorities helped as much as they could until it was rumoured the British would not intervene in the war, when it came. Then the French sources of supply dried up. After Britain joined in the hostilities the French provided the Military Aeronautics Directorate with engines and other equipment once again.

The highest authorities had received ample warning of this unhappy situation long before the war began, but not much was done to correct it. In 1913 Sir Charles Ottley, who had served as Secretary of the Committee of Imperial Defence until 1912, was invited to assemble a team of experts to look into the question of British aeroplane supply 'in case of emergency'. When his investigations were finished Ottley prepared what he called a 'very secret report' and sent it to Colonel Seely, on 15 December 1913. The report made clear that the British aircraft industry was deficient in the area of aircraft engine production and deficient also in the area of air-frame or fuselage production:[36]

SECRET

AEROPLANES
Sudden expansion of output in case
of emergency.

It is assumed that circumstances might arise necessitating a sudden and very large demand for new military 'Planes. Such a demand might rise to a need of 200 new 'Planes in 4 months; from the two great factories of Armstrong & Vickers.

Could such a demand be met? The limiting factor today is believed to be the engines. Enquiries I have set on foot lead me to believe that Messrs. Armstrong could obtain 50 Gnome, Rhone and Clerget engines in 4 months, always assuming that cash down was paid with the orders. . . . To get the other . . . engines in 4 months from Renault . . . would not I believe be possible, but my experts say that with the addition of engines from Benz, Argus, Maybach . . . (a) second 50 could be provided in 4 months. Of course all these difficulties will largely disappear so soon as a British system of Aeroplane Engine Manufacture is firmly established. . . .

Turning now to the question of the manufacture of the Fuselage – the present British Army Plane ('B.E.') is I believe a very good one. . . . It owes its superiority largely I believe to the extreme accuracy . . . and nicety of its various parts. This multiplicity of small parts implies a maximum amount of skilled hand labour, and this condition imposes a limitation upon the number of 'planes we could then turn out in a given time.

VIII

Despite these technical shortcomings the young men of the Flying Corps and the Naval Air Service did not hesitate to accept the challenges of war when it became obvious to them that hostilities were about to begin. We have seen how the naval pilots concentrated at their various air stations and then began the patrol work assigned to them by the Admiralty. In the case of the Flying Corps, action was taken in conformity with long-held plans.

It had been arranged that each squadron of the Flying Corps was to mobilize at its peace station. Each was to be ready to move on the fourth day. Then, the aeroplanes would fly to Dover in preparation for the flight across the Channel to France. The Royal Flying Corps possessed five aerodromes at this time. They were: Farnborough; Montrose in Scotland; Netheravon on Salisbury Plain; Gosport in Portsmouth harbour; and Dover. The squadrons set out for Dover on 8 August and all but one arrived by the 12th. By 13 August three squadrons of the Royal Flying Corps were ready to take their places in the Expeditionary Force's order of battle.[37]

However, the condition of the Corps was complicated by a startling fact. Its commanders hated each other.

The extent of these hatreds has not been noticed before but it is fair to say that Lieutenant-Colonel Sykes despised Major Trenchard and that Trenchard, in his turn, loathed Sykes. Sir David Henderson concealed his own feelings but they were strong ones and we will take notice of them later on in this account.

Sykes believed that Trenchard was a 'rival whose strength of character seemed an inadequate substitute for other imagined deficiencies, and whose loud voice seemed to betray a permanently vacant mind'. Trenchard looked upon Sykes as a 'most conceited and indecisive staff officer'.[38]

It was only on 3 August that Trenchard learned he would not go to France but that he would replace Sykes at Farnborough. On 7 August he was formally appointed Officer Commanding the Royal Flying Corps (Military Wing), with the temporary rank of Lieutenant-Colonel.

Years later, Lord Trenchard told of a bizarre incident that took place upon this change-over of command. He reported at once to Farnborough where Sykes handed him, 'with an air of solemn mystery', two keys and a confidential box. According to Trenchard, Sykes said the box contained 'detailed plans for dealing with any German attack on this country by airships'. The smaller of the two keys fitted this box and Trenchard slept with it under his pillow that night. In the morning the confidential box was gone from the office. When Trenchard found it after a search he discovered it was full of shoes and that it had always been used for that purpose, and for no other. Trenchard's account continued: '. . . I understood. Remembering Sykes's instructions, I went to the safe and opened it with the other key. There were no papers whatever inside; but I was already prepared for that'.[39]

After Sykes left Farnborough he motored to Dover with Sir David Henderson. They planned to watch the aeroplanes of their Corps fly off to France and to follow in a ship that would take them to Calais. When Sykes described these incidents in his memoirs, which he called *From Many Angles*, he twice claimed that in 1912 he had been given a verbal promise that he would command the Royal Flying Corps in the field in case war broke out. He wrote:

> As I have mentioned, I had received a verbal promise that I should command the R.F.C. in the field, and very probably had I held higher rank in the Army at the time the promise would have been kept. . . . If a more senior officer were needed Henderson was the best choice, and he was certainly a charming man to work under.[40]

From Many Angles was published in December 1942. When Lord Trenchard read the book he was outraged by these claims. The ancient animosities of August 1914 flared again. Trenchard at once prepared some notes of his own, dated 19 January 1943. In them he wrote as follows of Sir Frederick Sykes and his book:[41]

I have never read a book so egotistical and so 'smug' as this. . . .

. . . When the Air Force went to France it went under Sir David Henderson, who appointed Sykes as his staff officer. I do not believe for one moment that he had ever been promised command of the Flying Corps if the war should eventuate in the year 1912. This is not only untrue . . . but grotesque that anybody should promise in 1912 that in years to come he should lead the Flying Corps in the Field. . . .

There is no doubt whatever that David Henderson, with whom I sometimes disagreed in the past, was the one man more than any other who had the true picture of the Air Force of the future in these early days. His work has never been sufficiently recognized chiefly owing to his Staff Officer.

Sir Frederick Sykes died at the end of September 1954.

His obituary notice in *The Times* angered Lord Trenchard, yet again. Trenchard at once spoke to Marshal of the Royal Air Force Sir Edward Ellington about the obituary, and Ellington agreed to write a letter of protest to *The Times*. Ellington was a pioneer of the early days of the Royal Flying Corps. In his letter to *The Times*, published on 13 October 1954, he declared:

It is not in any way my intention to deprecate the debt Sykes's successors owe to him . . . but only to plead that the contribution of others should not be forgotten, especially that of Sir David Henderson, who was in command of the four squadrons which went to France in August, 1914.

Trenchard was old and ill by this time. He was unable to read or write but he was so annoyed that he dictated a letter to Henderson's widow:[42]

It is a long time since I wrote to you but I was rather perturbed the other day when I saw the obituary notices for Sir Frederick Sykes, and I talked to several people about them.

The obituary notice in The Times read as though he was the 'Father of the R.F.C.' and took it to France in 1914, whereas he was, as you know, a junior Staff Officer to Sir David Henderson. Sir David was the man whose influence at the War Office, and his great strength of character, made him the power in Air matters that he was.

I saw Sir Edward Ellington and he agreed with me about Sir David's part in the formation of the Royal Air Force. . . .

Unfortunately, I cannot see to read or write now, so you must excuse a dictated letter.

Trenchard was not alone in his hatred of Frederick Sykes. Sir David Henderson also mistrusted his Chief of Staff. In August 1914 Henderson worried that the younger man was scheming to take his place as the commander of the Royal Flying Corps in the field.[43]

Lady Henderson was so moved by the terrible events of August 1914 that she wrote down, and preserved, her own contemporary account of them. These jottings contained bitter comments about Sykes and we cannot doubt they reflected the opinions of her husband.[44]

Shortly after the Royal Flying Corps left for France, Lady Henderson had lunch with Bertie Fisher, one of her husband's staff officers. Fisher had been ordered to remain behind in the Directorate of Military Aeronautics in the War Office despite his eagerness to accompany his chief to the front. He had watched the squadrons fly off from Dover and had then bidden Henderson goodbye when the latter sailed for the continent. Lady Henderson, in her notes, wrote of Fisher:

> He also had a good deal to tell me about Sykes and the way he left Aldershot denuded of every man, aeroplane, mechanic, and typewriters!! How when the officer who took over from him went down there he was in despair how to start afresh . . . Colonel Brancker and Bertie say all he wanted was to make a fine show for himself on the other side of the water and that he didn't care a hang how he left things here. They are very disgusted with him. . . . How he dislikes Sykes and I get frightened and wonder if Sykes will cart D. He is hated by all soldiers for a mean narrow small minded man.

It seems reasonable to conclude it was no band of brothers that travelled to France when the headquarters and staff of the Royal Flying Corps, Military Wing, embarked for Calais on that August day.

IX

At 6.25 a.m. on 13 August the machines of Nos. 2, 3, and 4 Squadrons began to take off for France. The aeroplanes followed each other at two minute intervals and arrived at Amiens before noon. One flight of No. 4 Squadron remained at Dover to carry out patrol duties. It was a token contribution to the home air defence system. No. 5 Squadron was delayed and it moved somewhat later than the other three. Within a few days, however, 44 British aeroplanes were in France. Only one of them was armed with a machine-gun. Its pilot, L. A. Strange, claimed later that he had equipped his aeroplane with a Lewis gun, and that he had done so on his own initiative.[45]

Reconnaissance flights began on 19 August. A few days later, on 22 August, Lieutenant Strange was in action with his gun-equipped machine. He and his observer, Lieutenant L. da C. Penn-Gaskell, attempted to attack a German Albatross aeroplane but the gun and its mounting turned out to be so heavy they were unable to rise above 3500 feet. The Albatross escaped by flying at a higher altitude. Strange was at once ordered to discard the Lewis gun and mounting and to equip his observer with a rifle.[46]

Some of the aeroplanes and many of the seaplanes of the Royal Naval Air Service were fitted with engines that were much more powerful than those employed in the machines of the Flying Corps in the earliest days of the war.

Winston Churchill and his colleagues in the Admiralty Air Department had planned to employ their aircraft to carry torpedoes and to drop bombs and they recognized, from an early stage, that powerful engines were a requirement that could not be neglected if efficient performance in wartime was to be achieved.

Sir David Henderson and Lieutenant-Colonel Sykes had planned an entirely different air force. The Royal Flying Corps had been organized by them, from its earliest days, to carry out reconnaissance missions. Its B.E.2 aeroplane, for example, had been designed with this purpose in mind. The B.E.2 flew with a fair degree of stability and provided the observer in it with a clear view of the ground below. It was not equipped with an especially powerful engine because it could carry out its reconnaissance tasks, very successfully, without one.

In the earliest fighting in the air, which began in August 1914, the pilots of the Royal Flying Corps understood that their machines were not powerful enough to be fitted with machine-guns. In these circumstances they tried to destroy enemy aircraft in the air by dropping grenades and bombs on them, from above. These grenades and bombs weighed far less than a machine-gun and its mounting.

A few days after Lieutenant Strange's failure, Sir David Henderson realized the consequences of his own lack of foresight. Early in September he wired home as follows:

> There are no aeroplanes with the Royal Flying Corps really suitable for carrying machine-guns; grenades and bombs are therefore at present more suitable. If suitable aeroplanes are available, machine-guns are better undoubtedly. Request you endeavour to supply efficient fighting machines as soon as possible.[47]

X

The great German assault in the west which began in August 1914 had been planned with care for years past. The object of the German commanders was to deliver so smashing an initial blow that it would break their opponents' will to resist. The German idea was to crush France in a lightning campaign. When the French were beaten the German Armies would then turn eastward and deal with the Russians separately in their turn. This was the essence of the famous Schlieffen plan which governed all German military activity in the first days of the war.

The French line between Luxembourg and Switzerland was perhaps the most heavily fortified in Europe. It would be impossible to pierce it quickly enough for the German purpose. The solution to this problem was the invasion of Belgium. A success there would enable the German Armies to turn their enemy's flank at the same time that they avoided his main fortifications.

The attack began with a tremendous assault upon Liège which guarded the crossings of the river Meuse. The city was taken on the night of 5–6 August and by 14 August the way to the west was clear for the German forces. On 20 August the German First Army entered Brussels in triumph and then it began to push its advance units even further to the west.

These German successes produced an instant result in London.

There, Sefton Brancker was among the first to sound the alarm. On 24 August he wrote a minute for General Launcelot Kiggell, the Director of Military Training. In it he explained the significance of the new situation in clear terms. He told Kiggell:[48]

> Now that Germany has occupied Western Belgium, Dover and the Thames Defences are within reach of German Aeroplanes.

In these circumstances the orders sent out earlier to the anti-aircraft gun defences in Britain had to be altered. As we have seen, on 18 August the War Office had ordered that anti-aircraft fire should be directed at Zeppelin airships only. Aeroplanes and seaplanes were not to be fired on because they might be British machines engaged in the recently activated coastal patrols.

After consultation with Murray Sueter in the Admiralty Air Department, new orders were issued to the anti-aircraft gun defences. They were now instructed to fire at hostile aeroplanes. At the same time a 'confidential list showing the exact positions of anti-aircraft guns' was sent to Lieutenant-Colonel Trenchard at Farnborough. He was advised that his aeroplanes 'should avoid these localities as much as possible'.[49]

In the Admiralty there was an even stronger reaction when the Germans established themselves in Belgium. It was thought that the Germans would use Belgium as a base for aeroplanes and airships, also. It was reckoned the enemy was now in a position to launch a regular series of air attacks against London and the Thames estuary. In these circumstances the aircraft of the Royal Naval Air Service were redeployed. They were concentrated between the Humber and the Thames and frequent patrols were carried out to report the movements of any enemy aircraft. The Wash was patrolled especially heavily because it was thought to be the most likely landfall for a German airship attempting to attack London.[50]

Winston Churchill did not limit himself to these arrangements. On 25 August Wing Commander Samson was summoned to London and ordered to take his squadron to Ostend at once. According to the official history Ostend 'had been chosen to serve as an advanced base for reconnaissance'.[51] Samson's Eastchurch squadron was the best and most highly trained of all the naval air units and it was also the first to be sent abroad. It did carry out reconnaissance duties with a force of British marines in the Ostend area but more was involved in the move, as Samson's later account makes clear. In 1914 his squadron, as a result of practice, was capable of fighting in the air. He wrote:[52]

> In 1914 we did a lot of practice fighting in the air, and tried raids on Chatham and Magazines in the vicinity, two aeroplanes attacking and 6 or 8 forming a protective screen. We got valuable data from this, and there is no doubt that this work in peacetime made us, from Eastchurch, such a splendid squadron when we all went together to active service in France [sic] in 1914 (August 27th) when war broke out.

On 30 August the Marine Brigade and Samson's squadron were ordered to return to England. The aeroplanes flew back by way of Dunkirk. There, the entire squadron was delayed for a period of three days because one of the pilots had smashed his machine upon landing. Then, there was a change of plan.

On 1 September the Admiralty ordered Samson to remain at Dunkirk 'to operate against Zeppelins and enemy aeroplanes' and also to carry out any reconnaissance missions that were needed. The Admiralty's new policy was to 'deny the use of territory within a hundred miles of Dunkirk to German Zeppelins, and to attack by aeroplanes all airships found replenishing there. . . . The immunity of Portsmouth, Chatham, and London from dangerous aerial attack is clearly involved'.[53]

XI

On 25 August a Zeppelin attack had been made on Antwerp and a dozen people were killed there as a result. This helped to spur Winston Churchill to more intense activity even though Colonel Brancker, in the War Office, was contemptuous of the air threat to London at this time. In Brancker's opinion, the Germans were not yet ready to launch a serious air attack against the capital. Furthermore, he believed that 'aerial home defence by aircraft had not developed sufficiently to be worth talking about'.[54]

Winston Churchill entertained a different view of the situation. On 1 September he gave orders that the largest possible force of naval aeroplanes was to be stationed at Calais or Dunkirk. He told his subordinates in the Admiralty that reports had been received

Winston Churchill Takes Charge 317

that the Germans were planning to attack London and other places in Britain with Zeppelin airships. He believed such a course was 'extremely probable'.[55]

Churchill was well aware of the air-fighting capabilities of Commander Samson and his pilots. When they were reinforced he wanted them to use Dunkirk as a base to search the country for a hundred miles inland seeking out Zeppelins and destroying them before they could begin an attack against London or other British targets. On 1 September he ordered that 'all steps' should be taken to supply Samson with the pilots, aeroplanes, and equipment he would need for the task.[56]

These subjects were discussed by the Cabinet on the next day. H. H. Asquith and his colleagues took the threat of a Zeppelin attack against London quite seriously. On 2 September the Prime Minister reported to the King:[57]

> Mr Asquith, with his humble duty to your Majesty, has the honour to report that the Cabinet met yesterday and today.
> The subjects chiefly considered were as follows:
> (1) The importance of establishing without delay a naval air reconnoitring station at Dunkirk;
> (2) the precautions to be taken in London against bomb-throwing by Zeppelins. It was agreed that the measures to be taken should be concerted between the Army & Navy & communicated to the Home Office.

In this period of his life Asquith was in love with an aristocratic young woman named Venetia Stanley. In order to impress her he adopted the disgraceful practice of sending her letters which contained, among other things, exact details of the country's most important military secrets. He did this on a regular basis. Men in the field or at sea were risking and losing their lives daily while the British Prime Minister provided his friend with immediate accounts of their activities, accounts he never should have sent and she was in no way entitled to receive. Information about the perceived and developing threat from the air was included in these intimate budgets. On 2 September, the same day that he reported to the King, Asquith sent Venetia Stanley a 'short synopsis of topics' taken up by the Cabinet. He told her: 'We had our regular Cabinet this morning . . . and you can judge of the kind of thing that comes up by this short synopsis of topics (1) Naval air reconnoitring at Dunkirk (2) protection of London against bomb throwing from Zeppelins . . .'.[58] On the next day the Cabinet was informed that Churchill had ordered some naval aeroplanes to Antwerp to attack the Zeppelin which had been dropping bombs there. Asquith at once wrote to Venetia Stanley: 'We had a Cabinet this morning . . . nothing really interesting, except that Winston has sent 4 of his Aeroplanes to Antwerp to scuttle the Zeppelin that has been throwing bombs there . . .'.[59]

It was at this stage that a major change took place in Britain's defensive arrangements. On 3 September 1914 when they met at a Cabinet meeting Lord Kitchener asked Churchill 'whether I would accept, on behalf of the Admiralty, the responsibility for the aerial defence of Great Britain, as the War Office had no means of discharging it'.[60]

This was a remarkable development. As late as July 1914, as we have already seen, the War Office continued to demand 'sole responsibility' for the home air defences.[61]

Now, however, Kitchener and his air adviser, Colonel Brancker, wanted to send every military aeroplane they could spare to the front in France. Therefore, they gave up their earlier claim that the War Office should exercise sole jurisdiction in this area of the nation's defences.

After Churchill agreed to Kitchener's request the new arrangement was sanctioned by the Cabinet, on 3 September: 'By a Cabinet decision, the Naval wing of the Royal Flying Corps assumed the responsibility for taking the necessary steps for the defence against aerial attack of this country'.[62]

This decision of the War Office to surrender its control of the home air defences in September 1914 was curious because the soldiers there continued to cling to the old principle that their department, and not the Admiralty, should be in charge of them. They had not abandoned their original outlook in any way.

The situation was dominated by the intense rivalry that existed between the two services. Shortly after the change was made Colonel Brancker prepared a minute about it for his superiors. His paper was an exact reflection of the military point of view:[63]

> If the Army had the unrestricted use of the aeroplane industry, and the entire output of the Central Flying School, we would even now be prepared to carry out the aerial defence of Great Britain quite as well as the Admiralty.

Just previous to the declaration of war, the General Staff had practically agreed that 5 more aeroplane squadrons . . . were urgently necessary; but owing to lack of funds, the development of the R.F.C. had been painfully slow, and consequently on mobilization we were only able to provide an insufficient complement of aeroplanes for the Expeditionary Force.

The fact that the Admiralty were allowed to compete for land aeroplanes from the trade placed them in a position to undertake (somewhat inadequately I fear) the provision of aircraft for the aerial defence of Great Britain – but this is no reason why this principle should be accepted forever.

By the terms of the arrangement of September 1914 the War Office was not divorced from playing a role in the home air defence system. Whenever military aeroplanes were available they could be employed for the purpose. Certain vulnerable points and other places, also, were protected by anti-aircraft gun emplacements, manned by soldiers. On 5 September Asquith explained the new system to the King: '. . . the Admiralty are taking over the aerial defence of the country, & in consultation with the War Office & Office of Works will at once make plans for London by night . . . .'.[64]

At this time, Winston Churchill acted more generously than his military colleagues. He plunged into this new aspect of his work with energy and enthusiasm. On 5 September he explained his plans for 'the aerial defence of England' in a document he sent to Prince Louis of Battenberg, the First Sea Lord, and to Murray F. Sueter of the Admiralty Air Division. He believed that Dunkirk was a place of high importance in his arrangements. He now wanted a force of 'armed motor cars' to be sent there as a support for Commander Samson and his squadron of aeroplanes. He calculated that the two units, in combination, could keep the Germans away from the coast and allow the Admiralty to maintain 'aerial control over the area'.[65]

Churchill was convinced that his wisest course would be to attack the Zeppelins at their bases, even while they were still in their sheds. He planned to launch aeroplane attacks againnst them as soon as they could be arranged. This offensive spirit was a legacy he left to those in the Air Force who came after him. They forgot, however, that he came to the decision to attack partly because there was no defensive system in place in Britain that could deal with the air raids he expected the Germans to launch at any moment. Years later, in *The World Crisis*, he explained the situation as he saw it in September 1914:[66]

> As the Germans overran Belgium and all the Channel ports were exposed, the danger of air attacks upon Great Britain became most serious and real. Zeppelins had already cruised over Antwerp, and it was known that London was in range of the Zeppelin sheds at Düsseldorf and Cologne. To meet this danger there was nothing except the naval aeroplanes the Admiralty had been able to scrape and smuggle together. . . . I thereupon undertook to do what was possible with the wholly inadequate resources which were available. There were neither anti-aircraft guns nor searchlights, and though a few improvisations had been made, nearly a year must elapse before the efficient supplies necessary could be forthcoming. Meanwhile at any moment half a dozen Zeppelins might arrive to bomb London or, what was more serious, Chatham, Woolwich or Portsmouth. . . .
>
> . . . It was easy to order the necessary guns, searchlights, etc. and set on foot the organization which should produce and employ them. But it was no use sitting down and waiting for a year while these preparations were completing. Only offensive action could help us.

Churchill grasped that aircraft had provided the German enemy with a novel method of attacking every corner of the United Kingdom. He was convinced that such air attacks would be made, and in the very near future. In these circumstances he did not hesitate to respond to the new challenge, though he later had reason to regret his actions. In *The World Crisis* he wrote:

> Looking back with after-knowledge and increasing years, I seem to have been too ready to undertake tasks which were hazardous or even forlorn. Taking over responsibility for the air defence of Great Britain when resources were practically non-existent and formidable air attacks imminent, was from a personal point of view 'some love but little policy'.[67]

Nevertheless, the nation would soon have cause to be grateful for what he did at this time.

Notes and References

1. Introductory: Britain in the Air Age

1. For the interview see Alfred Gollin, *No Longer An Island, Britain and The Wright Brothers, 1902–1909* (London, 1984), ch. xi, hereafter cited as Gollin, *No Longer An Island*.
2. For these details see *Flight*, 8 May 1909.
3. For this point see Gollin, *No Longer An Island*, p. 460.
4. *Parliamentary Debates*, 5th Series, 2 August 1909, cols. 1564–1565. This important revelation has been neglected by British aeronautical historians.
5. Richard Burdon Haldane, *An Autobiography* (London, 1924), pp. 232–3, hereafter cited as Haldane, *Autobiography*.
6. The document is printed in the classic work of C. F. Snowden Gamble, *The Air Weapon* (Oxford, 1931), pp. 109–10, hereafter cited as Snowden Gamble, *The Air Weapon*.
7. The letter is printed in Marvin McFarland (ed.), *The Papers of Wilbur and Orville Wright* (New York, 1953), vol. II, p. 926, hereafter cited as McFarland, *The Papers of Wilbur and Orville Wright*.
8. This prize was increased later to the sum of £1000.
9. Enthusiasts of Rudolf Martin's type sometimes hoped to see a league of Continental Powers created. The object of such a league would be to defy Great Britain.
10. See Cab. 16/7. 'Report and Proceedings of A Sub-Committee of the Committee of Imperial Defence on Aerial Navigation', dated 28 January 1909 and marked 'SECRET'. For an analysis of this Sub-Committee in action see Percy B. Walker, *Early Aviation at Farnborough* (London, 1974), vol. II, ch. xii; this work will be cited hereafter as Walker, *Early Aviation at Farnborough*; and Gollin, *No Longer An Island*, ch. xiii. All further quotations in this and the next two sections of this chapter, unless otherwise specified, are drawn from this source.
11. For some comment on this idea see *The Times*, 21 December 1908.
12. For this meeting see Cab. 2/2. Committee of Imperial Defence, Minutes of 101st Meeting, 25 February 1909, marked 'SECRET'.
13. Walker, *Early Aviation at Farnborough*, vol. II, pp. 327–8.
14. For this and material in the preceding paragraph, see Gollin, *No Longer An Island*, pp. 429–30.
15. For the details of Bacon's resignation see Arthur Marder, *From The Dreadnought to Scapa Flow* (London, 1961), vol. I, pp. 191ff, hereafter cited as Marder, *From The Dreadnought to Scapa Flow*.
16. For these details and those in the preceding paragraphs see Gollin, *No Longer An Island*, pp. 465ff.

2. The Advisory Committee for Aeronautics

1. Haldane's remarkable knowledge of explosives resulted from his experiences as a lawyer. He took part in two great legal cases involving explosives and mastered a good deal of technical information in preparing for them.
2. Asquith sent Lord Rayleigh a letter of thanks for this consideration in his own hand, written from 10 Downing Street. See Rayleigh Papers, Asquith to Rayleigh, 24 June 1908.
3. See Griffith Brewer, *Fifty Years of Flying* (London, 1946), p. 113.
4. Haldane, *Autobiography*, pp. 232–3.
5. B. M. Add.MSS. (Northcliffe Papers) Haldane to Northcliffe, 4 May 1909.
6. Oddly, Snowden Gamble in his classic *The Air Weapon* at p. 108 attributes the statement of 5 May to Haldane. However, it was Asquith who spoke on that day and not his Secretary of State for War. The above account of his speech is based on the report in the *Daily Mail*, 6 May 1909 and on the report in the *Daily News* of the same date.
7. See the *Daily News*, 6 May 1909.
8. B.M. Add. MSS. (Northcliffe Papers) Northcliffe to Haldane, 9 May 1909.
9. The subjects Northcliffe here refers to were included in a list published in the *Daily Mail* for 6 May. The paper listed as questions to be dealt with by the scientists at Teddington the following: frictional resistance of planes in a current of air; determination of the centre of pressure of inclined planes in a current of air; the efficiency of small propellers and motors; the curvature and general form of an airship or aeroplane.
10. Northcliffe was very interested in the development of motors and motor-cars. In 1902 he wrote a book entitled *Motors and Motor Driving*. He used his newspapers to advance the cause of motoring in Britain from an early date. Royce, the partner of C. S. Rolls, and Napier were pioneers and competitors in this field.
11. Garvin Papers, Northcliffe to Garvin, 7 and 11 May 1909.
12. B.M.Add.MSS. (Northcliffe Papers), Northcliffe to Lee, 13 May 1909.
13. Ibid., Arthur Lee to Northcliffe, 13 May 1909.
14. Ibid., Haldane to Northcliffe, 18 May 1909.
15. This refers to the fate of Cody, Dunne, and later on, to that of Colonel Capper.
16. The reference here is to the appointment of Dr Ludwig Prandtl, mentioned earlier in this chapter.
17. From the account in the *Daily Telegraph* 21 May 1909.
18. Sir Walter Raleigh, *The War in The Air* (Oxford, 1922), vol. I, p. 160, hereafter cited as Raleigh, *The War in The Air*.
19. For these incidents see the excellent account of them in Dudley Sommer, *Haldane of Cloan* (London, 1960), pp. 338–9.
20. Halevy, *History of The English People*, vol. VI, p. 123.

3. The Phantom Airship Scare of 1909

1. Winston S. Churchill, *The World Crisis* (London, n.d.), vol. I, p. 24, hereafter cited as Churchill, *The World Crisis*.
2. This paragraph is based on a report in the *Manchester Guardian*, 20 May 1909.
3. *Parliamentary Debates*, 5th series, 17 May 1909, col. 6.
4. The supposed flights of mysterious airships flying over Britain in May 1909 is a theme neglected by most historians dealing with this period. E. L. Woodward in his great work *Great Britain and the German Navy* mentions these flights in a footnote, p. 235, n. 1. But he does not even refer to the Prime Minister's aeronautical announcement of 5 May in the House of Commons. Woodward incorrectly states that the debates on the Naval Estimates and the vote of censure of March 1909 caused the airship scare in May. See also A. J. A. Morris, *The Scaremongers* (London, 1984), p. 159, hereafter cited as Morris, *The Scaremongers*.
5. The material in this and the following paragraphs is based upon the account in the *Morning Post* for 20 May 1909. All the newspapers carried such accounts which were sent to the London journals by the Cardiff Correspondent of the Press Association.
6. Oron J. Hale, *Publicity and Diplomacy* (New York, 1940), p. 354.
7. Zara S. Steiner, *Britain and the Origins of the First World War* (New York, 1977), p. 168.
8. Garvin Papers, Northcliffe to Garvin, 22 May 1909. Two of Northcliffe's mother's relatives had married Germans and these are the 'German relations' mentioned in this letter.
9. *Parliamentary Debates*, 5th series, 24 May 1909, col. 812. It was discovered later that the rifles mentioned here were obsolete weapons which had been acquired by the Society of Miniature Rifle Clubs.

4. The Beginning of Air Power Politics in Britain

1. Airships of this period were classified into three categories. Rigids; non-rigids; and semi-rigids. The first of these possessed a complete framework which retained its shape even when the gas-bags were removed. The second, the non-rigid, had no framework at all and maintained its shape as a result of gas and air pressure. The semi-rigid was equipped with a girder which ran the length of the envelope. For a full account see Higham, *The British Rigid Airship*, p. xix.
2. The correspondence between Lee, Clément, and the War Office was published in the *Daily Mail*, 29 October 1910.
3. See Walker, *Early Aviation at Farnborough*, vol. II, p. 330; and Snowden Gamble, *The Air Weapon*, p. 111.
4. See the *Daily Mail*, 26 July 1909 for de Forest's letter and telegram.
5. For the incident see Randolph Churchill, *Winston S. Churchill* (London, 1967), vol. II, pp. 156ff.
6. Haldane Papers, R. B. H. to 'My dearest Mother', 3 August and 4 August 1909.
7. *Parliamentary Debates*, 5th Series, 2 August 1909, col. 1564. The debate

of 2 August is to be found in cols. 1564–1617 and all references to it reproduced above are taken from this source.
8. See Viscount Lee of Fareham, *A Good Innings and A Great Partnership* (Privately Printed, 1939), vol. I, pp. 399–400. The reader should be clear this reference is taken from Lee's privately printed book, in several volumes, and not from the one volume edition edited by Alan Clark.
9. Octave Chanute Papers, Chanute to Colonel Glassford, 27 April 1909.
10. Ibid., 3 August 1909.

5. The Forward March of Aeronautics

1. C. H. Gibbs-Smith, *Aviation* (London, 1970), p. 146.
2. From the account in the *Daily Mail*, 25 August 1909.
3. B.M.Add.MSS. (Northcliffe Papers), A. E. Widdows to Northcliffe, 15 September 1909.
4. Ibid., Northcliffe to Repington, 25 January 1915. The letter is mentioned in Koss, *Haldane*, p. 144.
5. Ibid., Northcliffe to Fisher, 5 February 1915. The letter is mentioned in Koss, *Haldane*, p. 144.
6. Ibid., Northcliffe to Balfour, 11 July 1915.
7. See Dudley Sommer, *Haldane*, p. 341.
8. Harald Penrose, *British Aviation: The Pioneer Years* (London, 1967), p. 199, hereafter cited as Penrose, *British Aviation: The Pioneer Years*.
9. See Air 1/1613/204/88/7. 'Ammunition For The Attack of Balloons' by H. R. A. Mallock, 6 July 1909.
10. This is Mallock's phrase, employed in the next section of his paper.
11. See the copy preserved in Lord Rayleigh's Papers: Memorandum by Sir G. Greenhill, 6 August 1909.
12. Rayleigh Papers, F. J. Selby to Lord Rayleigh, 29 October 1909. Selby was the Advisory Committee's Secretary.
13. See Higham, *The British Rigid Airship*, p. 36, n. 2.
14. See Air 1/729/176/4. Reports on German Dirigible Balloons.
15. These phrases are his own and are to be found in his *Who's Who* entry.
16. See Air 1/729/176/4/4. 'Dirigibles', marked 'Confidential', dated 10/6/09, and signed F. Trench. Col. G. S. The paper was sent to the War Office in London.
17. Air 1/1613/204/88/14. Trench to Capper, 3 September 1909, marked 'Confidential'.
18. Air 1/685/21/13/2243. F. Trench to War Office, 10 September 1909.
19. See Air 1/824/5/69. 'Aviation in Foreign Countries', 15 November 1909, 'Confidential'. For a discussion of these reports see also Neville Jones, *The Origins of Strategic Bombing* (London, 1973), p. 34, hereafter cited as Jones, *The Origins of Strategic Bombing*.
20. Air 1/729/176/5/62. 'Aeronautical Reports For 1910. General Staff, War Office', and marked 'Confidential'.
21. Cab. 2/2. CID 'Minutes of 107th Meeting, July 14, 1910. SECRET'.
22. Ibid.

23. For an account of the talk and subsequent discussion see the *Journal of the Royal United Service Institution*, May 1909, vol. LIII, No. 375, pp. 559–77.
24. Capper's talk was printed in *The Journal of the Royal Artillery*, vol. XXVI, 1909–10, pp. 433ff.
25. Col. 1613.
26. See Walker, *Early Aviation at Farnborough*, vol. II, p. 332.
27. Walker, *Early Aviation at Farnborough*, vol. I, p. 255.

6. The Origins of British Air Defence

1. Penrose, *British Aviation: The Pioneer Years*, p. 197.
2. For one of these see Alfred Gollin, *Proconsul in Politics* (London, 1964), p. 202.
3. W.O. 32/5545. Paget to The Secretary, War Office, 23 October 1909, and marked 'Secret'.
4. W.O.32/5545. J. S. Ewart, DMO to CGS, 8 November 1909.
5. Air 1/654/17/122/491. 'Report of a meeting held on 4th Jan. 1910 to discuss recent progress made by Foreign Governments in aerial navigation . . .' dated 7 January 1910 and marked 'Confidential'. All further quotations in this section, unless otherwise indicated, are drawn from this source.
6. W.O.32/5545. W. Graham Greene to The Secretary, War Office, 4 February 1910, marked 'Secret'.
7. W.O.32/5545. A. J. Murray to CIGS, 14 February 1910. This minute was addressed to Sir William Nicholson, now styled CIGS (Chief of the Imperial General Staff).
8. W.O.32/5545. J. S. Ewart to CIGS, 17 February 1910.
9. W.O.32/5545. Nicholson to MGO (Master-General of the Ordnance) 9 March 1910.
10. W.O.32/5545. R. H. Brade to The President, Home Ports Defence Committee, 29 March 1910, marked 'SECRET'.
11. Cab.13/1. 'The Defence of Magazines, Cordite Factories and Other Vulnerable points against Airship Attack', HPDC Memorandum No.12-M, 19 May 1910, marked 'SECRET'. All material reproduced in this section, unless otherwise specified, is drawn from this source.
12. See above Ch. 1, section VI.
13. Cab.4/3. CID 117-B. 'Aerial Navigation. Note by the Secretary'. 6 July 1910, marked 'SECRET'.
14. See above Chapter 1, section VI.
15. Cab. 4/3. CID 117-B. 'Aerial Navigation. Note by the Secretary'. 6 July 1910, marked 'SECRET'.
16. See above Chapter 1, section VI.
17. See Cab. 2/2. Committee of Imperial Defence. Minutes of 107th Meeting, 14 July 1910, marked 'SECRET'. All quotations reproduced above, with respect to this meeting, are drawn from this source.
18. See above Chapter 5, section III.
19. The *Morning Post*, 29 April 1910.

20. David Lloyd George, *War Memoirs* (London, n.d.,), vol. I, pp. 20–21, hereafter cited as Lloyd George, *War Memoirs*.
21. *Parliamentary Debates*, 5th Series, 7 March 1910, cols. 1185–6.
22. See Lord Montagu of Beaulieu, *'Aerial Machines and War'*, published for the Aldershot Military Society, 1910. All references to this meeting reproduced above are taken from this source.
23. *Parliamentary Debates*, 5th Series (House of Lords) 13 April 1910, cols 646–51.
24. Ibid., col. 651.
25. Ibid., col. 652.
26. An excellent account of the League in 1909 and 1910 is in N. Jones, *The Origins of Strategic Bombing*, pp. 28–9.
27. Ibid., p. 29. An example of how the League worked occurred immediately before the General Election of January 1910. Then, Stephen Marples, Secretary of the Aerial League, sent a letter to each candidate for Parliament enquiring if the candidate, when elected, would support a vote for funds for Aerial Defence. The letter explained that the League was a 'patriotic non-party organisation' and warned of Germany's 'Aerial Fleet'. For this letter see *Aeronautics*, January 1910, pp. 10–11.
28. The leaflet is preserved in Air 1/648/17/122/398.
29. See Air 1/648/17/122/398.
29. See Air 1/648/17/122/398.
30. See Graham Wallace, *Claude Grahame-White* (London, 1960), p. 52.

7. The Paris Conference and its Consequences

1. There is a good summary of the details and significance of this International Conference in S. W. Roskill (ed.), *Documents Relating to The Naval Air Service*, Volume I 1908–1918 (Navy Records Society, 1969), pp. 13ff., hereafter cited as Roskill, *The Naval Air Service*; and in Gamble, *The Air Weapon* pp. 146ff.
2. See Cab.18/26. 'Interim Report of the British Delegates to the International Conference on Aerial Navigation', 12 July 1910: 'The Strategical Aspects of the International Conference on Aerial Navigation'. Note by Secretary. Marked 'SECRET'.
3. Cab. 4/4/1. CID 137-B. 'Proceedings of the Committee of Imperial Defence with reference to the International Conference on Aerial Navigation 1910'. Appendix IV 'Report of the InterDepartmental Committee . . .' 11 October 1909 marked 'SECRET'.
4. For these details in this and the previous paragraph see Cab.18/26. 'Interim Report . . .' 12 July 1910, Note by Secretary.
5. Cab. 18/26. 'Interim Report . . .' 12 July 1910.
6. Cab. 18/26. 'Interim Report . . .' 12 July 1910.
7. Cab.18/26. 'Interim Report . . .' 12 July 1910.
8. For the letter see Cab. 2/2. Committee of Imperial Defence. Minutes of 106th Meeting, 14 June, 1910, 'SECRET'. Appendix. Gamble to Under-Secretary of State, Home Office, 10 June 1910.
9. For these details see Cab. 4/4/1. CID 137-B 'SECRET'. Explanatory

Notes and References 327

Note by the Secretary. The Standing Sub-Committee differed only slightly from the full Committee of Imperial Defence.
10. Lord Hankey, *The Supreme Command* (London, 1961), vol. I, p. 111, hereafter cited as Hankey, *The Supreme Command*.
11. Cab.18/26. 'The Strategical Aspect of Certain Proposals before the International Conference on Aerial Navigation'. Marked SECRET, dated 23 June 1910.
12. Contemporaries included Denmark in this list.
13. Hankey, *The Supreme Command*, vol. I, p. 111.
14. For this see Gamble, *The Air Weapon*, p. 149. For the attitude of Haldane and Churchill see Hankey, *The Supreme Command*, vol. I, p. 113.
15. Cab. 4/3 CID 118-B. 'SECRET'. 'The International Conference on Aerial Navigation'. Memorandum by the General Staff. 11 July 1910. All quotations reproduced above dealing with the Memorandum are taken from this source.
16. See Cab.18/26. 'Interim Report of the British Delegates to the International Conference on Aerial Navigation' dated 12 July 1910 and marked 'CONFIDENTIAL'. The paragraphs above are based on this document and all quotations in this part of Section III of this chapter are drawn from it.
17. For this see Cab. 4/4/1. CID 137-B 'SECRET'. 'Proceedings of the Committee of Imperial Defence with reference to the International Conference on Aerial Navigation 1910' dated December 1911. Appendix 1.
18. Cab. 4/4/1. CID 137-B. Appendix 6. W. Graham Greene to the Foreign Office 21 July 1910, Copy.
19. See D. W. Sweet, 'Great Britain and Germany, 1905–1911', p. 226 in F. Hinsley (ed.), *British Foreign Policy under Sir Edward Grey* (Cambridge, 1977).
20. Oron J. Hale, *Publicity and Diplomacy* (New York, 1940), p. 310.
21. Alfred Pribram, *England and The International Policy of the European Great Powers* (Oxford, 1931), p. 121.
22. See R. C. K. Ensor, *England 1870–1914* (Oxford, 1949), p. 569. The later British writer is D. W. Sweet, 'Great Britain and Germany, 1905–1911', p. 226. The American account is Sidney Fay, *The Origins of The World War* (New York, 1948), vol. I, p. 243.
23. Fritz Fischer, *World Power or Decline* (New York, 1974), pp. 6–7.
24. Air 1/2100/207/27. E. Grey to Sir F. Bertie (and others) 29 July 1910, marked 'Most Confidential'.
25. For the memorandum see Air 1/2100/207/27. Foreign Office Memorandum, dated 29 July 1910, marked 'Secret'.
26. See above, Section II of this chapter.
27. Cab.4/4/1. CID 137–B. 'SECRET'. Note by Secretary, 4 December 1911.
28. Ibid.
29. See above Chapter 1, Sections V and VI.
30. See Percy Walker, *Early Aviation at Farnborough*, vol. II, p. 331. For the 'Note' see Roskill, *The Naval Air Service*, pp. 23–5.

31. See Cab. 4/3. CID 119–B. Aerial Navigation. Note by Lord Esher, 6 October, 1910 marked 'SECRET'.
32. See Lieut.-Colonel Charles A Court Repington, *Vestigia* (New York, 1919), pp. 295ff. The second article mentioned appeared in *Blackwood's Magazine*, July 1910, pp. 3ff.
33. See Cab. 4/3. CID 119-B. 'Note' by Lord Esher dated 6 October 1910. Lord Esher began his 'Note' by reproducing this statement.
34. Walker, *Early Aviation at Farnborough*, vol. II, pp. 332–3.
35. For Sir Alexander's views see Chapter 5, Section VI, above.
36. Cab. 4/3. CID 119-B. 'SECRET' Note by Lord Esher. 6 October 1910.
37. See Cab. 4/4/1. CID 137-B. 'SECRET'. Note by Secretary, 4 December 1911. R. B. Haldane acted as chairman of the Sub-Committee.
38. Hankey, *The Supreme Command*, vol. I, p. 112.
39. Ibid., p. 112.
40. Ibid., p. 113.
41. See Air 1/2100/207/27. Sir F. Bertie to Sir Edward Grey, 12 November 1910, marked *'Confidential'*. This letter is the source for these quotations and those in the following four paragraphs, above, dealing with Bertie's initiative in Paris at this time.
42. For this see Gamble, *The Air Weapon*, pp. 146–7. In this connection see also Cab. 2/3. The Minutes of the 121st meeting of the Committee of Imperial Defence deal with the Aerial Navigation Act, 1911. Further, see *Parliamentary Debates 5th series* vol. VIII, House of Lords, cols 1020ff., 1 June 1911.
43. Hankey, *The Supreme Command*, vol. I, p. 113.
44. Stephen Roskill, *Hankey, Man of Secrets* (New York, 1970), vol. I, p. 105, hereafter cited as Roskill, *Hankey*.

8. The Agitation for a National Air Force

1. B.M.Add.MSS (Northcliffe Papers) Northcliffe to Marlowe, *n.d.* but written in June 1911.
2. See W.O.163/15. Decision taken at the 128th Meeting of the Army Council, 30 July 1910.
3. Raleigh, *The War in The Air*, p. 189.
4. Penrose, *British Aviation: The Pioneer Years*, p. 248.
5. It should be mentioned he gallantly rejoined when the war began and eventually commanded the 2nd/5th North Staffordshire Regiment in 1917.
6. See the *Daily Mail*, 4 February 1911. It should be mentioned that Sir Alexander had support for his opinion from sources outside the General Staff. See, for example, Harald Penrose, *British Aviation: The Pioneer Years*, pp. 295ff who quotes the opinion of one authority who wrote the aeroplane had not improved very much by this time but that the pilots and engines were better.
7. Gamble, *The Air Weapon*, p. 149. This presents a fine account of the early activities at Eastchurch.
8. For Sueter's remark see Harald Penrose, *British Aviation: The Pioneer*

Years, p. 344; see also Group Captain G. E. Livock, *To The Ends of The Air* (London, 1973), p. 43.
9. See Airl/724/76/2. 'History of Naval Aeroplanes and Seaplanes' by Colonel C. R. Samson, DSO, dated 30/12/18.
10. In March 1911 in order to strengthen their party in the House of Lords Asquith arranged that Haldane should be raised to the peerage as Viscount Haldane of Cloan.
11. See the account in the *Daily Mail*, 5 April 1911.
12. See B.M.Add.MSS (Northcliffe Papers) C. Grahame-White to Northcliffe, 27 May 1911; and Graham Wallace, *Claude Grahame-White*, p. 132.
13. See Graham Wallace, *Claude Grahame-White*, pp. 135ff.
14. McKenna Papers (MCKN 3/15 f.75), Lord John Hay to McKenna, 17 May 1911.
15. This account of the deputation of 23 May is taken from the *Morning Post*, 24 May 1911.
16. Parts of the speech are reproduced in R. C. K. Ensor, *England 1870–1914* (Oxford, 1949), p. 435.
17. See Major-General Sir Frederick Maurice, *Haldane* (London, 1937), vol. I, p. 283.
18. Michael Howard, *The Continental Commitment* (London, 1972), p. 47.
19. A. G. Gardiner, *Pillars of Society* (London, 1913), p. 61.
20. Lucy Masterman, *C. F. G. Masterman* (London, 1939), p. 231.
21. For the incident see Gollin, *No Longer An Island*, p. 431.
22. Randolph Churchill, *Winston S. Churchill* (London, 1967), vol. II, p. 698, hereafter cited as Randolph Churchill, *Winston S. Churchill*, vol. II.
23. See Air 1/724/76/2. *History of Naval Aeroplanes and Seaplanes* by Colonel C. R. Samson, DSO, dated 30/12/18.
24. See Cab.17/9 Minute signed by W. S. C. and dated 9.11.11, and Eddie Marsh to Seely, dated 'Dec. 13th'. See also S. W. Roskill (ed.), *Documents Relating to the Naval Air Service* (London, 1969), vol. I, p. 26, hereafter cited as Roskill, *Naval Air Service*.
25. McKenna Papers (MCKN 3/15 f.76), R. H. Bacon to McKenna, 28 September 1911.
26. See Air 1/2442/6/4. *Rigid Airships of Zeppelin Type*.
27. For this and the following exchanges see *Parliamentary Debates*, 5th Series, 30 October 1911, cols 559ff.
28. C. F. Snowden Gamble, *The Air Weapon*, p. 157.
29. Cab.16/16. C.I.D.139-B. 'Report and Proceedings of the Standing Sub-Committee of the Committee of Imperial Defence on Aerial Navigation. 1912', dated March 1912 and marked 'SECRET'.

9. The New Arrangement and its Lapses

1. The lecture and remarks were printed in *The Aeroplane*, 23 November 1911. This lecture was referred to in a debate in the House of Commons on 14 December 1911. See *Parliamentary Debates*, 5th Series, 14 December 1911, cols 2673–4.

2. In 1911 the Italians used aeroplanes against the Turks in their campaign in Tripoli.
3. For the exchange of 6 December 1911 see *Parliamentary Debates*, Fifth Series, House of Lords, 6 December 1911, cols 649ff.
4. Cab.16/16. CID 139-B. 'Report by the Technical Sub-Committee', 27 February 1912. In addition to Seely, the members were: Brigadier-General G. K. Scott Moncrief; Brigadier-General D. Henderson; Commander C. R. Samson, RN; Lieutenant R. Gregory, RN; and Mervyn O'Gorman. Rear-Admiral C. L. Ottley and Captain M. P. A. Hankey acted as Secretary and Assistant Secretary, respectively.
5. See the excellent account in Gamble, *The Air Weapon*, pp. 164ff. It may be mentioned here that the rivalry between the War Office and the Admiralty was so great that the Royal Flying Corps did not, from the first, function as originally planned. From an early period the Naval Wing went its own way; and before the War broke out in August 1914 the term Royal Flying Corps, Naval Wing, had been replaced by the title 'Royal Naval Air Service'. This alteration was the work of the Admiralty alone. It was not the result of any decision by the Cabinet.
6. The reader will notice that this was the day Colonel Seely's Technical Sub-Committee completed its Report which was, of course, secret.
7. For these exchanges of 4 March see *Parliamentary Debates*, 5th Series, 4 March 1911, cols 69ff.
8. The letter is reproduced in Gollin, *No Longer An Island*, p. 274.
9. Cab.17/19. Tullibardine to Seely, 22 January 1912, marked 'Private'.
10. The Blair Atholl Syndicate managed to get an order from the War Office for two Dunne machines in March 1913. Sales of the aeroplane were also made to French and American companies. Tullibardine's letter of 22 January 1912 was read out by Seely at a meeting of his Technical Sub-Committee, on 24 January 1912. For this see Cab.16/16. CID 139-B. 'Minutes of Meeting, January 24, 1912'.
11. See Parliamentary Debates, 5th Series, 6 March 1912, cols 407–8 for Lee's remarks on this occasion.
12. See Lord Lee of Fareham, *A Good Innings and A Great Partnership* (1939), vol. I, p. 445. The reader should note this remark is taken from the privately printed version preserved in the Lord Lee of Fareham Papers in the Courtauld Institute in London.
13. See the excellent account in Snowden Gamble, *The Air Weapon*, p. 169. He is very critical of the failure to establish a unified air service in this early period.
14. Cab.21/524/12/4. 'The Unified Air Force'. Undated but docketed 'Not to be circulated without reference to A(ir) M(inistry)'. The two services were called eventually the Royal Flying Corps and the Royal Naval Air Service.
15. See above, Chapter 4.
16. These were seaplanes equipped with floats.
17. *Parliamentary Debates*, Fifth Series, 12 March 1912, col. 1009.
18. See Cab. 14/1. 'Air Committee. Minutes of First Meeting', 31 July 1912.

19. Cab.16/16. CID 139-B. 'Report by the Technical Sub-Committee', 27 February 1912.
20. H. A. Jones, *Sir David Henderson, A Memoir* (London, n.d.), p. 10. See also H. A. Jones, 'Sir David Henderson, Father of the Royal Air Force', in The Journal of *The Royal Air Force College*, vol. xi, Spring 1931, pp. 6ff.
21. John Buchan, 'David Henderson', in *The Quarterly Review*, No. 470, January 1922, p. 131.
22. See Henderson Papers. AC 71/12, Box 8. H. A. Jones to Lady Henderson, 6 May 1929. In this letter Jones, the official air historian, reported Trenchard's remark to Henderson's widow. It should be noted in this connection that Trenchard himself was often called the 'Father of the Royal Air Force'.
23. Sir Frederick Sykes, *From Many Angles* (London, 1942), p. 66, hereafter cited as Sykes, *From Many Angles*.
24. Sykes, *From Many Angles*, p. 77.
25. Ibid., p. 91 and p. 95.
26. For this report see Sykes Papers. AC 73/35. 'Notes on Aviation in France. November 1911. by Captain F. H. Sykes, General Staff'. The paper was sent to The Secretary, The War Office. This paragraph and the following five paragraphs draw on material contained in this paper. Later, Sykes delivered two lectures on this subject which were published in *The Army Review*. In these lectures fighting in the air and air bombing are mentioned but the greatest emphasis is placed upon the importance of aerial reconnaissance. For the lectures see Sykes Papers. AC 73/75. 'Military Aviation', in *The Army Review* July 1913, pp. 127ff and 'Further Developments of Military Aviation', in *The Army Review*, April 1914, pp. 440ff. In his *From Many Angles*, p. 107, Sykes refers to these lectures and makes clear his opinion that the 'chief function of the air arm . . . would be reconnaissance, and information would have to be fought for. . .'.
27. Lord Hankey, *The Supreme Command*, vol. i, p. 110.
28. *Parliamentary Debates*, Fifth Series, 30 October 1911, col. 665.
29. Winston Churchill, *The World Crisis*, vol. i, p. 265.
30. Air 1/118/15/40/56. F. H. Sykes, Lieut. Colonel, Commanding Royal Flying Corps (MW) to DGMA (Director-General of Military Aeronautics), War Office, 16 December 1913.
31. Churchill, *The World Crisis*, vol. i, p. 265.
32. Air 1/2314/222/6. 'Aerial Defence' by W. S. C., Admiralty, 22 October 1914.

10. Home Air Defence and Lamp-posts

1. See above, Chapter 6.
2. There is an excellent account of this in Neville Jones, *The Origins of Strategic Bombing*, pp. 40ff. Jones also makes valid criticisms here of the opinions of the official historians, Sir Walter Raleigh and H. A. Jones. For the difficulties experienced by the Committee of Imperial

Defence in its relations with the service departments, see Nicholas d'Ombrain, *War Machinery and High Policy* (Oxford, 1973), *passim*.
3. See W. O. 32/5546. L. E. Kiggell, D. S. D. to Director of Fortifications and Works, 18 January 1911. This and the following paragraph are based on this document.
4. See Cab. 16/16. CID 139-B. Report by the Technical Sub-Committee. Memorandum by the General Staff on the Present Position of Military Aviation in the United Kingdom, and in France and Germany Respectively, 29 November 1911.
5. Cab. 13/1. Home Ports Defence Committee. 20-M. 'Decentralisation of Stores and Manufacturing Plant at Woolwich and in its Vicinity'. July 1912, marked 'SECRET'. Material in this and the following three paragraphs is based on this document.
6. Neville Jones, *The Origins of Strategic Bombing*, p. 40.
7. See above Ch. 6, section II where it is explained that the key to British defence policy was the maintenance of supremacy at sea by the Royal Navy. However, if a raiding force evaded the Fleets and crossed the coast it at once became the concern of the army. In accordance with this ancient tradition it was accepted that if enemy aircraft flew over the British Isles the responsibility for dealing with them would rest with the War Office, and not with the Admiralty.
8. See Cab. 17/20. O. Murray to The Secretary, Committee of Imperial Defence, 4 May 1912, marked 'Confidential'. This paragraph and the two following paragraphs are based on material contained in this document.
9. W.O. 32/5547. M. P. A. Hankey to The Secretary, War Office, Whitehall. 15 May 1912, marked 'Secret'.
10. W.O. 32/5547. Minute by Hadden dated 30/5/12.
11. W.O. 32/5547. Minute by Colonel R. Whigam, dated 10/6/12.
12. Mottistone Papers, J. Seely to Churchill, 11 December 1912. Churchill was very concerned about the vulnerability of magazines to air attack. See his Memorandum on the subject in Randolph S. Churchill, *Winston S. Churchill*, vol. II, *Companion, part 3* (London, 1969), p. 1876, hereafter cited as Randolph Churchill, *Companion*, part 3.
13. Air 1/626/17/42. Admiralty Air Department Paper 'Protection of British Dockyards From Aerial Attack'. 1913. 'Aerial Attack on Dockyards'.
14. See Air 1/762/204/4/179. 'Experiments. Shooting at Moving Aerial Targets (Kites) 1914'.
15. Neville Jones, *The Origins of Strategic Bombing*, p. 46.
16. See Air 1/2314/222/6. 'Aerial Defence' by W.S.C. dated 22 October 1914. Churchill was very angry with the War Office because of the paucity of these emplaced guns. The War Office, in the pre-war period, did plan to employ howitzers against hostile airships but Churchill was not satisfied by their proposals. If these guns were not in place he wondered how long it would take 'after the appearance of an enemy's airship' to bring the howitzers into the target area, mount them, and then open fire. See, in this connection, his minute dated 29 January 1913 in Adm. 1/8398. For Admiralty experiments

designed to discover if the gas in dirigibles could be set alight easily, see Roskill, *Naval Air Service*, p. 76.
17. For the appendix to Churchill's paper see Adm.1/8398.
18. See the account in H. A. Jones, *The War in The Air*, vol. III, pp. 72–3.
19. See Adm.1/8398. Document dated 29 January 1913 and initialled W.S.C. The War Office in the period after October 1913 experimented with bombs suspended from aeroplanes by a wire which could be 'fired electrically as a means of destroying dirigibles . . .'. For this see Air 1/757/204/4/101. Other experiments involved plans to drop light bombs on dirigibles. See, in this connection, Walter Raleigh, *The War in The Air*, vol. I, pp. 268ff.
20. H. A. Jones, *The War in The Air*, vol. III, pp. 73–4.
21. See Roskill, *Naval Air Service*, pp. 148ff. for an extensive account of this sub-Committee's first meeting.
22. Roskill, *Naval Air Service*, p. 153.
23. Air 1/2314/222/6. 'Aerial Defence' by W.S.C. dated 22 October 1914.
24. For his 'proposals' and the table see Air 1/511/16/3/59. Henderson to the Director of Military Training, 27 July 1914.
25. Basil Collier, *Heavenly Adventurer* (London, 1959), p. 33, hereafter cited as Collier, *Heavenly Adventurer*.
26. Brancker Papers, Unpublished Autobiography, holograph section. A slightly different version of this account is published in Norman Macmillan, *Sir Sefton Brancker* (London, 1935), p. 68, hereafter cited as Norman Macmillan, *Brancker*. Macmillan prepared Brancker's memoirs for publication and his version was published as this book; but his account occasionally differs slightly from the unpublished memoirs in the possession of the present writer.
27. Extracts from the minutes of this meeting are reproduced in Roskill, *Naval Air Service*, pp. 39ff.
28. See Cab. 16/16. Report and Proceedings of the Standing Sub-Committee of the Committee of Imperial Defence on Aerial Navigation, 1912. CID 139-B, marked 'SECRET'. Minutes of Meeting of 18 December 1911.
29. See Roskill, *Naval Air Service*, p. 40, and f.n.2.
30. See Cab. 16/17. CID 159-B. Report and Proceedings of the Technical Sub-Committee of the Committee of Imperial Defence on Aerial Navigation. Airships. 1912. Dated 6 August 1912 and marked 'SECRET'. See also Roskill, *Naval Air Service*, pp. 45ff.
31. Air 1/2471/6/37. Hugh Watson to Sueter, 9 December 1911.
32. Their report presented in July 1912 is Appendix VII of Cab. 16/17. CID 159-B. Report and Proceedings of the Technical Sub-Committee of The Committee of Imperial Defence on Aerial Navigation. Airships. 1912. Marked 'SECRET'. Appendix VII is entitled 'Report by Captain Sueter and Mr O'Gorman on Continental Airships'. All discussion of and references to this report, above, are drawn from this source, and are based on it.
33. In 1910 Goschen discussed the subject of Zeppelins with the German Emperor in Berlin. See C. H. D. Howard (ed.), *The Diary of Edward Goschen* (London, 1980), p. 203.

34. On 30 June 1912 Sueter reported to Admiral Beatty in London that he and O'Gorman had inspected the Parseval Works. 'We were shown everything', he wrote. He asked Beatty to report this to Winston Churchill. For this see Cab.17/20. Murray Sueter to Beatty, dated 30.6.12.
35. A brief account of the visit is in Murray F. Sueter, *Airmen or Noahs* (London, 1928), p. 132.
36. See Cab. 16/17. CID 159-B. Report and Proceedings of the Technical Sub-Committee of the Committee of Imperial Defence on Aerial Navigation. Airships. 1912. Marked 'SECRET'. The membership of this Sub-Committee included, in addition to Seely: Brigadier-General D. Henderson; Captain Sueter; Commander Oliver Schwann; and Mervyn O'Gorman. Hankey acted as Secretary. All further quotations from this report, and comment on it, in this section are based on this document. See also, Roskill, *Naval Air Service*, pp. 45ff.
37. For the list see C. F. Snowden Gamble, *The Story of A North Sea Air Station* (London, 1967), p. 24, hereafter cited as Gamble, *North Sea Air Station*, and Roskill, *Naval Air Service*, p. 59, and Gamble, *The Air Weapon*, p. 202.
38. Cab. 2/3. Committee of Imperial Defence. Minutes of 120th Meeting, 6 December 1912. Marked 'SECRET'. All quotations and comments with respect to this meeting, above, are drawn from this source. See also in this connection, Marder, *From The Dreadnought to Scapa Flow*, vol. I, pp. 339ff., and Roskill, *Naval Air Service*, pp. 65ff.
39. In his *The World Crisis*, vol. I, p. 265 Churchill claimed that he 'rated the Zeppelin much lower as a weapon of war than almost anyone else.' This was, perhaps, true of his attitude in 1913 but in 1912 he was seriously concerned by the menace of Zeppelin airships.
40. Cab. 2/2. Committee of Imperial Defence. Minutes of 119th Meeting. 1 August 1912, marked 'SECRET'.
41. There is a good account of the incident in Higham, *The British Rigid Airship*, pp. 63–4. He cites the work of the technical expert, Dr Douglas Robinson, to establish the fact that no Zeppelin flew over England at this time.
42. Air 1/2456/6/17/1. Murray F. Sueter to 3rd Sea Lord, 14/11/12.
43. Eastchurch is quite near Sheerness and was the nearest naval air station.
44. *The Aeroplane*, 28 November 1912.
45. For these exchanges see *Parliamentary Debates*, Fifth Series, 27 November 1912, cols 1243–5.
46. Cab. 2/3. Committee of Imperial Defence. Minutes of 120th Meeting, 6 December 1912, marked 'SECRET'.
47. E. B. Ashmore, *Air Defence* (London, 1929), p. 40.
48. These remarks were made in a conversation with the present writer at Cherkley on 13 April 1964. On that occasion Beaverbrook discussed the various phases of his public career.

11. The Air Panic of 1913
1. For the meeting see Roskill, *Naval Air Service*, p. 71.
2. See Air 1/657/17/122/563. 'Dirigible Airships in time of War', dated 7 December 1912 and signed by Hugh Watson, 'Captain and Naval Attaché', and by Alick Russell, 'Lieutenant Colonel and Military Attaché, marked 'Confidential'.
3. Air 1/657/17/122/563. Docket signed A. K. W., dated 2.2.13.
4. For Churchill's letter and Wilson's marginal comments see Cab. 4/5. CID 172-B. 'Aerial Navigation'. 'Letter from Mr Churchill to Admiral of the Fleet Sir A. K. Wilson, with marginal notes by Sir A. K. Wilson'. The letter is dated 3 February 1913 and the entire document is marked 'SECRET'. See also Roskill, *Naval Air Service*, pp. 73ff.
5. For this, see A. Temple Patterson, *The Jellicoe Papers* (London, 1967), p. 27.
6. See Roskill, *Naval Air Service*, for these comments at the meeting of 6 February, reproduced in this and the following two paragraphs.
7. See T. H. Hawkins, 'The Attack of Airships', in the *Journal of the Royal Artillery*, vol. xxxvii, 1911–12, pp. 302–6. Other articles of this general type, published at about the same time, were: Lieutenant J. W. Marsden, RGA, 'The Defence of a Fortress against Aerial Attack', in the *Journal of the Royal Artillery*, vol. xxxvii, pp. 384–6; and Major H. T. Belcher, RFA, 'A Method of Attacking Aircraft with Shrapnel Fire', in the *Journal of the Royal Artillery*, vol. xxxix, 1912–13, pp. 499–505.
8. Air 1/653/17/122/483. E. W. D. Ward to The Chairman, Home Ports Defence Committee, 3 January 1913, marked 'Secret'. A copy of this letter was sent to the Admiralty.
9. Cab. 13/1. 31-M. 'The Allotment of Anti-Aircraft Guns'. Memorandum by the Home Ports Defence Committee, dated 1 April 1913 and marked 'SECRET'. This and the following six paragraphs are based upon this document.
10. Air 1/653/17/122/483. E. W. D. Ward of the War Office to The Secretary to the Admiralty, 4 April 1913, marked 'SECRET'. This paragraph and the two following, above, are based on this letter.
11. See above, Chapter 3, Section IV.
12. George Dangerfield, *The Strange Death of Liberal England* (London, 1936), p. 114.
13. See, for example, the warm praise of Churchill's work for the Naval Wing in *The Aeroplane*, 27 February 1913.
14. Mottistone Papers, John Bernard Seely to 'Dear Prime Minister', 9 December 1912.
15. Royal Archives. Geo. V., F. 447. Seely to Stamfordham, 14 March 1913.
16. See the account in the *Daily Mail*, 20 March 1913.
17. *Parliamentary Debates*, Fifth Series, 19 March 1913, col. 1072.
18. Ibid., 24 March 1913, col. 1398.
19. Ibid., col. 1402.
20. Norman Macmillan, *Brancker*, pp. 33–4.
21. Asquith Papers, Esher to Mrs Asquith, 25 March 1914. A copy of this

letter was given to the present writer by Michael Brock, The Warden of Nuffield College, Oxford. See in this connection, Michael and Eleanor Brock (eds), *H. H. Asquith Letters to Venetia Stanley* (Oxford, 1982), p. 636.
22. A. J. P. Taylor, *The Struggle for Mastery in Europe* (Oxford, 1954), p. 496.
23. R. C. K. Ensor, *England, 1870–1914* (Oxford, 1949), p. 571. See also the opinion of a German scholar, V. R. Berghahn, *Germany and The Approach of War in 1914* (New York, 1973), p. 156, 'That the Army's proposals were . . . accepted by the civilian authorities with relatively little fuss demonstrates . . . how far the considerations of the General Staff had . . . come to dominate German policy-making. . .'. He adds the *Wehrbeitrag* 'could be dressed up as a 'sacrifice, a patriotic gift'. . .'.
24. See *The Times*, 1 April 1913.
25. See *The History of The Times, The 150th Anniversary and Beyond*, Part I (London, 1952), p. 99.
26. HLRO, Bonar Law Papers, B. L. 29/3/8. Northcliffe to Bonar Law, 8 April 1913.
27. Mottistone Papers, Northcliffe to Seely, 11 April 1913.
28. B.M.Add.MSS (Northcliffe Papers) Northcliffe to Marlowe, 12 April 1913.
29. This was the headline in the *Daily Mail* for 12 April.
30. *Parliamentary Debates*, Fifth Series, House of Lords, 29 April 1913, cols 337–45.
31. See the account in the *Daily Telegraph*, 6 May 1913.
32. See the *Daily News and Leader*, 6 May 1913.
33. Ibid.
34. See Chapter 4, above.

12. Preparing for War with Germany

1. The expert is Howard Roy Moon. See his excellent unpublished University of London PhD. thesis, 'The Invasion of the United Kingdom: Public controversy and Official Planning 1888–1918', July 1968, p. 433, hereafter cited as Moon, 'Invasion'. In addition to the high service leaders, Asquith, Balfour, Lloyd George, Churchill, Haldane, Lord Morley, McKenna, Sir E. Grey, Harcourt, Lord Esher, and the Marquess of Crewe served on this Standing Sub-Committee. Captain Hankey acted as Secretary.
2. For this and the assumption about Germany see Cab. 16/28A. 'Report and Proceedings of A Standing Sub-Committee of the Committee of Imperial Defence appointed by the Prime Minister to reconsider the question of Attack on the British Isles From Oversea', marked 'SECRET', dated 15 April 1914.
3. See above, Chapter 8.
4. The authority is Arthur Marder. See his *The Anatomy of British Sea Power* (New York, 1976), p. 80; and *From The Dreadnought to Scapa Flow*, vol. I., p. 345. See also Roskill, *Hankey*, vol. I, p. 130, who writes

Notes and References 337

of this enquiry of 1913–14 that 'one wonders why it was considered necessary'.
5. See Gollin, *No Longer An Island*, p. 327 for Balfour's arguments.
6. See Cab. 3/2. CID 62-A. Standing Sub-Committee of the Committee of Imperial Defence 'Attack on the British Isles from Oversea'. Report. Marked 'SECRET' and dated 15 April 1914.
7. For these points see Moon, 'Invasion', p. 429.
8. For this memorandum see Cab. 16/28B. O.A.6. 'The Influence of Aircraft on Attack from Oversea'. Memorandum by the Admiralty, dated 18 March 1913 and marked 'SECRET'. This and the five paragraphs following are based on this document.
9. For Lord Roberts's fear of such an air attack see Gollin, *No Longer An Island*, pp. 328–9.
10. See Cab. 16/28. A. Report and Proceedings of A Standing Sub-Committee of the Committee of Imperial Deference. . . . 'Attack on the British Isles From Oversea', dated 15 April 1914, marked 'SECRET'. The memorandum by Roberts, Repington, Scott, and Lovat is Appendix VI of the Report and is dated 30 March 1913. It is numbered O.A.-25.
11. See Cab. 16/28A. O.A.-57. Appendix XXIX of the Report: 'Proceedings of a Conference held at the Admiralty on the 19th November, 1913'. This conference was one of a series of meetings known as the 'High Level Bridge' and it is dealt with in that context in section v of this chapter, below.
12. See Cab.16/28.A. Report and Proceedings of A Standing Sub-Committee of the Committee of Imperial Defence, appointed by the Prime Minister to reconsider the question of 'Attack on the British Isles From Oversea', dated 15 April 1914 and marked 'SECRET'.
13. For these arguments see paragraphs 93 and 94 of the Report.
14. For this see paragraph 95 of the Report.
15. For this see paragraphs 103 and 104 of the Report.
16. Hankey, *Supreme Command*, vol. I, p. 84.
17. For Asquith's interest see Hankey, *Supreme Command*, vol. I, p. 84.
18. See Major-General J. E. B. Seely, *Adventure* (London, 1930), p. 140.
19. See Randolph Churchill, *Winston S. Churchill, Volume II, Companion, Part 3, 1911–1914* (London, 1969), pp. 1896–7.
20. Ibid., p. 1897. Sir John French was Chief of the Imperial General Staff and Prince Louis of Battenberg was the First Sea Lord.
21. Ibid., p. 1897. This letter is also in Roskill, *Naval Air Service*, pp. 119ff.
22. For the paper see Randolph Churchill, *Winston S. Churchill*, vol. II, pp. 613ff from which this and the following three paragraphs, above, are drawn.
23. For the details of this meeting see Cab. 18/27. 'Proceedings of Conferences held at the Admiralty and War Office between August 14th 1913 and May 22nd 1914'. 'High Level Bridge Conferences', marked 'SECRET'.
24. For the exchange set out in this paragraph and the next see Air 1/652/17/122/469 for Sueter's submission dated 29 August 1912, and attached minute sheets.

25. Cab. 18/27. 'High Level Bridge' Conferences. Proceedings of a Conference held at the War Office on October 15, 1913, marked 'SECRET'.
26. See Cab. 18/27. 'High Level Bridge' Conferences. Proceedings of a Conference held in the Admiralty Board Room, 19 November 1913. Marked 'SECRET'.
27. See above, Chapter 9.
28. For Churchill's minute see Raleigh, *The War in The Air*, vol. I, pp. 265ff, and Snowden Gamble, *The Story of a North Sea Air Station*, p. 61.
29. See Air 1/2104/207/34. 'Naval and Military Aeronautics – 1913'. Part I. 'General Progress of British Naval Aeronautics'. '(Corrected to 31st December 1913)'. Admiralty War Staff, Intelligence Division. Marked 'CONFIDENTIAL'. This and the following six paragraphs are based on this document.
30. This is the term employed in the report. For further developments in connection with this title, see below, Section VIII of this chapter. There is some discussion of the Military Wing's employment of aeroplanes in the next chapter, below.
31. See Roskill, *Naval Air Service*, p. 118.
32. For these trials see Snowden Gamble, *The Story of a North Sea Air Station*, p. 67.
33. For Churchill's remarks in the House of Commons on 17 March 1914 see *Parliamentary Debates*, 5th Series, 17 March 1914, cols. 1910ff. All remarks about the speech, above, are based on this source.
34. See Gamble, *The Story of a North Sea Air Station*, pp. 71–2.
35. See F. R. Scarlett, Inspecting Captain of Aircraft (acting) to The Commander-in-Chief, H. M. Ships & Vessels, THE NORE, 7 April 1914, marked 'SECRET', in Air 1/652/17/122/469. This and the following two paragraphs are based on this document.
36. For this minute dated 18 May 1914 see Roskill, *Naval Air Service*, pp. 138–9; and Randolph Churchill, *Companion Volume II*, Part 3, pp. 1916–17.
37. Seely's handling of the so-called 'Curragh incident' was the reason for his resignation.
38. See Hankey, *Supreme Command*, vol. I, who stresses the Admiralty acted as it did only after Seely resigned his place.
39. Adm. 116/1275. Minute signed W.S.C. dated 12 March.
40. Asquith MS. Winston S. Churchill to 'My dear Prime Minister', dated July, 1914 and marked 'Private'.
41. For the regulations and comment on them, see Snowden Gamble, *The Air Weapon*, pp. 266ff.
42. The present writer hopes to take up this subject, in a subsequent volume.
43. See 'The Defence of Localities Against Aerial Attack' by Colonel Louis Jackson in *The Journal of The Royal United Service Institution*, vol. LVIII, June 1914, No.436, pp. 701ff. Jackson was a fortification engineer. See also Snowden Gamble, *The Air Weapon*, pp. 262ff.
44. In the 1930s after Hitler announced the existence of a German air force the term 'knock-out blow' came into frequent use. In that period

it was expected that the Germans would begin the next war with a series of devastating air-raids against London, designed to break the morale and the war-will of its civilian inhabitants. This 'knock-out blow', it was thought, would force the British Government to seek peace even though its fleets at sea and its armies in the field had suffered no loss or defeat. These are exactly the themes Colonel Jackson touched upon in this lecture of 22 April 1914.
45. Holland's letter is reproduced at the end of the printed version of the lecture in *The Journal of The Royal United Service Institution*, mentioned above. For his exchanges with Jackson, all originally printed in *The Times*, see pp. 724ff of this edition of *The Journal*.
46. For this see B.M.Add.MSS.49712 (Balfour Papers) Fisher to Balfour, 22 November 1911; and 29 November 1912.
47. *The Times*, 5 June 1914.
48. There was a good deal of correspondence in the Press about Scott's letter in *The Times* of 5 June 1914. Six admirals disagreed with his arguments and so did a former Secretary of the Committee of Imperial Defence.
49. See Cab. 17/21. M. P. A. Hankey to Prime Minister, 16 June 1914.
50. Cab. 17/21. Terms of Reference dated 16 June 1914, marked 'SECRET'.
51. Cab. 17/21. Docket signed 'H.H.A.' dated 17 June 1914.
52. Sykes, *From Many Angles*, p. 111.

13. Winston Churchill Takes Charge

1. For these meetings see Roskill, *Naval Air Service*, pp. 148ff. It was at the meeting held on 21 July 1914 that General Robertson sought to reclaim for the War Office 'sole responsibility' for the home air defences. This so angered Churchill that he mentioned it prominently in his 'Aerial Defence' paper of October 1914. In this connection, see above Chapter 9, section vii.
2. Winston Churchill, *The World Crisis*, vol. i, p. 167.
3. See Air 1/2653. Holograph minute by W.S.C. dated 29.7. See also Churchill, *The World Crisis*, vol. i, p. 168.
4. A.O.P. is Admiral of Patrols.
5. For these details see Snowden Gamble, *Story of a North Sea Air Station*, pp. 86–7, and Christopher Cole and E. F. Cheesman, *The Air Defence of Britain, 1914–1918* (London, 1984), p. 6, hereafter cited as Cole and Cheesman, *Air Defence of Britain*.
6. Air 1/2653. Cypher telegram of 29.7.14.
7. Sykes, *From Many Angles*, p. 118.
8. W.O. 32/5258. 'Instructions issued re action to be taken against aircraft . . .'. The first of these is dated 27 July 1914.
9. Sykes, *From Many Angles*, p. 122. Sykes has been criticized for this attitude. Neville Jones in his excellent book, *The Origins of Strategic Bombing*, p. 51, points out that the Germans were capable of raiding London and the area of the Thames estuary and might have done so, successfully. In this connection see also Alfred Gollin, 'Anticipating

Air Attack – In Defence of Britain', in *Aerospace Historian*, vol. 23, No. 4, December 1976, pp. 197ff.
10. See 'A Possible Raid on London', in *Aeronautics*, September 1914.
11. Winston Churchill, *The World Crisis*, vol. I, p. 172.
12. The reader may recall that when the Home Ports Defence Committee produced its schedule of places to be protected by anti-aircraft guns, in April 1913, the Cleethorpes establishment was included in the priority list. See above, Chapter 11, Section III.
13. See Adm. 1/8368/31. Minute by E. Charlton, dated 18 April 1914.
14. See Adm. 1/8368/31. Minute by F.C.J. Tudor, DNO, 3 April 1914.
15. See above Chapter 11, Section III.
16. Adm. 1/8368/31. Minute by F. C. J. Tudor, dated 21 April 1914.
17. See Adm. 1/8368/31. B. B. Cubitt to the Secretary of the Admiralty, 15 July 1914.
18. See Air 1/648/17/122/385. Royal Naval Volunteer Reserve Anti-Aircraft Corps. Diaries and Notes compiled by Captain L. S. Stansfeld, RN, May, 1919.
19. See MEPO 2/1956. Home Office letter to Police, 18 August 1913.
20. MEPO 2/1956. R. H. Brade to General Officer Commanding, London District, 2 August 1914.
21. The scheme, undated is in MEPO 2/1956.
22. An early example of these riots was the disturbance at Keighley in Yorkshire on 29 and 30 August, 1914. See in this connection, H.O. 45/10944/257142. It is the report to the Home Secretary by the Acting Chief Constable of Yorkshire dealing with the Keighley rioting, dated 31 August 1914.
23. W. S. Churchill, *The World Crisis*, vol. I, p. 190.
24. Cole and Cheesman, *Air Defence of Britain*, p. vi.
25. Ibid., p. 6.
26. PRO. 30/57/72. Winston Churchill to Kitchener, 7 August 1914.
27. W. D. 32/5258. B. B. Cubitt to The General Officer, Commanding-in-Chief, Western Command, 18 August 1914, marked 'Secret'.
28. Winston Churchill, *The World Crisis*, vol. I, p. 211.
29. See above, Chapter 9, Section VII. The memorandum is in Air 1/2314/222/6. 'Aerial Defence' by W. S. C., Admiralty, 22 October 1914. The memorandum is printed in Roskill, *Naval Air Service*, pp. 181–2.
30. See Martin Gilbert, *Winston S. Churchill, Companion* vol. III, Part 2, p. 1238.
31. See Winston Churchill, *The World Crisis*, vol. I, p. 215.
32. Martin Gilbert, *Winston S. Churchill*, vol. III, p. 561.
33. For these figures see Penrose, *British Aviation*, p. 532.
34. Sykes, *From Many Angles*, p. 111.
35. Penrose, *British Aviation*, p. 532.
36. Mottistone Papers, Ottley to Seely, 15 Dec 1913 with report entitled 'Aeroplanes', dated 15 December 1913 and marked 'SECRET'.
37. No. 4 Squadron arrived at Dover after flying from the Eastchurch naval air station.
38. For these opinions see Andrew Boyle, *Trenchard* (London, 1962), p. 117, hereafter cited as Boyle, *Trenchard*.

39. Boyle, *Trenchard*, p. 115.
40. Sykes, *From Many Angles*, p. 122.
41. Trenchard Papers, IV/54/99. 'Note by Lord Trenchard on reading Sykes's book "From Many Angles". . .'. dated 19 January 1943.
42. Sir David Henderson Papers, AC71/12, Box 4. Trenchard to Lady Henderson, 7 October 1954.
43. Boyle, in his *Trenchard*, p. 116 does not recognize the extent of Henderson's hostility to Sykes. The relationship between Henderson and Sykes is misunderstood in Malcolm Cooper, *The Birth of Independent Air Power* (London, 1986), p. 22. Cooper argues that Henderson began to dislike Sykes only in 1915 but it is clear that he was angry with him in August 1914.
44. Sir David Henderson Papers, AC/71/12, Box 3. 'Three days of the war'. Notes made by Lady Henderson beginning on 3 August 1914. This and the paragraphs immediately following are based on this document.
45. Lt. Col. L.A. Strange, *Recollections of An Airman* (London, 1933), pp. 42–3.
46. See Raleigh, *The War in The Air*, vol. I, p. 328.
47. See Raleigh, *The War in The Air*, vol. I, p. 412.
48. W.O. 32/5258. Brancker to D.M.T., 24.8.14.
49. W.O.32/5258. L. E. Kiggell to The Officer Commanding RFC, Farnborough, Upavon, dated 27 August 1914 and marked 'CONFIDENTIAL'.
50. For these details see Raleigh, *The War in The Air*, vol. I, pp. 360–61.
51. Ibid., p. 371.
52. Air 1/724/76/2. 'History of Naval Aeroplanes and Seaplanes' by C. R. Samson, DSO, written after the war.
53. See Raleigh, *The War in The Air*, vol. I, pp. 375–6 and Martin Gilbert, *Winston S. Churchill, Companion*, vol. III, p. 75.
54. For Brancker's ideas see Macmillan, *Brancker*, p. 76.
55. For Churchill's directions on 1 September, see Martin Gilbert, *Winston S. Churchill, Companion*, vol. III, p. 75.
56. Brancker was full of contempt for Samson and his activities at Dunkirk. See Macmillan, *Brancker*, pp. 77ff. for his opinions.
57. Asquith MS. Asquith to the King, 2 September 1914.
58. See Martin Gilbert, *Winston S. Churchill, Companion*, vol. III, p. 76, and Michael and Eleanor Brock (eds), *H. H. Asquith, Letters to Venetia Stanley*, p. 215.
59. Martin Gilbert, *Winston S. Churchill, Companion*, vol. III, p. 78, and Michael and Eleanor Brock (eds), *H. H. Asquith, Letters to Venetia Stanley*, p. 218.
60. Churchill, *The World Crisis*, vol. I, p. 265.
61. See above, Chapter 9, Section VII.
62. W.O. 32/5260. The sentence is contained in an undated Report marked 'Confidential', prepared by Murray F. Sueter in October 1914. It is interesting that the term Royal Naval Air Service was not used in this context.
63. W.O. 32/5260. Minute by Brancker, dated 2 December 1914.

64. Asquith MS. Asquith to the King, 5 September 1914. See also Gilbert, *Winston S. Churchill, Companion*, vol. III, p. 87.
65. See Gilbert, *Winston S. Churchill, Companion*, vol. III, p. 88 for these plans.
66. Churchill, *The World Crisis*, vol. I, p. 265. The present writer hopes to take up the subject of air defence in the first World War in a subsequent volume in this series.
67. Ibid., p. 273.

Index

Abel-Musgrave, Dr C., 224
Admiralty, 81, 82–3, 97, 145–6, 158, 201
  air defence, 111–14, 205, 206, 207, 210–11, 213
  anti-aircraft guns, 207–8
  aviation policy, 21, 167–8
  'High Level Bridge', 269–73
  preparation for war, 292–5
  rivalry with War Office, 281–3, 300–2, 330 n5
  war strategy, 178–9, 263–4
Advisory Committee for Aeronautics, 27–33, 67, 78, 96–8, 107, 109, 121, 190
  Northcliffe's reaction to, 38–46
  press reactions to, 33–8
  success of, 46–7
Aerial Defence Committee, Navy League, 255
Aerial League of the British Empire, 6–7, 8, 46, 67, 105, 106, 129–31, 253
aerial navigation, free in principle, 137, 138–40
Aerial Navigation Bill 1910, 157–8
Aerial Navigation Sub-Committee, 14–19
Aero, The 106, 165
Aero Club of the United Kingdom, 8, 46, 67
aerodrome, Brooklands, 160–1
Aeronautical Society of Great Britain, 8, 17, 48, 67
Aeronautics, 173–4, 296
aeronautics
  Parliamentary debate, 76–85
  as political issue, 257–8
  press views, 85–6
  scientific research, 3, 25–6

Aeroplane, The, 187–8, 190, 223, 224, 226–7, 234, 238, 241, 248, 249–50, 296
aeroplane flight, English Channel, 63
aeroplanes, 10, 126
  advocacy of, 2–3
  against airships, 222
  air defence, 211–15, 237–8, 278–80, 293–5
  British Army, 183–4
  coastal patrols, 265–6
  cross-Channel flight, 68–71
  experiments with, 95–6
  flight trials, 11
  France, 159–60, 188
  Germany, 13, 31
  ground attack, 117
  military uses, 78–9, 105, 145, 148, 171, 175, 199–200
  naval, 275, 279–80, 286, 287
  need for, 243
  Northcliffe advocacy of, 92–4, 161
  numbers, 246–7, 249–50
  potential, 120, 256
  production, 307–9
  purchase, 3, 4
  scepticism over, 165, 187–8
  scouting, 277–8
  value of, 150–2
Agadir crisis, 175–7, 191, 263
air attacks
  legality, 286
  magazines, 110–11
  'nerve centres', 7–8
  possibility of, 283–6
  vulnerability, 54
Air Battalion, Royal Engineers, 163–5, 191

Index

air bombardment, 19, 79–80, 83–4
Air Committee, Committee of Imperial Defence, 158, 190, 196, 226
air corps, proposed, 153–4
air defence, 18–19, 96–7, 103, 111–23, 142, 187, 204–7, 227–8, 230–8, 259, 270–1, 304–5
  aeroplanes, 211–15, 237–8, 279–80, 293–5
  Churchill's responsibility, 318–20
  guns, 207–11
  improvement, 289
  inadequacy of, 200–3
  naval, 277, 278–80
  neglect, 196, 197
  War Office responsibility, 272–3
air displays, Hendon, 171–2, 194–5
air enthusiasts, 102–6, 174
air fleet, German, 251
air force
  Churchill on, 181–2
  separate, 80–1
Air Ministry, 196
air raids, London, 338 n44
air service, unified, 195–6
aircraft
  coastal patrols, 268–9
  Germany, 98–102
  invasion and, 265–7, 270–1
  military potential, 115–16
  naval duties, 272–3, 283
  night attacks, 79–80
  overflying, 137, 138–40, 146
  production, 228–9
  purchase, 3, 37
  reconnaissance, 123, 186–7
  scepticism over, 12, 15, 140
  scouting, 84–5, 268
aircraft engines, 307, 308, 313
Airship battalion, Prussian Army, 64
airship panic 1913, 238–44
airship stations, 220
airships
  air attack, 7
  attack, 7, 112–13, 117, 118
  bombardment, 162–3
  bombing, 102–3, 219, 232, 271
  categories, 323 n1
  Churchill's views on, 230, 231–3
  civilian targets, 83, 84
  Germany, 37–8, 106, 127, 154, 216–19, 225, 231–2, 242, 253, 265, 283–4
  invasion, 19, 100
  Lebaudy, 16, 17
  military uses, 145, 148
  naval, 19, 21, 31, 219–20, 279–80
  operations against, 275
  overflying, 138–40, 141, 143
  potential, 256
  reconnaissance, 100, 103
  Sheerness incident, 223–7
  transfer to Admiralty, 272
  see also dirigibles
Albatross aeroplanes, 313
Amery, Leo, 195–6
anti-aircraft guns, 99, 127, 200, 207–11, 221, 234–8, 276, 284, 295, 315
  Admiralty, 298–9
  War Office policy, 302–3
Antoinette monoplane, 69
Antwerp, 306, 316, 318
Arbuthnot, Major-General, H. T., 105–6, 253
Argyll, Duke of, 259
armament, aeroplanes, 294
armaments, expenditure, 50
armed motor cars, 319
army
  aeroplanes, 150
  home defence, 262
Army Aircraft Factory, 190
army aviators, 161
Army Bill, German, 250–3
army estimates, 244–50
Arthur of Connaught, Prince, 171
artillery, anti-aircraft, 103, 127
Ashmore, Major-General E. B., 228
Asquith, H. H., 14, 15, 23, 27, 32, 33, 41, 45, 47, 53, 54, 101, 120, 128, 144, 149, 150, 162, 171–2, 177, 178, 180, 181, 184–5, 200, 213, 215, 221, 222, 226, 230,

243, 263, 264, 269, 274, 288–9, 305–6, 317–18, 319
Australia, 51
Austria-Hungary, 60, 293
aviation, early pioneers, 1–2
aviation meeting, Reims, 89–92
Avro aeroplanes, 307

B.E. aeroplanes, 307, 309
B.E.2 aeroplanes, 313
Bacon, Admiral, 100
Bacon, Captain Reginald, 14, 15, 19, 21, 32, 36, 182, 321 n15
Baden-Powell, Major B. F. S., 5, 6, 7, 17, 58, 106
Balcarres, Lord, 84
Balfour, A. J., 4, 26–7, 34, 44–5, 62, 125, 172, 174, 255, 258, 262–3, 264
Balfour, Evelyn, 26–7
Balloon Factory 10, 17, 31, 65, 97, 107–8, 109, 117, 164, 167
balloon guns, 113
Balloon School, 152–3, 163, 164, 167
balloon troops, German, 99
ballooning, 8
balloons, 96
   German, 134–5
   reconnaissance, 99
Bannerman, Major Sir Alexander, 104, 152–3, 163, 165–6, 167, 170, 188, 194, 216, 328 n6
Barlow, Sir John, 62
Barrow-in-Furness, 21, 65, 97, 182, 236–7
Bathurst, Earl and Countess, 67
battleships, 303
   obsolescence, 286, 287
Beaverbrook, Lord 228–9, 334 n48
Belfield, Major-General H. E., 104
Belgium, 139, 149, 315, 320
Beresford, Admiral Lord Charles, 67
Berliner Neueste Nachrichten, 59, 61
Berliner Tageblatt, 57
Bertie, Sir Francis, 147, 155, 156–7
Bethell, Rear-Admiral Hon. A. E., 114, 144

Bethmann Hollweg, Theobald von, 250
Blackwood's Magazine, 151
Blair Atholl Aeroplane Syndicate, 193, 330 n10
Blériot aeroplanes, 95, 307
Blériot, Louis, 70–1, 72, 73, 74, 75, 80, 87, 90, 95, 151, 191, 199
'Blue Water' theory, 262, 263, 264
Boer War, 10, 26
'Bolt from the Blue' theory, 262
bombing, 16, 18, 19, 162–3, 171, 187, 256, 275–6
   airships, 219, 232
   London, 228, 295–9
Bonar Law, Andrew, 192, 254
Brade, Reginald, 300
Brancker, Colonel Sefton, 214–15, 292, 295, 307–8, 312, 315, 316, 318–19
Brewer, Griffith, 28
British Air Corps, 150
British Air Service, 3, 21–2, 24–5, 27, 106–8, 109, 153, 190
British Army, aeroplanes, 183–4
British Expeditionary Force, 123, 178, 197, 202, 207, 211, 212, 214, 268, 294–5
   passage to France, 302–3
British Isles, air defence, 18–19
Brooklands, 160–1, 198
Brun, General Jean, 151, 152, 156
Brunner Mond and Company, 83
Brussels, 314
Buccleugh, 7th Duke of, 47
Buchan, John, 198
Burke, Captain C. J., 186–7
Burnett, Sir David, 253

Cambon, Paul, 136
Campbell, Vice-Admiral Sir Charles, 103–4
Campbell-Bannerman, Sir Henry, 180, 263
Canada, 51
capital levy, armed forces, 250, 251
Capper, Colonel, J. E., 10–11, 17–18, 65, 99–100, 103, 104–5, 107, 108, 117, 120, 153, 163

Index

Cardiff, 55–6
cavalry, 199
  aeroplanes and, 152
Central Air Office 276–7
Central Flying School, 190, 196, 244, 281, 307, 318
Chalmers, Sir R., 185
Chamberlain, Joseph, 180
Channel Fleet, 303
channel flight, aircraft, 4
Channel ports, 320
Chanute, Octave, 87–8
Charlton, Rear-Admiral E. F. B., 297–8
Chatham, 212, 236, 280, 320
Chattenden, 209–10, 211, 235, 236, 237, 280
Churchill, Randolph, 181
Churchill, Winston, 18, 20–1, 75–6, 94, 101, 137, 140, 144, 185, 222, 244, 245, 267, 290, 313, 315, 318–20
  aeroplanes at Dunkirk, 316–17
  air defence, 202–3, 209, 210, 211, 212, 213, 220–1
  airships, 215–16, 230, 231–3
  criticisms of, 306–7
  First Lord of the Admiralty, 179–82
  on German navy, 50
  'High Level Bridge', 269–73
  on naval aircraft, 274–7
  preparation for war, 292–5
  relations with War Office, 303–5, 332 n16
  relationship with Kitchener, 300–1
  responsibility for air defence, 318–20
  separate naval air service, 281–3
  Sheerness incident, 224–5
Circuit of Britain, 160, 170
civil disturbances, 299–300
civilian targets, aircraft, 83, 84
Clacton, 54
Cleethorpes, Wireless Telegraph Station, 237, 297, 298
Clément-Bayard II, 66–7

Coalition Government, proposed, 125
coast defence, 212
Coast Defence Flotillas, 212
coastal patrols, 265–6, 268–9, 301–2
Cockburn, George, 90, 92, 93, 169
Cody, S. F., 11, 17, 107
Colchester, 56
Cologne, airship station, 101
Cologne Gazette, 239
Committee of Imperial Defence, 14, 18, 19–21, 31, 78, 101–2, 114, 116, 119, 120–2, 137–40, 144, 149, 150, 152, 153, 154, 158, 164, 169, 177–9, 180, 182, 184–5, 190, 204, 205, 213, 215, 220, 222, 226, 261, 262–3, 264, 266, 272
Connaught, Duke and Duchess of, 171
Conneau, Jean, 160
conscription, 98, 263, 266
Constitutional Conference, 124
cordite factories, 111, 114, 122, 204, 235, 236
Cox, Harold, 82–3
Crewe, Earl of, 34
cruiser squadrons, 303
Cubitt, B. B., 302
Curtiss, Glen, 90

Daily Chronicle, 288
Daily Mail, 4, 5–6, 13–14, 30–1, 35–8, 39, 45–6, 53, 54, 57–8, 59, 60, 61, 63, 65, 67, 68, 69, 71–3, 75, 76, 85–6, 90–1, 104, 131–3, 159, 161–3, 165, 166, 167, 170, 172, 175, 191–2, 224–5, 238–9, 254, 255, 284, 288
Daily News, 34, 54, 56, 57, 74, 86, 164, 288
Daily News and Leader, 259
Daily Telegraph, 33–4, 56, 73–4, 258
Dangerfield, George, 238
Dardanelles, 306
Darwin, Horace, 32
Davies, Lieutenant Richard, 181
Dawson, Geoffrey, 254

Index

Dawson, Lieutenant Trevor, 97
de Forest, Baron Arnold, 74–6
Denmark, 139, 149
Dernburg, Friedrich, 57
Desborough, Lord, 259
destroyer flotillas, 303
Dickson, Captain Bertram, 95
dirigible balloons, 96
dirigibles, 2, 3, 10, 11, 15, 17, 18, 31, 65, 78, 81, 167, 244
  danger of, 126–7
  Germany, 98–9, 266–7
  ground attack, 118
  military use, 151–2
  naval, 121, 182–3
  reconnaissance, 105
  see also airships
dispatch carrying, 171
Dixon, Norman, 12
dockyards, 111, 122, 204, 210–11, 220, 271, 280
Donop, Colonel von, 105
Dreadnought battleships, 171, 254
Du Cros, Arthur, 9, 32, 34, 65–6, 80–2, 86, 170, 195
Dunkirk, 316, 317, 318, 319
Dunlop Rubber Company, 9, 65–6
Dunne, Lieutenant J. W., 11, 17, 107, 193

Eastchurch aerodrome, 161, 168–9, 196, 202, 207, 223, 224
Eastchurch flying school, 281
Eastchurch naval air station, 280, 290, 296, 315, 316
Edward VII, King, 124, 146
Ellington, Marshal of the Royal Air Force Sir Edward, 311, 312
Ellison, General Gerald, 22
Elswick Ordnance Company Works, 236
endurance records, aircraft, 12–13
England, invasion, 5, 13, 17, 18, 58, 100
English Channel, 261–2
  aeroplane flight across, 68–71
Ensor, Sir Robert, 146, 251
entente, Anglo-French, 176
Escapette, torpedo destroyer, 70

Esher, Lord, 14, 16, 20, 118, 119, 120, 129, 130, 144, 150, 151, 152, 159, 180, 185, 250, 263
European crisis, 1914, 291–2
Ewart, Major-General J. Spencer, 14, 111, 115
Excellent, HMS, gunnery school, 287
expenditure, armaments, 50
Explosives Committee, 26, 29, 36

Farman aeroplanes, 95, 307
Farman, Henri, 1, 69, 90, 104
Farnborough, 108, 150, 152, 163, 220, 281, 310
Felixstowe, 290
Fez, 175
field intelligence, 197, 198
Fischer, Fritz, 146–7
Fisher, Admiral of the Fleet Lord, 15, 21, 93–4, 101, 171, 178, 182, 286, 287
Fisher, Bertie, 312
Fleischer, 99
Flight, 165, 225, 227–8, 238, 240–1, 249
flight duration, 90
flying height, 90
Foreign Office, 142, 154, 158
  reactions to Air Conference, 144–50
foreigners, supposed presence, 56
fortifications, against invasion, 262
France, 22
  aeroplanes, 14, 151–2, 154, 159–60, 188, 307
  Agadir crisis, 175–7
  airships, 283–4
  aviation pioneers, 1
Free, Egerton, 54
Free Trade, 51, 82, 180
French Air Service, 199
French Army, 178
French, General Sir John, 208–9, 270, 303
Fulton, Captain J. D. B., 95

Gamble, C. F. Snowden, 71, 184

Gamble, Rear-Admiral Sir
    Douglas, 136, 137, 142–4, 155
Gardiner, Alfred, 34
Garvin, J. L., 41, 60, 257–8
General Election, 1906, 50, 180
General Elections, 1910, 124
General Staff, memorandum on
    Air Conference, 140–2
Geneva Convention, 84
George V, King, 244–5, 289
German army, 198–9
German assault, 1914, 314–15
German navy, 50
Germans, in Britain, rumours
    about, 49
Germany, 22
  aeroplanes, 307
  Agadir crisis, 175–7
  air fleet, 130
  aircraft, 98–102
  airships, 24–5, 37–8, 106, 127,
    154, 216–19, 225, 231–2, 242,
    253, 265, 266–7, 283–4
  Army Bill, 250–3
  aviation pioneers, 1–2
  encirclement, 146–7
  fear of, 25, 138
  First Air Conference, 135–8
  Haldane and, 47–8
  menace of, 4–5
  possibility of invasion, 263–71
  reactions to Phantom Airship
    Scare, 59, 60
  war with, 177–9
Gerrard, Lieutenant, 169
Gibbs, Lieutenant L. D. L., 95
Gibbs-Smith, C. H., 89
Glassford, Colonel, 87
Glazebrook, Dr R. T., 27, 32, 96
Goschen, Sir Edward, 217
Gosse, Edmund, 94–5
Grahame-White, Claude, 131–2,
    133, 161, 170–1
Great Yarmouth, 290
Greene, W. Graham, 114–15, 145–6
Greenhill, Sir George, 32, 97
Gregory, Lieutenant, 169
Grey, C. G. 187–8, 190, 223, 238,
    241, 248, 249–50

Grey, Sir Edward, 101, 142, 143,
    146, 147, 148, 149, 157, 158,
    176, 177, 292–3
gunpowder factory, royal, 236
Gwynne, H. A., 48

Hadden, Major-General Sir
    Charles, 14, 22, 32, 36, 107,
    163, 185, 209
Hague Convention 1907, 286
Haig, General Sir Douglas, 199
Haldane, R. B., 4, 6, 14, 54–5, 67,
    101, 117, 137, 144, 164, 170,
    171, 174–5, 177, 185, 191, 200,
    246
  accusations of pro-German
    attitudes, 47–8
  Advisory Committee for
    Aeronautics, 27–33, 46–7
  advocacy of scientific research, 3
  aeronautics debate, 77–80, 82,
    83, 84–5
  on aircraft, 120–1
  British Air Service, 9–10, 11,
    21–3, 24–5, 106–8, 109, 153
  criticism of, 125–31, 161–2,
    166–7, 190
  defence of aviation policy, 188–9
  meets Aerial Defence
    Committee, 172
  and Phantom Airship Scare,
    61–3
  raised to peerage, 329 n10
  relations with Northcliffe, 38–46,
    92–5
  relations with press, 33–8, 85–6
  scientific study of aeronautics, 3,
    16, 17, 43–4, 66, 77, 78, 97
  as War Secretary, 263
  on war strategy, 179
Halevy, E., 48
Hamburger Nachrichten, 59
Hankey, Captain Maurice (later
    Lord Hankey), 116, 138,
    139–40, 155, 156, 158, 200, 207,
    208, 269–70, 288–9
harbours, air attack, 103, 104
Hardinge, Sir Charles, 144
Hardwicke, Earl of, 188

Harmsworth, Cecil, 9, 80
Hawkins, Major H. T., 234–5
Hay, Admiral of the Fleet Lord John, 172
height records, aircraft, 12
Heligoland Bight, 268
Henderson, Brigadier-General Sir David, 174, 185, 197–8, 200, 201, 213–14, 227, 271, 292, 295, 309, 310, 311–12, 313, 314, 341 n43
Henderson, Lady 312
Hendon, 132, 161
   air displays, 171–2, 194–5
Hendon aerodrome, 170, 171
Hermione, HMS, cruiser, 21
Hicks-Beach, Sir Michael (Viscount St Aldwyn), 189
'High Level Bridge', 269–73
Hitler, Adolf, 338 n44
Holland, 139, 149
Holland, Thomas, 285–6
Holt Thomas, George, 159–61
home air defence, 122–3, 200–3, 204–7, 227–8, 304–5
   aeroplanes, 211–15
   guns, 207–11
   naval, 277, 278–80
home defence army, 262
Home Ports Defence Committee, 113–14, 115, 116–20, 204–7, 235–8, 299
Home Rule, Ireland, 124, 291–2
Hopkins, Admiral Sir John, 259
House of Lords, 123, 174
Hunter-Weston, Colonel Aylmer, 187–8
hydrogen, plant, 99

Imperial Chemical Industries, 83
Imperial Maritime League, 56
Imperial Preference, 51–2, 180
insular position, British Isles, 141, 142
intelligence, 197, 198
intelligence services, information on foreign aviation, 121–2
International Conference on Aerial Navigation, First, 134–8, 326 n1
   adjourned, 154–7
   British reaction, 140–50
   international law, 285–6
   invasion, 3, 5, 13, 17, 18, 19, 49, 58, 100, 261–7
Ireland, Home Rule, 124, 291–2
Irish Nationalist Party, 124
Isle of Grain, 290
Italy, 60

Jackson, Admiral Henry, 155, 157
Jackson, Colonel Louis, 283–6
Japan, 198
Jellicoe, Admiral Sir John, 233, 239
Jones, Neville, 207
Joynson-Hicks, William, 224, 226, 236–7, 249

Keyes, Commodore Roger, 303
Kiderlen-Wächter, Alfred von, 176
Kiggell, General Launcelot, 315
Kitchener, Field-Marshal Lord, 214–15, 300–1, 303, 306, 318
Kriege, Dr Johannes, 136–7, 138
Krupp, 99

L1, Zeppelin, 223
laissez-faire, 82
Lambert, Comte de, 69
Lanchester, Dr F. W., 32, 39
Lanrezac, General, 303
Lansdowne, Henry Charles, 5th Marquis, 26, 172
Lapeyrère, Admiral Boué de, 156, 157
Larkhill, 161, 164, 165, 211
Latham, Hubert, 69–70
Lebaudy airship, 16, 17, 68
Lebaudy Brothers, 67–8
Ledeboer, John, 296–7
Lee, Arthur, 9, 32, 41–2, 44, 46, 66, 79–80, 86–7, 173, 194–5, 249
legality, air attacks, 286
Lehmann, Rudolf, 31
Liége, 314
Lloyd George Budget 1909, 25, 50–1, 52, 54, 82, 123, 174, 250

Lloyd George, David, 14, 16, 17, 90–2, 124–5, 177, 221, 241
Locke-King, Mr and Mrs, 160
Lodge Hill, 209, 211, 235, 236, 237
London, air attack, 8, 228, 284, 285, 286, 295–6, 301, 317, 318, 320, 338 n44
London aerodrome, 161
London Air Defence Area, 228
London–Manchester flight, 131–3
Longmore, Lieutenant, 169
looting, 299
Loreburn, Robert, 1st Earl, 200
Louis of Battenberg, Vice-Admiral Prince, 185, 270, 319
Lovat, Lord, 266
Lucas, Lord, 128–9
LZ4 airship, 2

McClean, Frank, 169
MacDonald, J. Ramsay, 172
Macdonogh, Colonel George, 155
machine guns, 294, 313, 314
MacInness, Major Duncan, 197
McKenna, Reginald, 14, 101, 121, 130, 131, 170, 172, 173, 175, 183
Mackenzie, J. E., 251, 252
magazines, air attack, 110–11, 235, 237
Mallock, H. R. A., 32, 96–7
Manchester Guardian, 56–7, 174–5, 252–3, 257, 259
Mansion House meeting, 253, 255, 256, 257, 258–60
Mansion House speech, Lloyd George, 177
manufactured products, tariffs, 52
March, 53
Marlowe, Thomas, 162, 255
Martin, Rudolf, 5, 6, 7, 13, 14, 58
Maxim guns, 275
Maxim, Sir Hiram, 18, 21
Mayfly dirigible, 121, 167, 182–3, 220
Metropolitan Police, 299–300
Meuse river, 314
Meyer, Horatio, 55

Military Aeronautics Directorate, 292, 308
Military Wing, Royal Flying Corps, 190, 197, 201–2, 204, 207, 211, 212, 244, 249, 273, 274, 289, 292
Milner, Alfred, 1st Viscount, 67
Molesworth, Major, 299
Mond, Alfred (later Lord Melchett), 83–4
Montagu of Beaulieu, Lord, 7–8, 34, 35, 126–9, 249, 256–7
Morley, John, 1st Viscount, 242
Morning Post, 34–5, 48, 54, 56, 65, 67–8, 73, 86, 90, 91–2, 132, 167, 240, 259–60
Morocco, 175
Motor Airship Study Society, 31
Murray, Brigadier-General A. J., 115, 144
Murray, Oswin, 208
Murray, Sir George, 121
Musgrave, Major Herbert, 207

National Aeronautical Defence Association, 259
National Airship Fund, 67–8, 86
National Government, proposed, 125
National Physical Laboratory, 27, 28–9, 30, 31, 36, 44–5, 46, 66, 78, 97, 108, 109, 121, 322 n9
National Service, 98, 263, 266
Naval Air Service, 275, 281–3
naval air stations, 212–13, 265–6, 276, 280, 283, 290
naval aviation, Churchill's interest in, 181
naval aviators, 161
  Royal Aero Club initiative, 168–9
naval building programme, 25
naval crises, 49–50
naval duties, aircraft, 272–3
Naval Flying School, 196
naval magazines, 111, 114, 122, 204, 209, 220, 221

Index

Naval Wing, Royal Flying Corps, 190, 196–7, 202, 204, 219, 244, 270, 272, 277, 279, 281, 330 n5
Navy League, Aerial Defence Committee, 255
Netheravon, 289
New Zealand, 51
Nicholson, Captain Stuart, 131
Nicholson, General Sir William, 11–12, 14, 15, 16, 19, 21, 30, 101, 111, 115–16, 123, 140, 144, 153, 166, 187, 188, 199, 208
night attacks, aircraft, 79–80
No. 1 Rigid Naval Airship, 167, 182–3, 220
North Sea, 268
Northampton, 53
Northcliffe, Alfred Charles, 1st Viscount, 9, 29–30, 46, 104, 159, 170, 171, 172
  and Advisory Committee for Aeronautics, 38–46
  and aeronautical developments, 4–6
  on aeroplanes, 92–4
  on air defence, 253–5
  airship panic, 238, 239
  attitude to Haldane, 94–5
  on aviation policy, 160–3
  cross-Channel flight, 68–71
  London–Manchester flight, 131–3
  on Phantom Airship Scare, 58–60, 61, 62–3

Observer, 41, 132, 166, 256, 257–8, 288
Official Secrets Act 1911, 177
O'Gorman, Mervyn, 107–8, 109, 117, 164, 167, 185, 194, 216–18, 219
oil storage tanks, 220, 237, 280
old age pensions, 227
Ostend, 315–16
Ottley, Rear-Admiral Sir Charles, 14, 116, 119–20, 308–9
overflying, aircraft, 137, 138–40, 141, 146

oxygen, plant, 99

Paget, Lieutenant-General Sir Arthur, 110–11
Panther, gunboat, 176
Parker, Sir Gilbert, 78, 80
Parliament Bill 1910, 174
Parliamentary Aerial Defence Committee, 8–9, 32, 42, 65, 79, 86–7, 170–1, 172–4, 195
Parseval airship works, 216, 217, 334 n34
Paulhan, Louis, 131–2, 133, 160, 161, 192
Penn-Gaskell, Lieutenant L. da C., 313
Petavel, Professor J. E., 32
Peterborough, 53
Phantom Airship Scare, 53–8, 323 n4
  Northcliffe on, 58–60
Pichon, Stephen, 156–7
pilots, training, 84
Plumstead, 206
political issue, aeronautics as, 258–60
Portsmouth, 211, 236, 320
Prandtl, Dr Ludwig, 26
production, aircraft, 228–9
propellers, airships, 97
Prussian Army, Airship battalion, 64
purchase, aircraft, 17, 22, 37
Purfleet, 206, 211, 236, 295
Pustau, Kapitän zur Zee von, 232

Queensborough, 224

raids, aeroplanes, 200
Raleigh, Sir Walter, 47
Rayleigh, John William Strutt, 3rd Baron, 26–7, 29, 32, 39, 43, 45, 78, 96, 107
reconnaissance, 99, 100, 103, 105, 123, 141, 151, 152, 171, 186–7, 197, 198, 199–200, 227, 313
Reims aviation meeting, 89–92, 160
Renault, Louis, 136, 137, 155–6, 157

Repington, Colonel Charles, 150–2, 159, 251–2, 253, 257, 266
Review of Reviews, 238, 242–3
rifles, aircraft, 294
riots, 299, 300
Roberts, Field-Marshal Lord, 67, 74, 93, 106, 129, 172, 255, 258, 263, 266
Robertson, General Sir William, 203, 213, 292, 339 n1
Rohne, General, 99
Rolls, Hon. C. S. 16–17, 20, 21, 22, 46, 57, 180
Rolls-Royce, 16
Rosebery, Lord, 253, 255, 258
Rothschild, Lord, 172
Royal Aero Club, 168–9, 227, 238
Royal Air Force, 196
Royal Aircraft Factory, 191, 281, 307
Royal Arsenal, Woolwich, 205, 206
Royal Engineers, 182
 Air battalion, 163–5, 191
Royal Flying Corps, 190–1, 198, 214, 227, 301, 302, 307
 departure for France, 313–14
 division, 281–3
 functions of Wings, 196–7
 Military Wing, 201–2, 204, 207, 211, 212, 244, 249, 273, 274, 281, 289, 292
 mobilization, 295
 naval Wing, 202, 204, 219, 244, 270, 272, 277, 279, 281, 330 n5
 tensions in command, 309–12
Royal Flying School, 282–3
Royal Naval Air Service, 282–3, 289–90, 293, 295, 301, 303–4, 307, 313, 315, 330 n5
Royal Naval Reserve Anti-Aircraft Corps, 299
Royal Navy, 178, 254, 262, 263, 274, 275, 332 n7
Royce, Henry, 39
Royse, S. W., 243
Russell, Alick, 217
Russia, 198

Salisbury, Marquis of, 129
Samson, Commander C. R. 168–9, 170, 181, 185, 202, 207, 223, 279, 280, 289–90, 315–16, 317, 319
Sandys, Captain George, 183, 192, 249
Scarlett, Captain F. R., 276, 279–80
Schlieffen plan, 314
school for aeronauts, Germany, 99
science
 in aeonautics, 25–6, 66, 77, 78, 80
 Haldane's emphasis on, 43–4
Scott, Admiral Sir Percy, 287–8
Scott, Sir Samuel, 266
scouting, aeroplanes, 84–5, 268, 277–8
sea power, 332 n7
seaplane ships, 277–8
seaplane stations, 220, 268–9, 273
seaplanes, 202, 277–8, 280, 286, 290, 304, 313
searchlights, 200
Secret Service Bureau, 102, 122
Seely, Colonel, J. E. B., 173, 175, 181, 183–4, 185, 191, 192, 193–4, 195, 196, 200–1, 205, 209–10, 217, 218, 220, 221, 222, 224, 226, 230, 233, 244–50, 254, 259, 267, 269, 270, 274, 280–1, 289, 308
Serbia, 293
Seymour, Admiral Sir Edward, 259
Shaw, Dr W. N., 32
Sheerness, 212, 237, 276–7, 279
Sheerness incident, 223–7, 238, 239
Shoeburyness, 211
Short, Oswald, 169
Short Seaplane, 169
Smith-Dorrien, Lieutenant-General Sir Horace, 126
social reforms, 50, 227, 241
Socialism, 124
South Africa, 51
speed, 90
Spender, J. A., 166
Spithead review, 289
Stamfordham, Lord, 244

Standing Sub-Committee,
   Committee of Imperial
   Defence, 137–8, 144, 149, 150,
   154, 184–5, 190, 261, 264,
   267–9
Stanley, Venetia, 200, 317–18
Stansfeld, Captain L. S., 299
Stead, Alfred, 241–3
Stead, W. T., 241, 242
Stone, Brigadier-General F. G., 234
Stone, Colonel G. F., 102–3, 105
Strange, Lieutenant L. A., 313, 314
strategic reconnaissance, 199
strikes, 177, 299
Sturdee, Rear-Admiral Sir F. C.
   Doveton, 183
submarines, 268, 286, 287, 303
Sudan, 179
Sueter, Admiral Sir Murray, 21, 32,
   168, 181, 182, 183, 202, 216,
   217–18, 219, 272, 293, 319
Sykes, Lieutenant-Colonel
   Frederick, 197, 198–200,
   201–2, 214, 227, 274, 289, 292,
   295, 307, 309, 310–12, 313,
   331 n26, 341 n43

tactical reconnaissance, 200
Tariff Reform, 51–2, 180
Taylor, A. J. P., 250
Technical Sub-Committee,
   Committee of Imperial
   Defence, 196–7, 205, 216, 217,
   218–20, 222, 226, 233
Territorial Army, 40
Thameshaven, 237
The Times, 38, 39, 74, 76, 85, 89–90,
   150, 239–40, 251–2, 253, 254,
   257, 260, 283, 287–8
training, pilots, 84
Trench, Colonel Frederick, 98–100
Trenchard, Lieutenant-Colonel
   H. M., 292, 309–12, 315
Tripoli, 188, 199
Troup, Sir Edward, 144
Tudor, Rear-Admiral Frederick,
   298, 299
Tullibardine, Marquess of, 192–4

United States Army, 84
United States Army Signal Corps, 12
Upavon, 281
Upnor, 236

Vickers, Sons and Maxim, 21, 97,
   182, 236–7
Viktoria Luise, airship, 217, 219, 232
Voisin aeroplanes, 40

Walker, Percy, 20, 71, 108
Waltham Abbey, 211, 236, 295
war, with Germany, 177–9
war flights, 279–80
War Office, 81, 82–3, 97, 158
   aeronautical policy, 25, 46–7
   air defence, 201–3, 205–7,
      211–12, 213–15, 272–3, 304–5,
      318–19
   anti-aircraft guns, 208–9
   attitude to aeroplanes, 78–9
   attitude to air attack, 111
   aviation policy, 21
   Churchill's relationship with,
      303–5
   'High Level Bridge', 269–73
   military aviation reports, 100–2
   preparation for war, 295–6
   rivalry with Admiralty, 281–3,
      300–2, 330 n5
   war strategy, 178–9, 263–4, 292
war squadrons 280
Ward, Sir Edward 22
Warde, Colonel Charles, 226
Ware, Fabian, 34
Wehrbeitrag, 251, 252
Wells, H. G., 17, 72–3, 129, 254,
   286
Westminster Gazette, 165–6
Whale Island, 287
White, Dundas, 82
Whittaker, W. E. de B., 226–7
Widdows, A. E., 92–3
Wildman-Lushington, Lieutenant,
   169
Wile, F. W., 5, 13–14, 58
Wilson, Admiral of the Fleet Sir
   Arthur Knyvett, 130, 131,

140, 144, 178–9, 221–2, 230–1, 232–3, 293
Wilson, General Henry, 178
wireless station, Admiralty roof, 297–8
wireless telegraphy 69, 141, 217, 276, 280
Wireless Telegraphy Station, Cleethorpes, 237, 297, 298
Wisbech, 53
Woolwich, 236, 320
   vulnerability to air attack, 205–6
Wormwood Scrubs, 132
Wright aeroplanes, 13, 14, 16, 17, 20–1, 22, 37, 39, 40, 44, 46–7, 180

Wright, Katharine, 100
Wright, Orville, 1, 10–11, 12–13, 18, 20, 22, 28, 33, 39–40, 68, 79, 93, 100, 253–4
Wright, Wilbur, 1, 4, 10–11, 12, 18, 20, 22, 28, 33, 34, 39–40, 44, 68, 69, 79, 84, 93, 165
Wyatt, Harold, 56

Zeppelin, Count Ferdinand von, 2, 5, 10, 63, 64–5, 102, 223, 224
Zeppelins, 5, 6, 13, 17, 24–5, 31, 40, 58, 64–5, 139–40, 151, 216, 217, 225, 232, 233, 258, 284–5, 286, 295, 296, 302, 315, 316, 318, 319, 320